T0140549

Studies in Computational Intelligence

Volume 598

Series editor

Janusz Kacprzyk, Polish Academy of Sciences, Warsaw, Poland
e-mail: kacprzyk@ibspan.waw.pl

About this Series

The series "Studies in Computational Intelligence" (SCI) publishes new developments and advances in the various areas of computational intelligence—quickly and with a high quality. The intent is to cover the theory, applications, and design methods of computational intelligence, as embedded in the fields of engineering, computer science, physics and life sciences, as well as the methodologies behind them. The series contains monographs, lecture notes and edited volumes in computational intelligence spanning the areas of neural networks, connectionist systems, genetic algorithms, evolutionary computation, artificial intelligence, cellular automata, self-organizing systems, soft computing, fuzzy systems, and hybrid intelligent systems. Of particular value to both the contributors and the readership are the short publication timeframe and the world-wide distribution, which enable both wide and rapid dissemination of research output.

More information about this series at http://www.springer.com/series/7092

Dariusz Barbucha · Ngoc Thanh Nguyen
John Batubara

Editors

New Trends in Intelligent Information and Database Systems

Springer

Editors
Dariusz Barbucha
Department of Information Systems
Gdynia Maritime University
Gdynia
Poland

Ngoc Thanh Nguyen
Department of Information Systems
Faculty of Computer Science
    and Management
Wrocław University of Technology
Wrocław
Poland

John Batubara
Graduate Program in Information
    Technology
Bina Nusantara University
Jakarta
Indonesia

ISSN 1860-949X   ISSN 1860-9503   (electronic)
Studies in Computational Intelligence
ISBN 978-3-319-36850-4   ISBN 978-3-319-16211-9   (eBook)
DOI 10.1007/978-3-319-16211-9

Printed on acid-free paper

Springer International Publishing AG Switzerland is part of Springer Science+Business Media (www.springer.com)

Preface

Intelligent information and database systems are two closely related subfields of modern computer science which have been known for over thirty years. They focus on the integration of artificial intelligence and classic database technologies to create the class of next generation information systems. This new generation systems mainly aims at providing end-users with intelligent behavior: simple and/or advanced learning, problem solving, coordination and cooperation, knowledge discovery from large data collections, reasoning about information under uncertain conditions, support users in their formulation of complex queries self-organization, etc. Such intelligent abilities implemented in classic information systems allow for making them autonomous and user oriented, in particular when advanced problems of information management are to be solved in the context of large, distributed and heterogeneous environments.

The volume focuses on new trends in intelligent information and database systems and discusses topics addressed to the foundations and principles of data, information, and knowledge models, methodologies for intelligent information and database systems analysis, design, and implementation, their validation, maintenance and evolution. They cover a broad spectrum of research topics discussed both from the practical and theoretical points of view such as: intelligent information retrieval, natural language processing, semantic web, social networks, machine learning, knowledge discovery, data mining, uncertainty management and reasoning under uncertainty, intelligent optimization techniques in information systems, security in databases systems, and multimedia data analysis. Intelligent information systems and their applications in business, medicine and industry, database systems applications, and intelligent internet systems are also presented and discussed in the book.

The volume consists of 38 chapters based on original works presented during the 7th Asian Conference on Intelligent Information and Database Systems (ACIIDS 2015) held on 23–25 March 2015 in Bali, Indonesia. The book is divided into six parts.

The first part entitled "Advanced Machine Learning and Data Mining" includes seven chapters which focus on new developments in machine learning and data mining strategies as well as algorithms devised for computer user verification, outlier detection, and image, video and motion data analysis.

The second part of the book entitled "Intelligent Computational Methods in Information Systems" consists of six chapters which present computational intelligence methods (fuzzy set, artificial neural networks, particle swarm optimization, etc.) and their applications to solve selected decision and optimization problems related to intelligent information management.

In the third part "Semantic Web, Social Networks and Recommendation Systems" which encompasses five chapters, the authors propose and analyze new methods and techniques for sentiment analysis, social user interaction, and measuring information quality of geosocial networks. They also present how social network analysis can be used in selected areas of business.

The fourth part entitled "Cloud Computing and Intelligent Internet Systems" contains eight chapters. Majority of the papers included in this part focus on cloud computing in context of intelligent information management (ubiquitous networking, intelligent cloud based systems, etc.), as well as biomedical, educational, and other industrial applications of it. The second group of papers included in this part refers to Internet of Things which contributes to improving the scope and quality of communication and computing services. Utilization of above mentioned concepts together with new system design paradigms (communication enabled applications, service oriented architecture, user centricity, content and context awareness, etc.) will lead to design and implement advanced internet systems.

The fifth part "Knowledge and Language Processing" comprises four chapters and refers to both theory and application of knowledge engineering and natural language processing in context of information and database systems. They present and discuss effective ways of knowledge representation and integration as well as implementation of cognitive agents' perceptual memory in integrated management information systems. They also refer to advanced methods devoted to language processing.

And finally, eight chapters included in the sixth part of the book entitled "Intelligent Information and Database Systems: Applications" present practical applications of considered group of systems in different fields of business and industry.

The book can be interested and useful for graduate and PhD students in computer science as well as for researchers and practitioners interested in merging of artificial intelligence and database technologies in order to create new class of intelligent information systems.

We would like to express our sincere thanks to Prof. Janusz Kacprzyk, the editor of this series, and Dr. Thomas Ditzinger from Springer for their interest and support for our project. Our thanks are due to all reviewers, which helped us to guarantee the highest quality of the chapters included in the book. Finally, we cordially thank all the authors for their valuable contributions to the content of this volume.

March 2015

Dariusz Barbucha
Ngoc Thanh Nguyen
John Batubara

Contents

Part I: Advanced Machine Learning and Data Mining

Part IV: Cloud Computing and Intelligent Internet Systems

Part V: Knowledge and Language Processing

X   Contents

Part VI: Intelligent Information and Database Systems: Applications

Part I

Advanced Machine Learning and Data Mining

Exploration and Visualization Approach for Outlier Detection on Log Files

Ibrahim Louhi[1,2], Lydia Boudjeloud-Assala[1], and Thomas Tamisier[2]

[1] LITA-EA 3097, Université de Lorraine, Metz, France
{ibrahim.louhi,lydia.boudjeloud-assala}@univ-lorraine.fr
[2] ISC, Centre de Recherche Public - Gabriel Lippmann, Belvaux, Luxembourg
{louhi,tamisier}@lippmann.lu

Abstract. We propose a novel clustering-based outlier detection approach for data streams. To deal with the data streams, we propose splitting the data into several windows. In each window, the data is divided into subspaces. First, a clustering algorithm is applied on one subspace. Based on the existing relations between the different subspaces, the obtained clusters can represent partitions on another subspace. Then the same clustering algorithm is applied on each partition separately in this second subspace. The process can be iterated on n subspaces. We perform tests on firewall logs data sets, we choose to test our approach with two subspaces and to visualize the results with neighborhood graphs in each window. A comparison is provided between the obtained results and the *MCOD* algorithm results. We can identify visually the outliers events and observe the evolution of the stream.

Keywords: data streams, subspaces exploration, outlier detection, visualization.

1 Introduction

Nowadays, vast volumes of data are produced across a wide range of different fields. This data is characterized, in addition to a large size, by the temporal aspect which is generally associated with a stream.

In many fields, there is a need to detect unusual events. Hawkins [6] defines an outlier as *an observation that deviates so much from other observations as to arouse suspicion that it was generated by a different mechanism.* Johnson [8] defines it as *an observation in a data set which appears to be inconsistent with the remainder of the data set.* An outlier is, then, an atypical and suspicious observation that differs from other observations and which was caused by different events from those that caused the other observations.

Outlier detection is the process of identifying the aberrant observations in a data set and it is an important task in various fields. Outlier detection methods are used in several applications where it may indicate a change in the system behavior, intrusion attempts or errors. Their identification can solve problems, correct errors or simply alert users to unusual behavior [7].

© Springer International Publishing Switzerland 2015   3
D. Barbucha et al. (eds.), *New Trends in Intelligent Information and Database Systems*,
Studies in Computational Intelligence 598, DOI: 10.1007/978-3-319-16211-9_1

Network traffic is one of the fields where outlier detection is very important to ensure system safety and proper functioning. The simple text formatted log files produced by network traffic analysis, are often characterized by a very high number of recorded events, which can cause much difficulty in processing them.

In this paper, we propose a novel outlier detection approach based on using clustering within data streams. The clustering algorithm can be applied on the data subspaces. The clustering is applied firstly on one subspace, the obtained clusters are considered to be representative of partitions in one of the other subspaces. The clustering is then applied on the partitioned subspace elements that belong to the same partition separately. The process can be repeated on n subspaces. Data is split into windows to allow the comparison of the changes between each window.

The clustering results are visualized using neighborhood graphs on each window. The aim is to be able to detect outliers on each window and to follow their evolution on time. We can consider the clusters that contain only one element as outliers, and the clusters that contain a very small number of elements as suspicious [12]. The results are compared with those obtained with *MCOD* algorithm [10] using the *Cadral* tool [11]. *Cadral* is a software platform that contains a visual analytics module for knowledge discovery and data mining.

Tests are performed on data published as part of the VAST challenge (2012) [13], including the log files of the *Money Bank* network activity, recorded by the firewall Cisco Adaptive Security Appliance 5510.

This paper is organized as follows, we first present the related works, then we describe our clustering approach for outlier detection in data streams, and finally we present our experimental results and the conclusion.

2   Related Work

Outlier detection methods can be divided into three different approaches [4]:

- Supervised approaches: where a set of training data is used. The training set contains normal event instances and some outliers. Learning rules are then used on the new data in order to classify them as normal events or outliers.
- Semi-supervised approaches: where the training set contains only normal event instances. Any new data which does not match with this set is considered as outlier.
- Unsupervised approaches: are used when no training set and no information about the data exist. These approaches assume that only a small part of the data represents the outliers and they try to identify them based on that.

We are interested in unsupervised approaches where, generally, the training data or the complementary information about data are not available. Within the unsupervised approaches we can find distance-based methods [9]. They usually try to find the local distance between objects and they can handle large size sets of data [2] [12]. They assume that the outliers are far from their closest neighbors [4] and they measure the distance between each instance and the center of the nearest cluster. However, these methods can not deal with data sets that contain both dense and sparse regions [12].

Breuning [3] proposed a density based outlier detection, where an outlier is designated according to the degree to which the element is isolated from its neighbors. The local outlier factor *(LOF)* is used to determine the local density of each element neighbors and capture the element isolation degree.

Classical outlier detection approaches must be adapted for temporal data sets, to be able to deal with the stream evolution and to have real-time results. *MCOD* Algorithm (Micro-cluster-based Continuous Outlier Detection) [10] is an event-based approach for the outlier detection in data streams. For each new element, the algorithm fixes a time point in the future to check if the element is became an outlier instead of checking each element continuously. An element is outlier if it had less then a predefined number of neighbors.

Among the stream clustering approaches we find also methods based on Partitioning Representatives. The new elements are assigned to the closest representative cluster. *Stream* [5] is an algorithm that uses a k-median based clustering. The data stream is split into small windows that contain a predefined number of elements. The algorithm tries to find a set of k-medians for each window and associate each item to the nearest median. The purpose is to minimize the squared sum of distances between the elements and their associated median. An hierarchical clustering is applied across the medians : when a predefined number of representatives is reached, the clustering is applied on the representatives set. The limit of *Stream* algorithm is its insensitivity to the stream's evolution [1].

In our approach, we consider the clusters obtained in a subspace as representatives items of another subspace's partitions. Each element in a subspace can represent several elements in a second subspace based on the existing relations between the subspaces.

3   Proposed Approach

The approach that we propose is an outlier detection algorithm using a clustering based on the neighborhood. We use the relations between the data subspaces to process the data stream. The stream is split into windows with a predefined size.

As a first step, we preprocess the data elements to extract the distance tables in an initial subspace. We then apply the clustering to this first subspace to get the clusters. Using the existing relations between the subspaces, we cross the subspaces elements by replacing each element in the first subspace with the elements that it represent in another subspace. The clusters obtained in the first subspace are representative items of the second subspace partitions. We extract the distance table of each partition from the data, and then, we apply the clustering on each partition elements separately instead of the entire subspace elements. The process can be iterated on several subspaces (the second subspace clusters can represent partitions on a third partition, ...). Clusters that contain only one element are considered as outliers, and clusters that contain only two elements are considered as suspicious.

Algorithm 1. Outlier Detection

Require: Distance Table : $(DisSubSpace_i)$;
Ensure: Clusters.
  BEGIN
  for $i = 1$ to $n - 1$ **do**
    Build a neighborhood graph $G = (V, E, p)$ from $DisSubSpace_i$
    for Each cluster of $SubSpace_i$ **do**
      replace each element by the corresponding elements on $SubSpace_{i+1}$ based on the relations between the subspaces
    end for
    for Each partition of $SubSpace_{i+1}$ **do**
      Create the distance table $(DisSubSpace_{i+1})$
      Build a neighborhood graph $G' = (V', E', p')$ from $DisSubSpace_{i+1}$
    end for
  end for
  END.

The clustering approach is based on neighborhood graphs. The subspaces elements are represented by the nodes, and the edges between nodes have the distance as weight. The neighborhood structure is built by considering two elements to be neighbors if their distance is less than a threshold. This neighborhood graph allows us to get clusters on each subspace:

Algorithm 2. Neighborhood graph

Require: Distance Table: $(DisSubSpace)$; Threshold.
Ensure: Clusters.
  BEGIN
  Construct the graph of related neighbors.
  while There is a node not visited **do**
    Sort the edges of the graph in ascending order.
    Delete the edges with the weight greater than the threshold
    Delete the edge with the smallest weight of the graph (or the subgraph).
    for For each node which was connected by the deleted edge **do**
      Mark the node as "visited" and find its nearest neighbors.
      for Each neighbor **do**
        Find its nearest neighbors.
      end for
    end for
  end while
  END.

The major advantage of this approach is that instead of comparing each pair of elements in the entire space (all dimensions), we use only subsets of dimensions (subspaces). The existing relations between the subspaces allow the clusters obtained within a subspace to represent partitions of another subspace. We compare only the subspace elements that belong to the same partition.

4   Experimentation

We are interested in the outlier detection and the observation of their evolution in log files. We choose this type of data due to the very high number of events and the large number of dimensions that can be useful to extract significant knowledge.

We test our approach with the log files published for the VAST challenge 2 of 2012 [13] recorded by the firewall Cisco Adaptive Security Appliance 5510 which contains network event records. Log files are text files that contain the recorded events history where each line represents an event. Events are listed in a chronological order. We choose to split the stream into ten windows.

For each window, we choose to apply our approach on two subspaces which are extracted from this data: *destination ports* (the first subspace) and *IP sources* (the second subspace). We choose these two subspaces because the detection of a suspicious behavior (an attack for example) will be easier if we focus on the log events sources and on the fact that an attack will be focused on a specific port of the network machines. The outliers on the second subspace (*IP addresses*) represent the suspicious machines with an atypical behavior.

The main information that we are interested from the data are:

- *IP* addresses which designate machines (we consider them as the first space)
- *Ports* used by each machine (we consider them as the second space)

We next extract statistical data about the use of each subspace elements. The statistical data represent the attributes of the subspaces elements. The neighborhood-based clustering uses the Khi^2 distance measure (χ^2). We choose the χ^2 distance because the elements have observed values as dimensions.

We apply the neighborhood-based clustering on the first subspace (*ports*). As shown in the left of the figures below (Fig. 1 to Fig. 10), for each window, we get clusters of *ports*. We choose the use frequency of each *ports* by each *IP address* as the relation between the two subspaces. One *IP address* is represented by its most used *port*. In other words, each element from the first subspace can represent several elements from the second subspace.

The *ports* clusters are the representative items of the *IP address* partitions. A crossed table of the two subspaces, which contains the *ports* frequency use by each *IP address*, allows to get the partitions on the second subspace. Each of the *ports* is replaced by the *IP addresses* that it represents.

After replacing the obtained clusters by the corresponding elements in the second subspace, we can apply the neighborhood-based clustering on each partition of *IP addresses* separately to obtain the results shown in the right of the figures below (Fig. 1 to Fig. 10). The position of the *IP addresses* subspace corresponds to the position of their representatives clusters on the *ports* subspace.

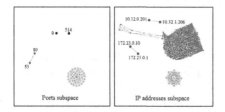
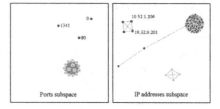

Fig. 1. The first window   **Fig. 2.** The second window

On the *ports* subspace in Fig. 1, we obtain a cluster with a large number of elements, and four elements that are suspected of being outliers: *80, 53, 514 and 0.* On the subspace of *IP addresses*, the *ports 80, 53* do not represent any element, we note that four elements are considered as outliers. The *ports 80, 53* do not represent any element because they are not used as destination *ports* on this window. The *IP addresses* detected as outliers have very aberrant values compared to the other IP addresses contained into the same partition.

In Fig. 2, on the *ports* subspace, *the ports : 53 and 514* are not used in this window. However, *the ports: 0 and 80* are still considered as outliers in addition to a new *port: 1341* which is used for the first time. In the subspace of *IP addresses*: *10.32.0.201* and *10.32.1.206* are integrated in a same cluster with two other *IP addresses*. Two of the *IP addresses* detected as outliers in Fig. 1 disappear in this window because they are not used, it can be a proven that they were true outliers. However, for the two other *IP addresses* which are integrated on a new cluster, the user can consider that they were not outliers on the first window.

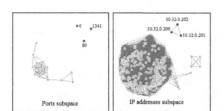
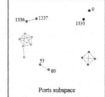

Fig. 3. The third window   **Fig. 4.** The forth window

In the third window (Fig. 3), the same outliers as the previous window (Fig. 2) are detected in the ports subspace. One of the suspicious *IP addresses* from the previous windows is once again regrouped with two other different *IP addresses*. Therefore these two new elements can be considered suspicious.

In the fourth window (Fig. 4), *the ports: 0, 53, 80* are still detected as outliers in addition to the *the port: 1333* and two other suspicious *ports*. These new elements were not used in the previous windows (Fig. 1, Fig. 2 and Fig. 3). Only the *port: 0* and the cluster on the bottom are representatives items. The others

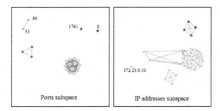

Fig. 5. The fifth window   **Fig. 6.** The sixth window

ports do not represent any element because they are not used as *destination ports*. On the *IP addresses* subspace, an element detected as an outlier in the first window is reappeared here in addition to a new *IP address*.

In Fig. 5, a new port is detected as an outlier (the *port: 1761*), this port is used for the first time in this window. Just like the previous window (Fig. 4), only the *port: 0* and the cluster on the bottom are representatives items. The other *ports* do not represent any element because they are not used here. In the second subspace, one of the previous outliers (Fig. 4) is not used in this window *(10.32.0.100)*, and the other is still detected as an outlier.

In Fig. 6, a port used for the first time is detected as an outlier *(the port: 1429)*, in addition to the usual outliers *(0, 80)*. The *ports: 1761, 53* are not detected because they are not used in this window. The *port: 1429* do not represent any *IP address* in this window because it is not used as a *destination port*. Two new *IP addresses* are detected as outliers and one of the suspicious *IP addresses* from the third window (Fig. 3) reappears.

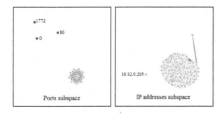

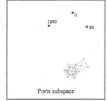 

Fig. 7. The seventh window   **Fig. 8.** The eighth window

In Fig. 7, one new port is detected as outlier (the *port: 1772*) because it is used for the first time in this window, in addition to two outliers detected on the previous window (Fig. 6). Only the *(ports: 0, 80)* and the cluster on the bottom of the *ports* subspace are representatives items of the elements on the *IP addresses* subspace. *The port: 1772* is not used as a destination *port*. One new *IP address* is detected as an outlier.

A new outlier is detected on the *ports* subspace in the eighth window (Fig. 8) as an outlier (the *port: 1890*). This element is used for the first time on this window. The *port: 1890* does not represent any element from the IP addresses subspace. An *IP address* detected as an outlier in the forth window (Fig. 4) reappears here in addition to a new outlier.

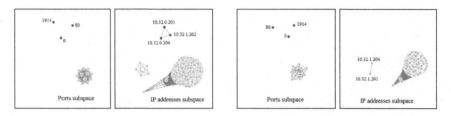

Fig. 9. The ninth window   **Fig. 10.** The tenth window

In Fig. 9, a new outlier is detected on *the ports* subspace (the *port: 1890*). This element is used for the first time in this window. In the *IP addresses* subspace, one new element is grouped in the same cluster with one of the outliers of the previous window (Fig. 9), and with another suspicious element which had appeared in the three first windows (Fig. 1, Fig. 2 and Fig. 3).

In the last window (Fig. 10), the same outliers as the previous window (Fig. 9) in the *ports* subspace are detected. The *IP address* that had appeared on the previous window (Fig. 9) is grouped with another new suspicious element.

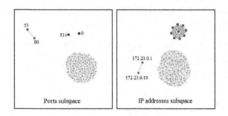

Fig. 11. The entire data

To compare the results obtained in the windows, we apply our approach on the entire regrouped data set, the obtained results are illustrated in Fig. 11. In the *ports* subspace, four outliers are detected: *(0, 514, 53, 80)*, these elements are detected as outliers in different windows. In the second subspace, two *IP addresses* are designed as outliers. The first *IP address* is detected as a suspicious element only in the first window (Fig. 1). The second is detected as suspicious in the first, the forth and the fifth windows.

We conclude that our approach can illustrate the evolution of the data stream. We can visualize the detected outliers and the stream changes over time.

To validate the obtained results, we apply the *MCOD* algorithm on the same data set. The outliers detected by *MCOD* are on the table below (Table 1):

Table 1. Results obtained by the *MCOD* algorithm and our approach

Windows	Ports	IP addresses
First	**514, 53, 0, 80,**	**172.23.0.1, 172.23.0.10**, 172.23.109.
Second	**80, 0, 1341**	172.23.41.1, 172.23.32.2, 172.23.109.
Third	**80, 0**, 1346	172.23.11.2, 172.23.27.2
Forth	**80, 53**, 1351, 1316	172.23.25.2, 172.23.17.9
Fifth	**1761, 80**, 1356, 1365	172.23.109., 172.28.29.6, 172.23.76.4
Sixth	**1429, 80**, 1386, 1394	172.23.14.2, 172.23.0.13
Seventh	**1772, 80**, 1411	172.23.93.2, 172.23.79.1
Eighth	**80, 0**, 1890	172.23.101., 172.23.87.2, 172.23.87.3, 172.23.118
Ninth	**80, 0, 1914**, 1479	172.23.86.2, 172.23.121
Tenth	**80, 0**, 2233, 1515	172.23.118., 172.23.122.
All Data	**514, 53, 0, 80**	**172.23.0.1, 172.23.0.10**

We note that our approach detects the same outliers detected by the *MCOD* algorithm on the entire regrouped data set. However, when the *MCOD* algorithm is applied to the windows, the results are different. The same outliers detected by both our approach and the *MCOD* algorithm are on bold in the Table 1.

In the first, the second and the eighth windows, the same outliers are detected in *the ports* subspace. For the other windows, the *MCOD* algorithm detects almost the same outliers as those detected by our approach, except for a few differences. In the IP addresses subspace, the results are completely different.

The differences on the *IP addresses* subspaces are due to the fact that *MCOD* does not take into account the partitions, contrary to our approach which detect the outliers between the elements that have the same representatives separately.

The similar obtained results on the entire data set by both the *MCOD* algorithm and our approach prove that our approach can detect the outliers. In addition to that, our approach allows the visualization of the outliers (*MCOD* returns just numeric results) and the clusters changes over time (*MCOD* is designed to process the entire data stream and do not split it into windows).

5 Conclusion

We propose a novel outlier detection approach that uses clustering within data streams. The data set is divided on subspaces, and based on the relations between the subspaces, one element can represent many elements in another subspace.

A clustering is applied on a first subspace, the obtained clusters are considered as representative items of partitions in one of the other subspaces. Then the clustering is applied to the partitioned subspace, but only to each partition element separately. We can iterate this process on n subspaces.

Our experiment was performed on the data published in the VAST Challenge 2 of 2012 [13] composed of log files containing the Cisco Adaptive Security

Appliance 5510 network traffic firewall records. We applied our approach to detect machines with an atypical behavior on the network.

We chose to divide the data into two subspaces. We applied a neighborhood-based clustering on the first subspace. The obtained clusters are representative items of partitions in the second subspace. Then, the clustering is applied to the second subspace partitions separately.

Resulting evaluations are encouraging, we are able to identify visually the same outliers detected by the *MCOD* algorithm. A numeric verification of the data set shows us that the outliers detected have atypical behavior.

As future work, we plan to use a visualization that allows an interaction between the user and the clustering algorithm. We intend also to test our approach with different data sets, and to iterate the approach on several subspaces.

References

1. Aggarwal, C.: A Survey of Stream Clustering Algorithms. Data Clustering: Algorithms and Applications, pp. 229–253. CRC Press (2013)
2. Angiulli, F., Pizzuti, C.: Fast Outlier Detection in High Dimensional Spaces. In: 6th European Conference on Principles of Data Mining and Knowledge Discovery, pp. 15–26. Springer, London (2002)
3. Breunig, M., Kriegel, H.P.: NG, R.T., Sander, J.: LOF: Identifying Density-Based Local Outliers. In: 2000 ACM SIGMOD International Conference on Management of Data, pp. 93–104. ACM, Texas (2000)
4. Chandola, V., Banerjee, A., Kumar, V.: Anomaly Detection: A Survey. ACM Comput 15, 15:1–15:58 (2009)
5. Guha, S., Mishra, N., Motwani, R., O'Callaghan, L.: Clustering Data Streams. In: 41st Annual Symposium on Foundations of Computer Science, pp. 359–366. IEEE Computer Society, Washington DC (2000)
6. Hawkins, D.M.: Identification of Outliers. Chapman and Hall, New York (1980)
7. Hodge, V.J., Austin, J.: A Survey of Outlier Detection Methodologies. Artif. Intell. Rev. 22, 85–126 (2004)
8. Johnson, R.A., Wichern, D.W.: Applied Multivariate Statistical Analysis. Prentice-Hall Inc., New Jersey (1992)
9. Knorr, E.M., Ng, R.T., Tucakov, V.: Distance-based outliers: algorithms and applications. The VLDB Journal 8, 237–253 (2000)
10. Kontaki, M., Gounaris, A., Papadopoulos, A.N., Tsichlas, K., Manolopoulos, Y.: Continuous Monitoring of Distance-based Outliers over Data Streams. In: 2011 IEEE 27th International Conference on Data Engineering, pp. 135–146. IEEE Computer Society, Washington (2011)
11. Pinheiro, P., Didry, Y., Parisot, O., Tamisier, T.: Traitement Visuel et Interactif dans le Logiciel Cadral. In: Atelier Visualisation D'informations, Interaction et fouille de Donnes, GT-VIF, pp. 33–44. EGC, Rennes (2014)
12. Ramaswamy, S., Rastogi, R., Shim, K.: Efficient Algorithms for Mining Outliers from Large Data Sets. SIGMOD Rec 19, 427–438 (2000)
13. VAST Challenge (2012), http://www.vacommunity.org/VAST+Challenge+2012

Object Segmentation under Varying Illumination Effects

Dini Pratiwi and Iman H. Kartowisastro

Computer Engineering Department, Bina Nusantara University
Jakarta, Indonesia
dini.holmes@gmail.com, imanhk@binus.edu

Abstract. Image segmentation is one of the medium levels of image processing. In this paper, illumination is the issue of segmenting various objects. This work proposed and evaluated two methods for image segmentation. First, we proposed region growing based method to find region that represented object of interest. Second, we proposed edge suppressing based method by the use of analyzing tensor. The experiment demonstrated that region growing method has limitation in image segmentation for object color variation. The system shows effectiveness of the method by implementing edge suppressing method in three different objects as foreground object, and four different objects as background object. Using edge suppressing, this system performed successfully under different illumination intensities.

Keywords: Segmentation, illumination, color, region growing, edge suppressing.

1   Introduction

Image processing plays an important role for industrial and service robot. In the areas of robot navigation, image processing has a main role in recognizing environments [6], [11]. Robust visual tracking are important in robotic application such as mobile robots [16]. A lot of works related to image processing was done in visual servoing for object tracking [5], [10]. However, object recognition and detection require a strong image segmentation [18]. Therefore, image segmentation plays an important role in image processing for object recognition and detection. [8] introduced color image segmentation using seed-fill algorithm based on YUV color space to recognize robot soccer. This method can quickly complete image recognition under condition of ensuring the accuracy. Image segmentation is a step of image processing to separate object of interest from the background. Many methods for image segmentation were proposed, such as threshold [4], [9], and edge detection [1], [13], [14]. Robot vision needs an excellent sensor to provide the information of the environment. However, variation in light illumination in environment can influence the image [17]. Therefore, the problem of image segmentation is the issue of illumination. Uneven illumination has impact on the image segmentation result. Top Hat transformation and watershed algorithm combine with homomorphic filtering can solve image segmentation with uneven illumination particle image problem. This filter works in frequency domain and Top-hat transformation can removes a large

© Springer International Publishing Switzerland 2015   13
D. Barbucha et al. (eds.), *New Trends in Intelligent Information and Database Systems,*
Studies in Computational Intelligence 598, DOI: 10.1007/978-3-319-16211-9_2

area of background. This method gives a satisfactory result and has some limit effects to the severe adhesion particle of image [19]. [12] proposed a segmentation method with two criteria: single point light source and general illumination condition for image segmentation undernear light source. [15] introduced Illumination Segmentation Model (ISM) to detect edge for non-zero autoregressive coefficient and perform segmentation. An image segmentation technique by analyzing shades and reflection of scene's object are able to separate novel object in complex environment under illumination at different wavelengths [3].

2   Design Method

This paper focused on segmentation to separate object of interest from the background under different illumination intensities. We present two algorithms for image segmentation: region growing and edge suppressing. The proposed model started with image formation model. An imagecharacterized by two components, that are the amount of illumination incident $i(x, y)$ and $r(x, y)$ illumination reflectance [7], that is :

$$f(x, y) = i(x, y)r(x, y) \qquad (1)$$

From (1), the condition of illumination reflectance affects pixel intensities of an image captured by a camera. Therefore, illumination plays an important role and this paper examined the effect of illumination with respect to the result of segmentation. Besides, the color of an object is the factor the image formed. Therefore, various objects were used for this experiment. This system consists of two main parts, image segmentation and image enhancement. The amount of illumination reflectance by an object was captured by a camera in image acquisition process and the computer received it as an input image. The computer process an input image with segmentation method to separate object of interest from the background. Region growing is one of the region based segmentation method, it starts from seed point $S(x, y)$ and expands itself into a region with the same characteristics until it finds a boundary to the adjacent region with different characteristics region. Region growing can give the information on a region that represents a point of an object H(x,y), based on homogeneity test, that is :

$$H(x, y) = \text{TRUE, if } |f(x, y) - S(x, y)| \le T \qquad (2)$$

Where H(x,y) can be obtained if the pixel intensities $f(x, y)$ is reduced by a seed value of less than or equal T as a threshold value. Image segmentation was implemented using region growing approach in two different objects with a contrast color as shown in Fig. 1. A blue can was used as a foreground object, and the other object as a background object. Using region growing, we can separate object of interest from the background as shown in Fig. 1. Further, the region growing method was tested on images with respect to object color variation.

(a)   (b)

Fig. 1. (a) Original image contained a can as an object of interest and a yellow box as an object background (b) A can was separated from the background after segmentation process using region growing approach

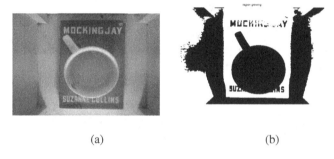

(a)   (b)

Fig. 2. (a) Original image before segmentation (b) Image segmentation result using region growing

Fig.2 shows respectively an original image and the corresponding experimental image segmentation result using region growing method for various colors of objects. Region growing method gave bad result of image segmentation against varying color object. Therefore, we proposed another method for image segmentation with edge suppressing. Edge suppressing proposed cross projection tensor derived from local edge structure in an image to suppress edge [2]. This method consists of two different images, those are image A and image B. Image A was captured as a foreground image consisted of an object of interest, and image B was captured as background image. Gradient of image convoluted with Gaussian kernel to produce smooth structure tensor G_σ is defined as:

$$G_\sigma = (\nabla I \nabla I^T) * K_\sigma \tag{3}$$

Where $\nabla I \nabla I^T$ denotes gradient of image at each pixel, gradient image transposed and K_σ denotes Gaussian kernel of variance sigma, that is:

$$K_\sigma = \frac{1}{2\pi\sigma^2} e^{-\frac{x^2+y^2}{2\sigma^2}} \tag{4}$$

Where x, y denotes pixel component in horizontal and vertical coordinate. Smooth structure tensor can be decomposed as:

$$G_\sigma = [v_1 \quad v_2] \begin{bmatrix} \lambda_1 & 0 \\ 0 & \lambda_2 \end{bmatrix} \begin{bmatrix} v_1^T \\ v_2^T \end{bmatrix} \tag{5}$$

Where λ and v are eigen value and eigen vector, respectively. The cross projection tensor D^B is obtained by the rules: if there is an edge in B ($\lambda_1^B > 0$), remove that edge by setting $\mu_1 = 0$ $\mu_2 = 1$. By selecting eigen vector v and eigen value μ_1 and μ_2, then the projection tensor can be stated as :

$$D^{self} = [v_1 \quad v_2] \begin{bmatrix} 0 & 0 \\ 0 & 1 \end{bmatrix} \begin{bmatrix} v_1^T \\ v_2^T \end{bmatrix} \tag{6}$$

$$u_1 = v_1^B, u_2 = v_2^B \tag{7}$$

We removed edges by using affine transformation for gradient of image A with cross projection tensor D^B, that is :

$$\nabla A' = D^B \cdot \nabla A \tag{8}$$

Reconstruction from $\nabla A'$ gives image A' as a result of this process. It contains an object of interest with free from the background.

3   Experimental Result

Having analyzed the study result mentioned previously, we implemented the system to several different illumination conditions and different objects. In the experiments, various images consisted of background object and object of interest were taken by using a web camera. We tested edge suppressing method under 12 illumination conditions with 10 lux until 1100 lux. The illumination was built using a light source (a lamp) located 27 cm from the scene. Background objects are represented by a book, an orange box, a lighter. In order to test the performance of the method, we executed the simulation programs with the use of MATLAB. Using edge suppressing, a mug as foreground object can be separated from background object shown in Fig. 3. Fig. 3 shows the corresponding segmentation result of the image a book as an object background is removed. Experiments are performed in complex condition as shown in Fig. 3 and give the successful percentage result in Fig. 8. The edge suppressing method is compared with canny edge detection method for image segmentation. Fig.6 shows that canny edge detection method failed to separate foreground object from background object. For the same illumination condition we used a mobile key (a security bank transaction device) for foreground object in Fig.4. Segmentation result in Fig. 4 shows error in boundary of mobile key as foreground object, because of color similarities between foreground object and background object.

The edge suppressing method was also evaluated on a long but thin object (a pen). Similar result of error was obtained in the boundary due to the color similarities between a pen and the corresponding background object in certain area as shown in Fig. 5.

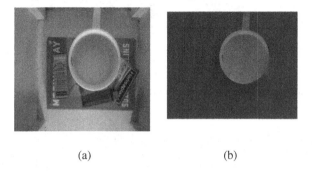

(a)   (b)

Fig. 3. (a) Original image before segmentation (b) Image segmentation result using edge suppressing

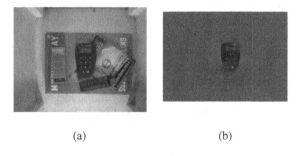

(a)   (b)

Fig. 4. (a) Original image before segmentation (b) Image segmentation result edge suppressing

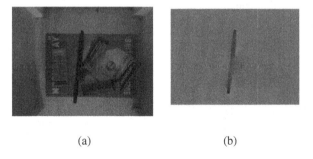

(a)   (b)

Fig. 5. (a) Original image before segmentation (b) Image segmentation result edge suppressing

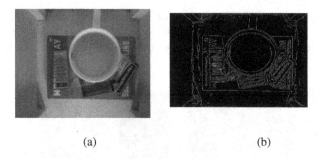

(a)   (b)

Fig. 6. (a) Original image before segmentation (b) Image segmentation result canny edge

The segmentation successful rate is defined as follows,

Successful rate = (sum of segmented pixels / total pixels that represent a perfect object   (9)
image)x 100 %

We evaluated edge suppressing method in two different background objects condition. First condition used one object as background, while second condition used four different objects as background.   Fig. 7 and Fig. 8 show the successful rate percentage of segmentation in different illumination intensities for various objects with different types of shapes. The segmentation was performed with a successful rate of higher than 90% in low to high illumination intensity (501 lux to 1100 lux) with a mug object as shown in Fig.7.   This method gives an effective and consistent result under 801 lux to 1100 lux of intensity with a pen as foreground object as shown in Fig. 7.   From the result of Fig.7, we consider the variation of color in mobile key under low light intensity effect segmentation.   Edge suppressing method perform better in larger object as shown in Fig. 7 a mug has the highest successful rate. Foreground object which has similarity color with background object is treated as part of background and gives the mobile key successful result lower than mug.

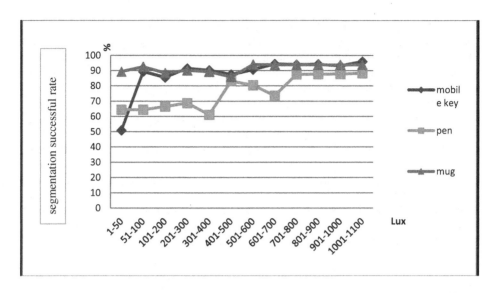

Fig. 7. Segmentation successful rate result using edge suppressing method in first condition

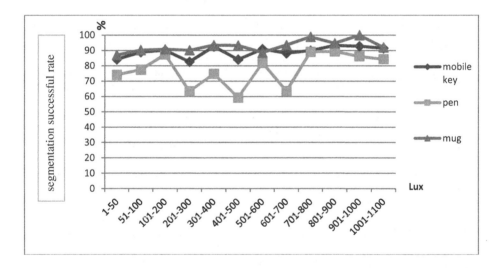

Fig. 8. Segmentation successful rate result using edge suppressing method in second condition

4   Conclusion

This paper presented image segmentation with problem issue in illumination effects. We introduce image segmentation method using region growing and edge suppressing techniques. Region growing and edge suppressing method were implemented in the system under varying illumination condition for various color objects. The result shows that edge suppressing method reached a satisfactory result for three different object of interest with high illumination intensity (501 lux to 1100 lux) using edge suppressing method. In low illumination intensity (10 lux to 500 lux), segmentation gives the best result for an image as a mug as an object. With a pen object, segmentation was performed with a successful rate of higher than 60% in low illumination intensity (10 lux – 500 lux). The experiment using region growing method shows limitless for image segmentation in various color objects. For various objects condition, region growing method failed for object segmentation, while edge suppressing method give better result for segmentation. Based on the experiment result, edge suppressing technique can be applied for image segmentation in real condition under varying illumination. Edge suppressing method gives consistent result in high intensities (801 lux to 110 lux) for three various foreground objects. Potential application for object segmentation using edge suppressing method can be used for pick and place application for various types of parts to be selected.

References

1. Acharjya, P.P.A., Das, R., Ghoshal, D.: Study and Comparison of Different Edge Detectors for Image Segmentation. Global Journal of Computer Science and Technology Graphic and Vision 12, 29–32 (2012)
2. Agrawal, A., Raskar, R., Chellapa, R.: Edge Suppressing by Gradient Field Transformation using Cross-Projection Tensor. In: IEEE Conference on Computer Vision and Pattern Recognition, vol. 2, pp. 2301–2308 (2006)
3. Almaddah, A., Mae, Y., Ohara, K., Takubo, T.: Arai: Visual and Physical Segmentation of Novel Objects. In: IEEE/RSJ International Conference on Robots and System, pp. 807–812 (2011)
4. Beevi, Y., Natarajan, S.: An Efficient Video Segmentation Algorithm with Real Time Adaptive Threshold Technique. In: International Journal of Signal Processing, Image Processing and Pattern Recognition, vol. 61, pp. 304–311. Springer, Heidelberg (2009)
5. Cretual, A., Chaumette, F., Bouthemy, P.: Complex Object Tracking by Visual Servoing Based on 2D Image Motion. In: Proceedings of the IAPR International Conference on Pattern Recognition, Australia, pp. 1251–1254 (1998)
6. DeSouuza, G.N., Kak, A.C.: Vision for Mobile Robot Navigation: A Survey. In: IEEE Transactions on Pattern Analysis and Machine Intelligence, pp. 237–267 (2002)
7. Gonzales, R.C., Woods, W.E.: Digital Image Processing. Prentice Hall, New Jersey (2001)
8. Hai-Bo, L., Yo-Mei, W., Yu-Jie, D.: Fast Recognition Based on Color Image Segmentation in Mobile Robot. In: Proceedings of the Third International Symposium on Computer Science and Computational Technology, vol. 2(1), pp. 1–4 (2010)
9. Kaushal, M., Singh, A., Singh, B.: Adaptive Thresholding for Edge Detection in Gray Scale Image. International Journal of Engineering Science and Technology, 2077–2082 (2010)
10. Li, P., Chaumette, F., Tahri, O.: A Shape Tracking Algorithm for Visual Servoing. In: IEEE Int. Conf. on Robotics and Automation, pp. 2847–2852 (2004)
11. Mata, M., Armingol, J.M., Escalera, A., Salichs, M.A.: Learning Visual Landmarks for Mobile Robot Navigation. In: Proceedings of the 15th World congress of The International Federation of Automatic Control, pp. 1–55 (2002)

12. Okabe, T., Sato, Y.: Effects of Image Segmentation for Approximating Object Appearance Under Near Lighting. In: Narayanan, P.J., Nayar, S.K., Shum, H.-Y. (eds.) ACCV 2006. LNCS, vol. 3851, pp. 764–775. Springer, Heidelberg (2006)
13. Shrivakshan, C.C.: A Comparison of Various Edge Detection Techniques used in Image Processing. International Journal of Computer Science Issues, 269–276 (2012)
14. Suji, G.E., Lakshmi, Y.V.S., Jiji, G.W.: Comparative Study on Image Segmentation Algorithms. International Journal of Advanced Computer Research 3(3), 400–405 (2013)
15. Stainvas, I., Lowe, D.: A Generative Model for Seperating Illumination and Reflectance from Images. Journal of Machine Learning Research, 1499–1519 (2003)
16. Stolkin, R., Florescu, I., Morgan, B., Kocherov, B.: Efficient Visual Servoing with ABCShift Tracking Algorithm. IEEE Transaction, 3219–3224 (2008)
17. Tu, K.Y.: Analysis of Camera's Images Influence by Varying Light Illumination for Design of Color Segmentation. Journal of Information Science and Engineering, 1885–1899 (2009)
18. Wang, L., Shi, J., Song, G., Shen, I.: Object Detection Combining Recognition and Segmentation. In: Proceeding of the 8th Asian conference on Computer Vision, pp. 189–199 (2007)
19. Wen-Cheng, W., Xiao-Jun, C.: A Segmentation Method for Uneven Illumination Particle Image. Research Journal of Applied Science, Engineering, and Technology, 1284–1289 (2013)

Review of Feature Selection Algorithms for Breast Cancer Ultrasound Image

Kesari Verma[1], Bikesh Kumar Singh[2], Priyanka Tripathi[3], and A.S. Thoke[4]

[1] Department of Computer Applications
National Institute of Technology, Raipur India
kverma.mca@nitrr.ac.in
[2] Department of Biomedical Engineering,
National Institute of Technology Raipur
bksingh.bme@nitrr.ac.in
[3] Department of Computer Application,
National Institute of Technology Raipur
ptripathi@nitttrbpl.ac.in
[4] Department of Electrical Engineering,
National Institute of Technology Raipur
asthoke.ele@nitrr.ac.in

Abstract. Correct classification of patterns from images is one of the challenging tasks and has become the focus of much research in areas of machine learning and computer vision in recent era. Images are described by many variables like shape, texture, color and spectral for practical model building. Hundreds or thousands of features are extracted from images, with each one containing only a small amount of information. The selection of optimal and relevant features is very important for correct classification and identification of benign and malignant tumors in breast cancer dataset. In this paper we analyzed different feature selection algorithms like best first search, chi-square test, gain ratio, information gain, recursive feature elimination and random forest for our dataset. We also proposed a ranking technique to all the selected features based on the score given by different feature selection algorithms.

Keywords: Feature Selection, Random Forest, Ranking of features, important feature selection.

1 Introduction

Correct classification of patterns in breast cancer is one of the important research issues of current era. Cancer is the second leading cause of death in developed countries and the third leading cause of death in developing countries. Ultrasound facility is one of the economic ways for early detection and screening.

Pathological tests are the most reliable and most traditional methods for disease detection. Computation in pathology data creates a revolution in the field of biological data. Fuchs and Buhmann[1] has given a concise definition for computational pathology.

© Springer International Publishing Switzerland 2015
D. Barbucha et al. (eds.), *New Trends in Intelligent Information and Database Systems,*
Studies in Computational Intelligence 598, DOI: 10.1007/978-3-319-16211-9_3

Computational pathology investigates a complete probabilistic treatment of scientific and clinical workflows in general pathology, i.e. it combines experimental design, statistical pattern recognition and survival analysis within an unified framework to answer scientific and clinical question in pathology [1]. In this paper we used ultrasound images dataset instead of using pathological data. Computer aided design helps to predict the correct class of cancer without an expert. The layout of breast cancer classification techniques is shown in Fig. 1. The figure consists of three parts. Part one shows the pathological investigation of tumor by medical experts in terms of size, randomness, ratio of height and width, these feature extraction requires human experts and it is time consuming process. Part two shows different computer aided techniques that used to extract the important features that contribute in image analysis and classification task. Our approach is to map the pathological features with computer aided features extracted from images. Part III is set of classification techniques that can be applied in order to accurately classify the images based on extracted feature from part II. The proportion of the total number of predictions that were correct (supervised learning) is called accuracy of classifier. Important recent problems in medical diagnosis is that images containing many input variables (hundreds or thousands), with each one containing only a small amount of information, identification of most important features is still challenging. A single feature selection algorithm will then have accuracy only slightly better than a random choice of classes, it is not able to select all the important feature that are contributing in classification process. Combining the random features can produce improved accuracy [10]. The feature selection techniques are applied from part II to Part III, in order to reduce computational time and space in memory.

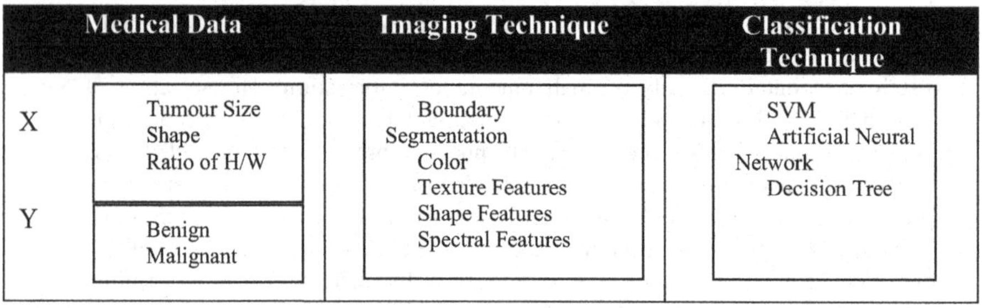

Medical Data		Imaging Technique	Classification Technique
X	Tumour Size Shape Ratio of H/W	Boundary Segmentation Color Texture Features Shape Features Spectral Features	SVM Artificial Neural Network Decision Tree
Y	Benign Malignant		

Fig. 1. Layout of Breast cancer Classification

1.1   Dataset Description

In this study we collected Ultrasound images data from J.N.N. Govt Hospital Raipur of Chhattisgarh, India. The images are labeled by medical professionals to train the model. Example of a benign and a malignant tumor is shown in Fig. 2. The geometric features of the images are the height, width, ratio of height and width, closeness of boundary.

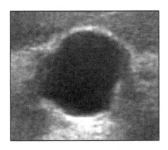 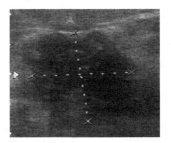

Fig. 2. (a) Benign   (b) Malignant

2   Features

Features are defined as a function of one or more measurements, the values of some quantifiable property of an object, computed so that it quantifies some significant characteristics of the object. A set of features that helps the model to recognize the pattern is called class label. The feature set may contain set of irrelevant features. The irrelevant input features will induce great computational cost. Feature subset selection is the process of identifying and removing as much irrelevant and redundant information as possible. The reduction of dimensionality the data and may allow learning algorithms to operate faster, accurately more effectively. The objective of this paper is find the most relevant features in order to discriminate different class of tumors in dataset. Table 1. shows different features extracted from ultrasound images.

Table 1. Extracted Features from Ultrasound Images

Particular	Number of Features
First Order Statistics (FOS) (features 1-5) [mean,variance,median,mode,skewness,kurtosis,energy,entropy]	06
Haralick Spatial Gray Level Dependence Matrices [2]	26
Gray Level Difference Statistics (GLDS) [contrast, entropy, energy, mean]	04
Neighbourhood Gray Tone Difference Matrix (NGTDM)	05
Statistical Feature Matrix (SFM) [feat, coarse, cont, period, roughness]	04
Laws Texture Energy Measures (TEM)	06
Fractal Dimension Texture Analysis (FDTA)	04
Shape (area, perimeter, perimeter^2/area)	04
Spectral Features	379
Law Features	06
Others	13
Total	457

2.1   Feature Selection Techniques

In previous section we described all 457 features extracted from ultrasound image. The feature selection process is divided into two categories feature ranking and subset selection which is shown in Fig. 3.

Feature Ranking. Kohavi and John [4] proposed variable ranking method for ranking the features based on their importance. Algorithm 1. demonstrate the layout of the feature ranking algorithm.

```
Algorithm 1. Ranking features

Input : S ← set of features
Output : N ← Top n ranked features
Method
    1. Features ← Evaluation_criteria(D) // Evaluation
       criteria   on   that   basis   the   features   are
       evaluated.
    2. Rank_features ← sort_descending(Features)

Return(top n ranked_features)
```

Feature
Selection

Fig. 3. Layout of different Feature Selection Techniques

Information gains, Gain Ratio, Best First search algorithm, Chi-Square test are some specific techniques that are widely using for feature selection purpose. The details are given as below.

Information Gain. This technique is based on decision tree induction ID3 [5] it uses information gain as its attribute selection measure. This measure is based on pioneering work by Claude Shannon from information theory. If p_i represents the number of times tuples occurred in data D. This attribute minimize the information needed to classify the tuples in the resulting partitions. The information gain is represented by equation 1.

$$\text{inf}(D) = -\sum_{i=1}^{m} p_i \log_2(p_i) \qquad (1)$$

Splitting attribute measures, that define information needed to exact classify the data is defined by equation 2.

$$\text{inf}_A(D) = \sum_{j=1}^{v} \frac{|D_j|}{|D|} \times \text{inf}(D_j) \qquad (2)$$

Information gain is difference between original information and information after splitting is defined in equation 3.

$$Gain(A) = \text{inf}(D) - \text{inf}_A(D) \qquad (3)$$

In this technique the features which have highest information will be ranked high otherwise low. Using Quinlan C4.5 algorithm [4] the attribute that are in the higher level of the tree are considered for further classification and these features have more importance.

Gain Ratio. C4.5 [5] a successor of ID3[6] uses, an extension to information gain known as gain ration. It applies a kind of normalization to information gain using split information defined in equation 5.

$$splitInfo(D) = -\sum_{j=1}^{v} \frac{|D_j|}{|D|} \times \log_2 \frac{(|D_j|)}{(\lfloor D \rfloor)} \qquad (4)$$

The gain ratio can be defined by equation 5. Intrinsic information: entropy of distribution of instances into branches by using equation 4.

$$GainRatio(S,A) = \frac{Gain(S,A)}{IntrinsicInfo(S,A)} \qquad (5)$$

Random Forest Filter. Breiman et. al [7] has proposed random forest algorithm, it is an ensemble approach that work as form of nearest neighbor predictor. The goal of ensemble methods is to combine the predictions of several base estimators built with a given learning algorithm in order to improve generalizability / robustness over a single estimator [15]. Ensembles are divide-and-conquer tree based approach used to improve performance of classifier. The ensemble method is that a group of weak learners that group together and work as a strong learner to take the decision for unknown attributes. Giger's et. al. [16] developed a method for automated feature analysis and classification of malignant and benign clustered for micro calcifications data.

Best First Search. Best first search [14] is an Artificial Intelligence search technique which allows backtracking in search path. It is a hill climbing, best first search through the search space by making change in current subsets.

2.2   Feature Subset Selection

In this approach subsets of features are selected, subset feature selection is an exhaustive search process. If data contain N initial features there exist 2^N possible subsets. Selection of features from 2^N possible subsets is an exhaustive search process that is call heuristic search algorithm.

Subsets of features are selected and analyzed their classification accuracy, if it is increasing, that feature is selected otherwise rejected, a new set of feature are participate in evaluation process.

Many feature selection routines used a wrapper approach [4] to find appropriate variables such that an algorithm that searches the feature space repeatedly fits the model with different predictor sets. The best predictor set is determined by some measure of performance. The objective of each of these search routines could converge to an optimal set of predictors. The layout of subset feature selection method is shown in Algorithm 2.

Algorithm 2: Subset feature selection

S ← All subsets {}
For each subset $s \in S$
        Evaluates (s)
Return {subset}

2.3   Recursive Feature Elimination [3]

Recursive feature elimination method is based on the concept that the features are eliminated recursively till the optimal set of features are not selected from the whole set. Random forest, forward subset selection, backword subset selection algorithm using caret, Boruta are the well-known techniques in R [8].

3   Experimental Study

For experimental evaluation of the proposed detection feature, we make use of our own created database of 188 ultrasound breast cancer images. The images were never-compressed gray scale images with resolution of 300x 340 and 90horizontal and vertical DPI. All feature extraction experiments were performed in MATLAB 2012 in windows environment with 4 GB RAM and 500GB Hard disk. For feature selection process we used R statistical package using caret [8], Borutha[9]. The selected features are with different feature selection algorithms is shown in Table 2.

Table 2. Selected Features using different techniques

Technique	Feature	
Information Gain	A50 100.00,A97 75.28 A137 69.66,A26 37.08 A93 22.47	Information from highest to lowest
Random Forest Rank	A33 6.325152, A53 -5.902789 A38 5.706619, A30- 5.491 A20 5.394731, A19-5.195 A50 5.162865, A42-5.158 A52 5.051218, A44-4.914 A34 4.883433, A171-4.714	Ranking of Features from top to bottom
Ranking features by Chi-Square	A141 A140 A171 A149 A163 A154 A137 A155 A158 A168 A169 A174 A175 A157 A172 A20 A55 A148 A152 A180 A165 A178 A156 A170 A177	
BestFirst Search	A33 A161 A170 A193	Exhaustive Search process stat with one feature and continue till the optimal features are not selected
Ranking by gain ratio	A163 100.000,A26 52.563 A33 39.641,A50 28.109 A44 23.424,A91 12.740 A8 11.594,A1 8.281	% of contribution from top to bottom
Random forest Feature selection measure [Borutha]	A20 A30 A33 A34 A37 A38 A42 A44 A50 A52 A53 A55 A171	
Recursive Feature Elimination	A50, A52, A39, A53, A17, A34, A35, A47, A49, A16, A32, A10	

In Table 2 it shows that information gain attribute selection measure selected 5 attribute A50, A137, A26 and A93. All the attributes are ranked based on information gain from higher to lower. Random forest rank selection algorithm ranked all feature from 6.32 to 4.71. Random forest rank algorithm and decision tree information gain selected 95% features from texture categories and 5% from spectral feature categories. Chi-square algorithm selected 23 important features from total 457feaures. Best first search is an exhaustive search algorithm that selected four features as important category. In Chi–Square and Best first search algorithm most of the features from spectral and very few features from texture categories. Gain ratio algorithm selected 8 features in important out of 7 are texture feature and one Spectral feature. Recursive feature elimination process selected all texture feature (12) are important feature but algorithm found spectral features are not important for classification. For Random forest feature selection 12/13 feature from texture and one feature selected from spectral category.

Based on Table 2 all features scores are evaluated which is shown in table 3. Table contain 0 and 1 value. 1 represents voted by algorithm 1 represent not voted by any of the algorithm. In experiment we found that Attribute A50 got highest ranked, six feature selection algorithm voted feature A40. We can conclude that Attribute 50 is most relevant feature for our dataset. Similarly other features are ranked attribute 20 has the lowest rank 2.

Table 3. Ranked Feature based on score of different feature selection algorithms

Features	Information Gain	Random Forest Rank	Chi Square	Best First Search	Gain Ratio	Random Forest	RFE	Rank
A50	1	1	0	1	1	1	1	6
A20	0	1	0	1	0	1	1	4
A33	0	1	1	0	1	1	0	4
A38	0	1	0	0	0	1	1	3
A42	0	1	0	0	0	1	1	3
A44	0	1	0	0	1	1	0	3
A52	0	1	0	0	0	1	1	3
A53	0	1	0	0	0	1	1	3
A55	0	0	0	1	0	1	1	3
A171	0	1	0	1	0	1	0	3
A19	0	1	0	0	0	0	1	2
	1	0	0	0	1	0	0	2
	0	1	0	0	0	1	0	2

All the selected features were evaluated using support vector classifier. The Classification accuracy using different kernel is shown in Table 4. The performance for kernel radial is highest in compared to other kernel so we selected radial kernel for measuring classification accuracy using selected features.

Table 4. Classification Accuracy using kernel gamma = .001 cost =10

Kernel	svm	tune svm
Radial Kernel	82.75	86.20
Polynomial	65.51	62.06
Gaussian	48.27	86.20
Sigmoid	62.06	79.31

3.1   Accuracy

Accuracy is the overall correctness of the model and is calculated as the sum of correct classifications (quantity) divided by the total number of classifications [9]. Other parameters that also important for correct prediction and selection of classifier are precision, recall/sensitivity, specificity and F-measrue. Experimental result with all feature selected algorithm, its accuracy, accuracy after tuning svm parameters, True positive (TP), False Positive (FP), False Negative (FN), True Negative (TN), recall, precision, sensitivity and F- measures results are shown in Table 5.

Table 5. Accuracy Measurement parameters

Precision	$\dfrac{TP}{TP+FP}$
Recall/Sensitivity	$\dfrac{TP}{TP+FN}$
Specificity	$\dfrac{TP}{TN+FP}$
F-Measure	$2*\dfrac{\Pr ecision*\operatorname{Re}call}{\Pr ecision+\operatorname{Re}call}$

The classification accuracy for support vector machine is shown in first row, which is using all 457 feature. The accuracy and other parameter are also shown in respective rows. Principal component feature selection measure (dimension reduction) algorithm performs worst using svm classifier for our dataset. We applied our scored feature table that is named as hybrid feature in last column of the table it accuracy is as same as original svm classifier as well as f-measure value is also equal to original classifier value. Our hybrid feature selector selected 43 features as important for classification task. We can conclude that the instead of taking 457 feature only 43 features are highly contributing most relevant feature for breast cancer classification using ultrasound images. The experimental results are shown in Table 6.

Table 6. Experimental results

Feature selection Technique	Selected Attributes	Accuracy	After Tuning	Recall	Sensitivity	Precision	F-Measures
SVM	457	82.75	86.2	0.77	1	0.73	0.87
C5.0 Ranking	5	79.31	75.86	0.68	0.92	0.6	0.78
Random Forest	25	79.31	82.75	0.76	0.92	0.733	0.83
Chi Square Ranking	25	79.31	79.31	0.75	0.85	0.733	0.79
Breath First Search	4	75.86	72.41	0.65	0.92	0.533	0.76
Borutha	13	75.86	82.75	0.81	0.85	0.533	0.83
RFE	14	82.75	82.75	0.8	0.85	0.8	0.82
PCA	2 PC	68.96	51.72		0		NA
Hybrid		86.2	86.2	0.77	0.2	1	0.87

4   Conclusion

In this paper different feature selection algorithm were analyzed for ultrasound breast cancer images. We extracted 457 feature features from images of texture and spectral categories. We performed the experimentation with all 457 features using support vector machine using 10 fold cross validation. After tuning the parameter we achieved 86.2069% of classification accuracy. We applied feature selection algorithms and svm classifier using 10 cross fold validation to all 7 feature selection algorithms and principal component for dimension reduction. We created a score matrix based upon all selected features and voting of all algorithms. Based on the score we arranged the features in the descending order, with this technique 43 feature were selected and svm classification technique was applied for classification purpose. We achieved the same accuracy as we got for all 457 features. It also reduced the computational time and memory for classification purpose.

References

1. Thomas, J.F., Buhmann, M.J.: Computational pathology: Challenges and promises for tissue analysis. Computerized Medical Imaging and Graphics 35(7-8), 515–530 (2011)
2. Haralick, R.M., Shanmugam, K., Dinstein, M.I.: Texture Feature for Image Classification. IEEE Transaction on Systems, Man and Cybernetics 3(6), 610–619 (1973)
3. (July 2014), `http://en.wikibooks.org/wiki/Data_Mining_Algorithms_In_R/Dimensionality_Reduction/Feature_Selection`
4. Kohavi, J.G.: Wrappers for feature subset selection. Artificial Intelligence 97(1-2), 273–324 (1997), doi:10.1016/S0004-3702(97)00043-X
5. Quinlan, J.R.: C4.5: Programs for Machine Learning. Machine Learning, vol. 16, pp. 235–240. Academic Kluwer Academic Publishers, Boston (1994)
6. Quinlan, J.R.: Induction of decision trees. Machine Learning 1(1), 81–106 (1986)
7. Kursa, M.B., Rudnicki, W.R.: Feature Selection with the Boruta Package. Journal of Statistical Software 36(11), 1–13 (2010)
8. R Statistical Package (July, 2014), `http://CRAN.R-project.org/package=varSelRF`
9. Svetnik, V., Liaw, A., Tong, C., Wang, T.: Application of breiman's random forest to modeling structure-activity relationships of pharmaceutical molecules. In: Roli, F., Kittler, J., Windeatt, T. (eds.) MCS 2004. LNCS, vol. 3077, pp. 334–343. Springer, Heidelberg (2004)
10. Breiman, L.: Random Forests. Machine Learning 45, 5–32 (2001)
11. Davis, J., Goadrich, M.: The Relationship Between Precision-Recall and ROC Curves. Technical report #1551, University of Wisconsin Madison (January 2006)
12. Tan, P.N., Kumar, V., Steinbach, M.: Introduction to Data Mining. Pearson education, 321321367th edn. Addison-Wesley (2005) ISBN : 0321321367
13. Hall, M.A.: Correlation-based Feature Selection for Machine Learning. Ph.D. thesis in Computer Science. University of Waikato, Hamilton, New Zealand (1999)
14. Rich, E., Knight, K.: Artificial Intelligence. McGraw-Hill (1991)
15. Ensemble method (Ocober, 2014), `http://scikit-learn.org/stable/modules/ensemble.html#b2001`
16. Gruszauskas, N.P., Drukker, K., Giger, M.L., Chang, R.F., Sennett, C.A., Moon, W.K., Pesce, L., Breast, U.S.: computer-aided diagnosis system: robustness across urban populations in South Korea and the United States. Radiolog 253, 661–671 (2009)

Recognition of Human Gestures Represented by Depth Camera Motion Sequences

Adam Świtoński[1,2], Bartosz Piórkowski[2], Henryk Josiński[2],
Konrad Wojciechowski[1], and Aldona Drabik[1]

[1] Polish-Japanese Academy of Information Technology,
Aleja Legionów 2, Bytom, Poland
{aswitonski,kwojciechowski}@pjatk.edu.pl
[2] Silesian University of Technology, ul. Akademicka 16, 44-100 Gliwice, Poland
{adam.switonski,henryk.josinski}@polsl.pl

Abstract. The method of gesture recognition useful in construction of hand controlled user interfaces is proposed in the paper. To validate the method, a database containing seven selected gestures, performed by six subjects is collected. In acquisition Microsoft Kinect device was used. In the first stage of introduced method, linear dimensionality reduction in respect to depth images of human silhouettes is carried out. Furthermore, the proper recognition is based on Dynamic Time Warping technique. To assess general features of human gestures across subjects, different strategies of dividing captured database into training and testing sets are taken into consideration. The obtained classification results are satisfactory. The proposed method allows to recognized gestures with almost 100% precision in case of training set containing data of classified subjects and with 75% accuracy otherwise.

Keywords: motion capture, motion analysis, times series classification, artificial intelligence, dimensionality reduction, dynamic time warping, depth imaging.

1 Introduction

Human-computer interaction is gaining more importance in construction of computer systems and their user interfaces. Quite new possibilities in this field are brought by still developing motion acquisition techniques, which allow to recognize gestures made by a human. Best precision of measurements of human motions is obtained by optical marker based motion capture systems. However such an acquisition has serious limitations related to requirements of mocap laboratory, markers which have to be attached on a body and calibration stage. In contrary there is a motion capture based on depth imaging. It gives worse precision of the measurements, but only shortcoming of the acquisition process corresponds to the limited distance between a human and capturing device.

Gestures can be defined as expressive, meaningful body motions involving physical movements of the fingers, hands, arms, head, face or other body parts

© Springer International Publishing Switzerland 2015   33
D. Barbucha et al. (eds.), *New Trends in Intelligent Information and Database Systems*,
Studies in Computational Intelligence 598, DOI: 10.1007/978-3-319-16211-9_4

[6]. The problem of human gesture recognition is broadly studied in the world literature for years. Below only the selected approaches which utilize Kinect depth imaging are described.

In [3] histogram based features of 14x14 grid placed on the determined foreground are computed. What is more simple differences between subsequent frames called motion profiles are calculated. The proper classification is carried out by multiclass Support Vector Machine. The proposed method is strongly simplified, because it does not consider the time domain and recognition corresponds only to single frames. This leads to worse accuracy, since there are some poses which are very similar in different gestures. To discriminate them analysis of whole sequence of poses is necessary. The similar approach is introduced in [8]. The static gestures are recognized by Finger-Earth Mover's Distance and template matching.

In [7] Neural Network, Support Vector Machine, Decision Trees and Naive Bayes classifiers are trained on the basis of parameters of skeleton model determined by Kinect software rather than raw depth images. Once again only single poses are recognized. The obtained classification accuracy is very high. in the best case it has even 100% of correctly classified instances, because very simple set of only three static poses - stand, sit down and lie down - is taken into consideration.

Whole motion sequences are analyzed for instance in [13], which deals with the problem of hand gesture recognition. In a preliminary stage palm is detected on every motion frame and its tangent angle at different time instants is calculated. Tangent angle relates to hand movement across images axes. The final classification is carried by Hidden Markow Models (HMM) - in the training phase single model for every class is determined and in the recognition stage the model with greatest probability is searched. Another approach based on HMM, but this time in respect to features extracted for four joints angles of the skeleton model: left elbow yaw and roll, left shoulder yaw and pitch - is presented in [5]. Prior to HMM modeling data is segmented by K-Means clustering into thirty groups and HMM in further processing takes into consideration only cluster centroids. Very naive method in which a multidimensional, skeleton model based motion sequences are transformed into single feature vector is described in [2]. To recognize gestures supervised classification is carried out. However such a method has the strong limitations related to the specified number of frames of the recordings and it does not considers possible local shifts between subsequent gesture phases.

There are also examples of Dynamic Time Warping (DTW) applications to Kinect data. In [1] DTW is used to align motion sequences to a median length, reference one. Further Gaussian Mixture Models (GMM) are utilized to construct feature vectors of subsequent time instants. In the final stage DTW once again is carried out with soft-distance based on the probabilities of a point belonging to G components of GMM. In [4] an approach of recognition of hand written digits is proposed. It consists of following steps: hand detection of depth motion sequences, feature extraction and DTW classification.

In the paper a strict machine learning based method for gesture recognition which analyzes whole motion sequences is introduced. In the preliminary stage dimensionality of extracted monochromatic silhouettes representing depth data is reduced. In the classification stage dynamic normalization of the time domains of compared motion sequences is carried out by Dynamic Time Warping (DTW) transform.

2   Gesture Database

For the preliminary validation of proposed method database containing data of six subjects and seven selected types of gestures is collected. In the acquisition a Kinect camera was used and raw data consist of sequences of low resolution (160x120) depth images. The chosen types of gestures visualized in subsequent rows of Fig. 1 are related to possible control actions used in construction of user interfaces. They are as follows: arm zoom, hand spinning, hand wave, sliding down, sliding forward, sliding inward and wheel steering. Hand spinning and waving are performed by left upper limb. In total there are 196 different instances of recordings wherein complete gesture is performed only once.

Fig. 1. Types of gestures

Fig. 2. Bounding box detection

To remove the influence of a location of a human on the recognition performed, the bounding box is determined as presented in Fig. 2. It is based on geometric center of silhouette depth image in which background points are labeled by zero value.

The captured database is available online at `http://as.pjwstk.edu.pl/aciids2015`.

3   Classification Method

As described above, there are two main stages of the proposed classification method - dimensionality reduction of silhouettes appearing in subsequent time instants and proper classification of reduced motion sequences. On the basis of our previous experiences on pose classification [11] and diagnosis of gait abnormalities [12], classical linear Principal Component Analysis (PCA) technique was chosen for a dimensionality reduction. Because of satisfactory results obtained in human gait identification challenge problem by classifier which compares dynamically scaled motion sequences [10], in proper recognition Dynamic Time Warping transform is applied.

3.1   Principal Component Analysis

The PCA method determines linear independent combinations Y of the input attributes X with the greatest variances, which are called principal components. New base of an input space is established. It turns out, that the base created by the eigenvectors v^T ordered according to corresponding eigenvalues of the covariance attributes matrix satisfies the demands [14].

$$Y = \begin{bmatrix} y_1 \\ y_2 \\ \cdots \\ y_n \end{bmatrix} = V^T \cdot X = \begin{bmatrix} v_1^T \\ v_2^T \\ \cdots \\ v_n^T \end{bmatrix} \cdot \begin{bmatrix} x_1 \\ x_2 \\ \cdots \\ x_n \end{bmatrix} \tag{1}$$

The variance of resultant attributes is denoted by corresponding eigenvalues λ_i, thus in dimensionality reduction only specified number of the first principal components is considered. What is more to evaluate the transformation, the variance cover vC(k) of the first k PCA attributes in respect to dimensionality n of the input space can be calculated. It assesses how the input space is explained by PCA features and it is expressed by the ratio constructed on the basis of proper eigenvalues:

$$vC(k) = \frac{\sum\limits_{i=1}^{k} \lambda_i}{\sum\limits_{i=1}^{n} \lambda_i} \qquad (2)$$

3.2 Dynamic Time Warping

Dynamic Time Warping firstly applied in spoken word recognition [9] scales time domains of analyzed pair of motion sequences. It tries to match most similar subsequent time instants. The transformation is represented by a warping path which indicates corresponding times instants of both sequences. The path is assessed by its cost - aggregated total distance between matched silhouettes. As the result a warping path with minimal cost is determined and the cost can be stated as a dissimilarity between whole compared motions. During DTW computation, similarity and cost matrices are calculated. The first one contains data of distances between every pair of silhouettes of compared sequences which is approximated by Euclidean metric applied to reduced PCA space. The specified cell c(n,m) of cost matrix corresponds to warping path with minimal cost for time series reduced to the first n and m time instants. What is more DTW is monotonic transformation, which means moving backward in time domain is not allowed. So, to determine specified value c(n,m) the minimum of three previous possible neighbors c(n-1,m), c(n,m-1) and c(n-1, m-1) is found and it is increased by a proper value of the similarity matrix. Thus, it is sufficient to carry on calculation in a proper order for subsequent rows or columns and the last cell of cost matrix corresponds to wanted cost of a complete warping path.

In case when motion dissimilarity is approximated, nearest neighbor classification scheme [14] can be applied - an object is being assigned to the class most common among its k nearest neighbors.

Example DTW paths determined with corresponding similarity matrices for 32 dimensional PCA space and input data reduced by 80x110 bounding box are visualized in Fig. 3. In case when the same gestures are compared (Fig. 3a and 3b) the path is matched in the narrow, dark region of more similar silhouettes, which is located in the left-upper quarter of similarity matrices. The right-lower

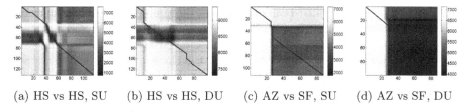

(a) HS vs HS, SU   (b) HS vs HS, DU   (c) AZ vs SF, SU   (d) AZ vs SF, DU

Fig. 3. Example DTW similarity matrices and DTW paths (HS-Hand spining, AZ-arm zoom, SL-sliding forward, SU-single user, DF-different users)

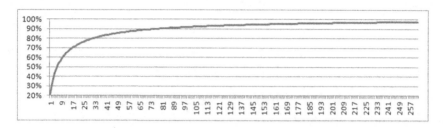

Fig. 4. Aggregated, total variance cover of the first n principal components

quarter is related to standing, final pose because duration of the recordings is longer than gestures activity. That is a reason why it quite uniformly distributed. There is no such an observation for different gestures in Fig. 3c and 3d. Data of separate users (Fig. 3a and 3b) by default are discriminated by visible greater scale, especially if the same gestures are analyzed as for instance in Fig. 3a.

4   Results and Conclusions

In Fig. 4 the aggregated, total variance cover of the first n principal components reflected by the horizontal axis is presented. The single attribute is sufficient to store over than 20% percent of variance, to preserve 50% only six components are required, 60% - 10, 70% - 19, 80% - 38 and 90% - 114. The results seem to be very satisfactory, though further components are noticeably less informative and even 512 of them is insufficient to represent 99% of variance. It is very naive to expect that only few attributes are able to preserve all details of movements. However a main assessment of the dimensionality reduction is related to an obtained accuracy of classification based on reduced silhouette spaces.

In Fig 5 3D trajectories of the first three principal components for four chosen gestures labeled by different colors are shown. In case of data of a single user trajectories are clearly separated in visualized reduced space. It is more difficult to notice such a discriminative features for trajectories of different users, however it seems still to be possible to partially recognize gestures. Surely it is difficult to state final conclusions on the basis of such a visualization, because charts contain only small part of collected data, thus they are strongly simplified.

The proper classification results are presented in Tab 1 and Tab. 2. The validation experiments are iterated across different dimensionalities of reduced spaces, two different sizes of applied bounding box in the preprocessing stage and also for complete silhouette images not reduced by a bounding box detection. What is more, there are two strategies of dividing collected dataset into the training and testing sets. In the first one further called S1 the classical leave one out method is utilized, which means that training set also contains data of classified user. It is a less convenient approach in respect to practical deployments, however we think it to be still acceptable. To directly reproduce such an implementation, a prior to appropriate system running a preliminary calibration stage in which user has to perform complete set of exemplary gestures is required. In the second strategy S2 there are six iterations for every user of collected database. In

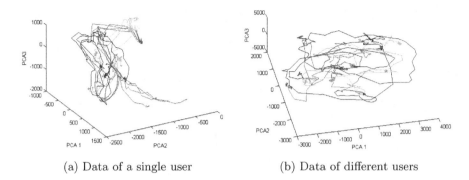

(a) Data of a single user   (b) Data of different users

Fig. 5. PCA 3D trajectories for arms zoom, hand wave, hand spinning and sliding forward gestures

a single iteration step, all data of specified user are moved to a testing set and remaining data are considered to be a training part. It means gestures are recognized on the basis of exemplary instances of other users, captured for instance in production stage of the system, thus no any extra calibration is required.

Table 1. Classification results for subsequent dimensionalities and in respect to different applied sizes of bounding box and training sets

Dimensionality	No Bounding Box		Boudning Box120x120		Boudning Box80x110	
	S1	S2	S1	S2	S1	S2
1	55,61	15,31	74,49	16,33	76,02	16,33
2	65,31	14,80	87,76	34,18	88,78	34,18
4	78,57	20,92	96,94	59,69	96,94	59,69
8	85,20	28,06	97,96	71,94	97,45	69,90
16	86,22	32,14	99,49	70,92	99,49	72,45
32	89,29	37,76	99,49	71,94	99,49	73,47
64	89,29	38,27	99,49	69,39	99,49	71,43
128	89,80	37,24	99,49	69,90	99,49	70,92
256	89,80	38,27	99,49	68,88	99,49	71,94
512	89,80	38,27	99,49	69,39	99,49	71,94

For the strategy S1 very high 99.49% accuracy of recognition, expressed by percentage of correctly classified instances of a testing set, is obtained. It means only a single missclassified gesture of 196. In such a case dimensionality reduction is also very efficient. Only the first principal component allows to classify with over than 75% precision and best results are achieved for 16 dimensional spaces. Thus, general conclusion can be stated - it is possible to precisely recognize gestures on the basis of reduced sequences of human silhouettes by machine learning in case when training data contains exemplary instances of gestures

Table 2. Classification results for subsequent users

Subject	No Bounding Box		Boudning Box120x120		Boudning Box 80x110	
	DIM 16	DIM 32	DIM 16	DIM 32	DIM 16	DIM 32
1	30,30	24,24	81,82	78,79	84,85	78,79
2	17,65	26,47	29,41	32,35	32,35	32,35
3	35,48	41,94	96,77	93,55	100,00	100, 00
4	35,48	45,16	83,87	87,10	80,65	87,10
5	8,57	17,14	57,14	57,14	57,14	60,00
6	68,75	75,00	81,25	87,50	84,38	87,50

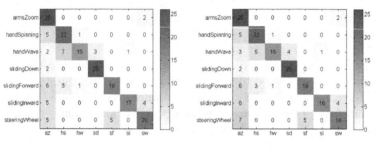

(a) Bounding Box 80x110   (b) Bounding Box 120x120

Fig. 6. Confusion matrices for 32 dimensional PCA spaces and S2 strategy

of a classified user. As it is expected it is much more difficult otherwise. It is caused by variations in movements across users for the interpreted individually gestures and different anthropometric features and postures of a human body which makes the recognition to be more challenging. Much more representative training set with greater number of users would probably minimize the influence of the first reason and classification based on the preprocessed model based skeleton data instead of raw silhouette images - the second one. Though obtained accuracy 73.48% for the strategy S2 in respect to considered seven classes is surely promising. Another observation is related to bounding box detection, which noticeably improves the performance of the classification. It is so, because bounding box removes from an input data their dependency on strict human location and it initially reduces the dimensionality of the input space, which makes further processing to be more efficient. What is a bit surprising smaller bounding box of the size 80x110 gives bit better results, though in some cases of very tall users or side stretched hands of T-pose silhouettes are cropped too much. Once again it can be explained in the same way - the drawback is balanced by lower dimensionality of input space.

As shown in Table 2, there is a user with id 2, which has vary poor classification accuracy of about 30%. In contrary the gestures of user 3 obtain 100% precision.

Table 3. Sensitivity and specificity of gestures classes for 32 dimensional PCA spaces

	Boudning Box 80x110		Bounding Box 120x120	
	Sensitivity	Specificity	Sensitivity	Specificity
arm zoom	92,86	85,12	92,86	82,74
hand spinning	78,57	94,05	78,57	95,24
hand wave	53,57	98,81	53,57	98,81
sliding down	92,59	98,22	92,59	97,63
sliding forward	65,52	97,01	65,52	97,01
sliding inward	65,38	99,41	61,54	99,41
steering wheel	66,67	96,39	60,00	96,39

If all the instances of users 2 and 3 are removed from the testing set, the average results are very alike.

To investigate similarities between gestures and their discriminative features confusion matrices, sensitivity and specificity [14] of subsequent classes are determined and presented in Fig. 6 and in Table 3 respectively. Arm zoom has a lot of false positives which are distributed quit uniformly across other gestures, but it has only two false negatives. It means there are many gestures instances which are badly recognized as arm zoom in contrary to only two missclassified arm zoom instances. It is analogous with hand wave which in some cases is badly recognized as to be hand spinning, but hand spinning is never recognized as to be hand wave. Acceptable level of sensitivity rate is obtained only for arm zoom and sliding down, but the first one has noticeably worse specificity.

Summarizing, the obtained results are quit promising. In case of strategy S1, which requires precalibration stage of every new user, the proposed method is very precise. For much more convenient in respect to practical implementation and usage strategy S2, the further improvements have to be proposed before deployments. As described above it is expected that model based data would obtain better accuracy of recognition. Collected database is multimodal, it also contains parameters of assumed skeleton model determined by Kinect software. Thus, stated hypothesis will be verified in the next step. The linear PCA transformation seems to be sufficient to efficiently reduce dimensionality of the silhouette spaces, however it is still possible that some other nonlinear techniques would outperform it. What is more the proposed method is going to be verified on the basis of chosen challenge gesture database with greater number of instances and classes, as for instance Chalearn (http://gesture.chalearn.org).

Acknowledgments. The work was supported by The Polish National Science Centre and The Polish National Centre of Research and Development on the basis of decision number DEC-2011/01/B/ST6/06988 and project UOD-DEM-1-183/001.

References

1. Bautista, M.Á., Hernández-Vela, A., Ponce, V., Perez-Sala, X., Baró, X., Pujol, O., Angulo, C., Escalera, S.: Probability-based dynamic time warping for gesture recognition on RGB-D data. In: Jiang, X., Bellon, O.R.P., Goldgof, D., Oishi, T. (eds.) WDIA 2012. LNCS, vol. 7854, pp. 126–135. Springer, Heidelberg (2013)
2. Bhattacharya, S., Czejdo, B., Perez, N.: Gesture classification with machine learning using kinect sensor data. In: Third International Conference on Emerging Applications of Information Technology, Kolkata, pp. 348–351 (2012)
3. Biwas, K.K., Basu, S.K.: Gesture recognition using microsoft kinect. In: 5th International Conference on Automation, Robotics and Applications, Wellington, pp. 100–103 (2011)
4. Doliotis, P., Stefan, A., McMurrough, C., Eckhard, D., Athitsos, V.: Comparing gesture recognition accuracy using color and depth information. In: Betke, M., Maglogiannis, I., Pantziou, G.E. (eds.) PETRA, p. 20. ACM (2011)
5. Gu, Y., Do, H., Ou, Y., Sheng, W.: Human gesture recognition through a kinect sensor. In: IEEE International Conference on Robotics and Biometrics, Guangzhou, pp. 1379–1385 (2012)
6. Mitra, S., Acharya, T.: Gesture recognition: A survey. IEEE Transactions on Systems, Man, and Cybernetics - part C: Applications and Reviews 37, 311–324 (2007)
7. Patsadu, O., Nukoolkit, C., Watanapa, B.: Human gesture recognition using kinect camera. In: Ninth International Joint Conference on Computer Science and Software Engineering (JCSSE), Bangkok, pp. 28–32 (2012)
8. Ren, Z., Meng, J., Yuan, J., Zhang, Z.: Robust hand gesture recognition with kinect sensor. In: 19th ACM International Conference on Multimedia, New York, pp. 759–760 (2011)
9. Sakoe, H., Chuba, S.: Dynamic Programming Algorithm Optimization for Spoken Word Recognition. IEEE Transactions on Acoustics, Speech and Signal Processing 8, 43–49 (1978)
10. Switonski, A., Joinski, H., Zghidi, H., Wojciechowski, K.: Selection of pose configuration parameters of motion capture data based on dynamic time warping. In: 12th International Conference of Numerical Analysis and Applied Mathematics, Rodos (2014)
11. Świtoński, A., Josiński, H., Jędrasiak, K., Polański, A., Wojciechowski, K.: Classification of poses and movement phases. In: Bolc, L., Tadeusiewicz, R., Chmielewski, L.J., Wojciechowski, K. (eds.) ICCVG 2010, Part I. LNCS, vol. 6374, pp. 193–200. Springer, Heidelberg (2010)
12. Switonski, A., Joisski, H., Jedrasiak, K., Polanski, A., Wojciechowski, K.: Diagnosis of the motion pathologies based on a reduced kinematical data of a gait. Electrical Review 57, 173–176 (2011)
13. Wang, Y., Yang, C., Wu, X., Xu, S., Li, H.: Kinect based dynamic hand gesture recognition algorithm research. In: 4th International Conference on Intelligent Human-Machine Systems and Cybernetics, Nanchang, pp. 274–279 (2012)
14. Witten, I., Frank, E., Hall, M.: Practical Machine Learning Tools and Techniques. Morgan Kaufmann, Boston, Massachusetts (2011)

Segmentation of MRI Data to Extract the Blood Vessels Based on Fuzzy Thresholding

Jan Kubicek[1], Marek Penhaker[1], Karolina Pavelova[1], Ali Selamat[2],
Radovan Hudak[3], and Jaroslav Majernik[4]

[1] VSB–Technical University of Ostrava, FEECS, K450
17. listopadu 15, 708 33, Ostrava–Poruba, Czech Republic
{jan.kubicek,marek.penhaker,karolina.pavelova.st}@vsb.cz
[2] UTM-IRDA Center of Excellence, UTM and and Faculty of Computing, University
Teknologi Malaysia, 81310 UTM Johor Bahru, Johor, Malaysia
aselamat@fsksm.utm.my
[3] Department of Instrumental and Biomedical Engineering,
Faculty of Mechanical Engineering, Technical University of Košice,
Letná 9, Košice, 042 00, Slovakia
radovan.hudak@tuke.sk
[4] Department of Medical Informatics, Faculty of Medicine,
Pavol Jozef Šafarik University in Košice, Slovakia
jaroslav.majernik@upjs.sk

Abstract. The article discusses the design of appropriate methodology of segmentation and visualization of MRI data to extract the blood vessels. The main objective of the proposed algorithm is effective separation individual vessels and adjacent structures. In clinical practice, it is necessary to assess the progress of the blood vessels in order to assess the condition of the vascular system. For physician who performs diagnosis is much more rewarding to perform analysis of an image that contains only vascular elements. The proposed method of image segmentation can effectively separate the individual blood vessels from surrounding tissue structures. The output of this analysis is the color coding of the input image data to distinguish contrasting behavior of individual vessels that are at the forefront of our concerns, the structures that we need in the picture.

Keywords: Image segmentation, MRA, Fuzzy tresholding, Color mapping.

1 Introduction

Current methods for noninvasive imaging of the vascular system include Doppler ultrasonography, CT angiography (CTA) and MR angiography (MRA). MR angiography provides examination with absence of risks that are associated with exposure to ionizing radiation. The main disadvantages of this diagnostic method may include higher cost of examinations and difficult accessibility. A major problem is the inability perform examination of people with pacemakers, implanted defibrillators, because these devices are not exclusively from non-magnetic material. MRA offers

© Springer International Publishing Switzerland 2015   43
D. Barbucha et al. (eds.), *New Trends in Intelligent Information and Database Systems,*
Studies in Computational Intelligence 598, DOI: 10.1007/978-3-319-16211-9_5

contrasting view of blood vessels without contrast fluid. Currently native MRA examination is intended for intracranial arterial and venous circulation. Currently, it is one of the most elaborate techniques, which gives very good results [6, 7].

2   MR Angiography (MRA)

Method angiography allows examination of the blood vessels by imaging methods. This method is normally considered non-invasive examination method. This method allows perform examination of different body parts (e.g.. Limbs or brain). This method can be divided into angiography with administration of contrast fluid and without its use [10].

This contrast fluid is to enhance the contrast of the individual structures. For example it is possible to highlight pathology, or dynamic processes. These contrast fluids normally contain paramagnetic chelates. Subsequent contrast-enhanced examination is appropriately timed to contrast agent was present mainly in the arterial blood stream, or in the venous system [11].

The resulting MRA image is given by subtraction image without contrast and image with contrast. Non-contrast MRA is used to display the bloodstream is only flowing blood in the investigated area. Signal of surrounding stationary tissues is intentionally suppressed [8, 9, 10, 11].

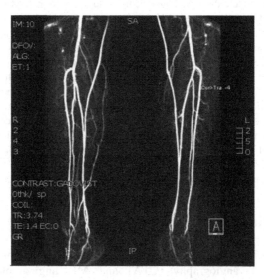

Fig. 1. MRA of lower limbs [9]

3   The Proposed Segmentation Algorithm

Segmentation methods play a key role in the processing of medical imaging data. We often solve the task of extracting parts of the image that exhibit specific properties. Image data from the MRA are represented by gray-levels.

Detection of blood vessels was solved by more approaches, in particular on the basis of edge operators. With edge detection is possible to find various objects in the image, but it is difficult to identify them. Significant benefit of the proposed solution is recognition of individual image elements based on the color spectrum and clearly distinguish the image structures.

The main objective of our analysis is the separation of gray levels in the output sets, so in order to create a contrasting color map, which represents the individual tissue structures. For this purpose fuzzy classifier was used, which has sufficient sensitivity to even slight contrasting shades of gray were distinguished. The histogram of the input image is divided into K different regions, which is the first key stage segmentation procedure, because there we specify what the difference image, we are able to separate from each other. We assume that we have a number of recognizable areas:

$$k = \{n\}_{n=1}^{K} \tag{1}$$

Subsequently, histogram normalization is solved in order to eliminate outliers and scaling of the image. The resulting histogram is restricted to the interval [0, 1].

$$\bar{h}(i) \rightarrow [0;1] \tag{2}$$

Number of output areas is determined by the number of distinguishable peaks of the histogram. The number of local extremes (N_{max}) can be defined by:

$$K = N_{max} \tag{3}$$

The final step of data preprocessing is image smoothing of lowpass filter, which is used to remove unwanted high frequency noise [3].

For the classification of each pixel in the output sets is used fuzzy logic rules. There it is necessary to first define the membership function for each output region R_n.

For the n - th classifier is used expression $\mu_n(f(x))$. There are many kinds of membership functions, but not all are suitable classifiers. The optimum properties have function of triangular shape (TS). Based on these facts, we have used for our algorithm pseudo trapeziodal shape membership function (PTS).

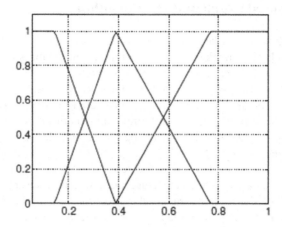

Fig. 2. The shape of the PTS function [4]

Individual blood vessels is segmented in successive iteration steps. The first step of the iteration is defined by:

$$S = \max_n \mu_n(f(x)) \tag{4}$$

The output of this procedure is to determine the maximum individual classification sets. This output is often insufficient, because the image data are suppressed noise that is transmitted to the output of segmentation, which degrades the quality of the entire segmentation process. For this reason, we have to evaluate the competence of pixels intended use of the operator, which is invariant to noise. For each channel of segmentation, we used the median, which is invariant to noise. The output of the segmentation procedure using the median is defined as [1], [3], [4]:

$$S = med(\mu_n(f(x))) \tag{5}$$

3.1   Structure of the Proposed Algorithm

The structure of the proposed algorithm segmentation of MRA data to extract the blood vessels can be summarized in three basic steps:

Data preprocessing - there is performed histogram normalization and the definition of each output regions. The number of regions is determined by the number of distinguishable peaks of the histogram. This phase determines how deep we are able to perform the final segmentation and how many objects we are able to distinguish from the input data.

Process of Filtration – for filtration is used low-pass filter. This phase is in the structure of the proposed methodology optional. It is used especially when the image data are suppressed by high frequency noise. Noise has a negative impact on the quality of the segmented images, after removing noise we are able to achieve a smooth contour segmentation.

The process of segmentation - in this key phase of segmentation individual pixels are grouped into output sets on the basis of their properties using the membership function. The aim is the separation of shades gray that represent the blood vessels and the removal of remaining tissues that belong to the background of images.

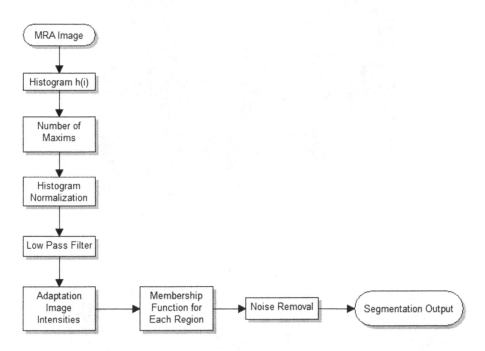

Fig. 3. The structure of the proposed segmentation procedure

4   Testing of the Proposed Algorithm

For testing of the algorithm, we used a series of MRA images 60 patients. The main requirement is approximation of vessels by the color model that separates the bloodstream from the background and subsequent filtering to display only vascular structures. Benefit of the proposed solution is the choice of output sets, which allows in-depth analysis even tissues that are poorly recognized. In the first part of the analysis we performed mapping of carotid arteries in order to detect the main trunk of the carotid artery.

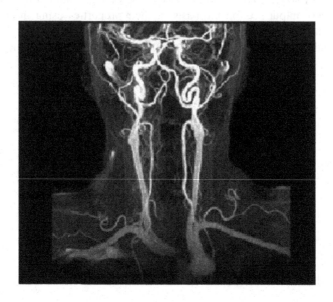

Fig. 4. MRA image of carotid arteries

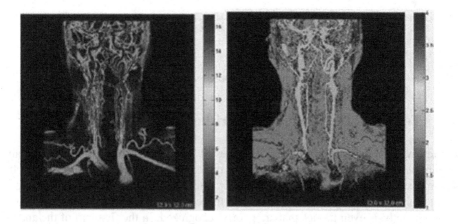

Fig. 5. Evaluation of the number output sets - 18 (maximum) on the left and 4 on the right

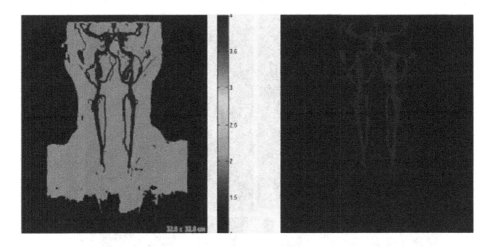

Fig. 6. Filtered image of carotid arteries -18 sets (left) and thresholded image (right)

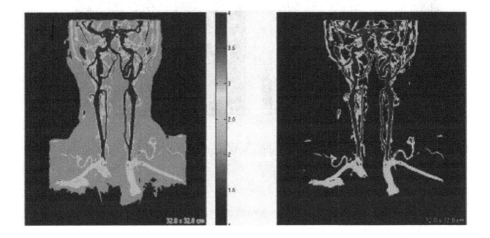

Fig. 7. Filtered image of carotid arteries – 4 sets (left) and tresholded image (right)

We used two segmentation outputs for 4 and 18 output sets. From this view, it appears that it is not always desirable to select the maximum number of output sets as the optimal solution. When comparing the number of output regions can be seen that the filter 18 suppresses the set portion of the structure that includes the blood stream and provides to the physician significant diagnostic information.

In the following part of the analysis, we focused on MRA sequences that generate areas of tibia and pelvis. There is the same requirement, and that separation of the bloodstream from the background.

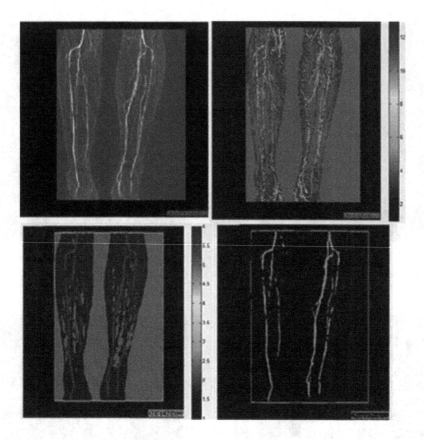

Fig. 8. MRA image of tibia (top left), 13 output sets (top right), filtered image (bottom left) and tresholded image (bottom right)

When comparing the images of the carotid arteries and tibia is at first sight that the tibia images are suppressed by more noise. This negative image component is transferred to the segmentation result and acts as an artifact. These image series have lower resolution, which is reflected on the resulting color map that is not fluent in some places.

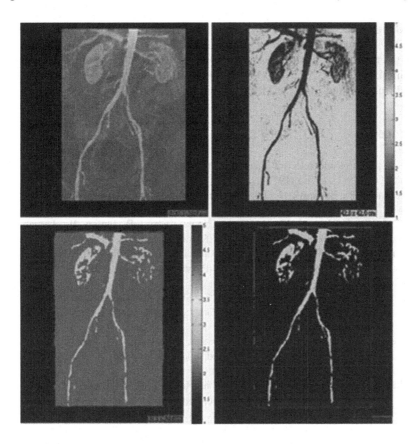

Fig. 9. MRA image of the pelvic bottom (top left), 5 output sets (top right), filtered image (bottom left), thresholded image (bottom right)

5   Conclusion

The main goal, which is described in the article is the development of a segmentation method that is able to reliably separate the blood vessels from the other structures. The proposed methodology is developed primarily for image data from MRA. The main requirement is only the detection of blood vessels, without the adjacent structures. This task is relatively easily implemented, because segmentation based on fuzzy logic is able to sharply demarcate the different tissue structures and choice of appropriate thresholding can be displayed only vessels that are at the forefront of our concerns. Another significant advantage is the high sensitivity of the algorithm even in places where there is only small changes in contrast. In a future time, we would like to focus on the calculation torturozity of vascular system from segmented data.

This parameter is very important because we get information about the curvature of the vessels, which corresponds with a vascular condition.

Acknowledgment. The work and the contributions were supported by the project SP2015/179 'Biomedicínské inženýrské systémy XI', and This paper has been elaborated in the framework of the project „Support research and development in the Moravian-Silesian Region 2013 DT 1 - International research teams" (RRC/05/2013). Financed from the budget of the Moravian-Silesian Region.

References

1. Otsu, N.: A threshold selection method from gray-scale histogram. IEEE Trans. on Sys., Man and Cyb. 9(1), 62–66 (1979)
2. Szczepaniak, P., Kacprzyk, J.: Fuzzy systems in medicine. Physica-Verlag, New York (2000)
3. Ville, D.V.D., Nachtegael, M., Der Weken, D.V., Kerre, E.E., Philips, W., Lemahieu, I.: Noisereduction by fuzzy image filtering. IEEE Trans. Fuzzy Sys. 11(4) (2003)
4. Fernández, S., et al.: Soft tresholding for medical image segmentation. IEEE EMBS (2010)
5. Bezdek, J.C., Pal, S.K.: Fuzzy Models for Pattern Recognition. IEEE Press, New York (1992)
6. Falcao, A.X., Udupa, J.K., Samarasekera, S., Sharma, S.: User-steered image segmentation paradigms: live wire and live lane. Graphical Models Image Process 60, 233–260 (1998)
7. Eckstein, F., Tieschky, M., Faber, S., Englmeier, K.H., Reiser, M.: Functional analysis of articular cartilage deformation, recovery, and fluid flow following dynamic exercise in vivo, pp. 419–424. Anat Embryol, Berl (1999)
8. McWalter, E.J., Wirth, W., Siebert, M., Eisenhart-Rothe, R.M., Hudelmaier, M., Wilson, D.R., et al.: Use of novel interactive input devices for segmentation of articular cartilage from magnetic resonance images, Osteoarthritis Cartilage, pp. 48–53 (2005)
9. Graichen, H., Al Shamari, D., Hinterwimmer, S., Eisenhart-Rothe, R., Vogl, T., Eckstein, F.: Accuracy of quantitative magnetic resonance imaging in the detection of ex vivo focal cartilage defects, Ann Rheum Dis, pp. 1120–1125 (2005)
10. Schmid, M., Conforto, S., Camomilla, V., Cappozzo, A., Alessio, T.D.: The sensitivity of posturographic parameters to acquisition settings. Medical Engineering & Physics 24(9), 623–631 (2002)
11. Severini, G., Conforto, S., Schmid, M., Alessio, T.: D': Novel formulation of a double threshold algorithm for the estimation of muscle activation intervals designed for variable SNR environments. Journal of Electromyography and Kinesiology 22(6), 878–885 (2012)

Precision of Gait Indices Approximation by Kinect Based Motion Acquisition

Agnieszka Michalczuk[1], Damian Pęszor[1], Henryk Josiński[2], Adam Świtoński[2], Romualda Mucha[3], and Konrad Wojciechowski[1,2]

[1] Polish-Japanese Academy of Information Technology,
Aleja Legionów 2, 41-902 Bytom, Poland
{amichalczuk,dpeszor,kwojciechowski}@pjwstk.edu.pl
[2] Silesian University of Technology, Institute of Informatics,
Akademicka 16, 44-100 Gliwice, Poland
{Henryk.Josinski,Adam.Switonski,Konrad.Wojciechowski}@polsl.pl
[3] Medical University of Silesia, Batorego 15, 41-902 Bytom, Poland
romam28@wp.pl

Abstract. Step length constitutes one of the important gait indices. The research described in the present paper was focused on the method of determination of the step length by means of the Kinect device. Gait sequences recorded in the Human Motion Laboratory of the Polish-Japanese Academy of Information Technology using the Vicon system played the role of the reference data. The group of six subjects participated in the experiments. Conclusions from the comparative analysis of the results of both approaches summarize the paper.

Keywords: step indices, motion capture, Kinect, gait.

1 Introduction

The gait is one of the fundamental elements of human motor skills. It depends on age, gender, coordination predisposition and construction of the human body. Exercise of motor skills of the human body leads to increase mobility, which is fundamen-tal for athletes and for people with the musculoskeletal system disorders. Mobility improvement is often associated with supervision specialist who controls the correct course of exercises. Recording enables data analysis of their correctness and error correction. One of the most effective data recording method is motion capture technique. It is based on the cameras working in the near infrared light. Cameras trace the cloud of points defining the position of an actor in the 3D space by recording the light reflected from the markers. Markers are placed in the anthropometric points on the human body. Cameras operate in high frequencies (e.g. 100 Hz) and in high resolution. In order to correctly identify a single marker it needs to be visible by two cameras at least. Thus, system efficiency depends on the number of cameras. In addition, system is designed for permanent installation in the premises, since its portability is troublesome.

© Springer International Publishing Switzerland 2015
D. Barbucha et al. (eds.), *New Trends in Intelligent Information and Database Systems*,
Studies in Computational Intelligence 598, DOI: 10.1007/978-3-319-16211-9_6

Another most popular method of digital recording is the use of video cameras. These cameras are relatively inexpensive and their sizes allow recording video in almost any place. Using a single camera limits the visibility of the subject because of occlusions. Another limitation is the noise in the recorded image. During recording an interesting object, the camera also records its neighborhood. This makes it difficult to extract the object from the background, especially when they are recorded in a similar color scheme.

Combining both methods of digital recording: motion capture system and video cameras constitutes motion sensor for Xbox 360 produced by Microsoft named Kinect. It comprises an emitter and a camera recording the reflected light to determine the distance of the objects in the camera range. Motion sensor is also equipped with a standard RGB camera, which allows easy acquisition of the scene in colors scheme proper for human. Although Kinect controller is quite handy device it is in fact stationary device and each position change requires new calibration.

Analysis of mobility of the human body is dependent on the method of registering the object. The technique of motion capture provides information about the location of the markers in 3D space and joint angles in the sagittal, frontal and transverse planes, among others. In [1] capture motion technique was used to determine the indices of gait for patients with Parkinson's disease. On the basis of the position of the markers indices such as the step length or wrist trajectory were determined for each patient. The second group of indices – the arm deflexion angle and the Decomposition Index which requires computation of time interval when selected joints are in motion – is based on the Euler angles. In addition to publications where the authors focus on the analysis of a single person, capture motion technique is also used to analyze a group of people. In [2] authors considered a human identification problem based on joints movement expressed by the Euler angles.

Digital recording using the video camera doesn't provide accurate output data hence their analysis is difficult. Depending on considered issue, authors put up with some inaccuracy, or they try to project objects from 2D to 3D space. In [3] the authors apply the particle swarm optimization algorithm which creates a model of human embedded in three-dimensional space by synchronization of images from four video cameras. Data processed in this way were recorded for 22 people and gave 90% success rate in their identification. For the purpose of person identification the authors used the distance between the ankles, the person's height and the angles in selected joints. Digital recording of objects using the Kinect sensor allows research of human gait identification. In [4] the authors explored the problem of classification of gender gait quality. Data were collected using the Kinect sensor. In the data preprocessing authors used tracking motion method to get a human skeleton in 3D space. Application of 3D version of the Discrete Wavelet Transform allowed obtaining 83.75% effectiveness in the problem of classification.

In [5] the authors present a method which detects phases of the gait cycle (stride): stance and swing. The authors compared the parameters obtained from the 3D virtual skeleton from the Kinect sensor and from the in-shoe pressure sensors and a gyro-scope attached to the wrist via straps. The average difference between the duration measured by the pressure sensor and the duration measured by the regression model is less than 1%.

Step length, step width and gait variability, which can provide information about mobility and balance of human, have particular importance in the gait analysis. Lack of reproducibility during gait and step length are associated with the fall risk [6] which, in particular for older people, may lead to permanent disability. Analyzing spatial gait variability parameters authors registered the biggest correlation between the step length and the level of medications. This may indicate that the step length is an important factor that reflects the loss of mobility in humans causing a fall risk. In addition, elevated serum anticholinergic activity has been associated with significant slowing of both gait speed and simple response time.

In [7] authors compared two web-cameras and two Kinects with Vicon. Estimation of the stride length was better with web-cameras than with Kinect. On the other hand, the Kinect sensors were better than video cameras in estimation of stride time and its velocity.

A Kinect sensor as a non-invasive and relatively cheap equipment which, compared to other image recording devices, could be used at home where user's behavior is natural which leads to objective and reliable data. However, data collected by a Kinect sensor may be more or less accurate. Thus, laboratory studies should identify causes of inaccurate measurements which could be, among others, location of the sensor, method of computation of the step length.

This work presents the analysis of step length parameter, based on the methods of digital recording that can be used at home. For this purpose, a Kinect sensor was used to record digital data. To verify the accuracy of recorded data, we used an additional source of recordings based on the motion capture technique.

2   Dataset

The recordings occurred in the Human Motion Laboratory of the Polish-Japanese Academy of Information Technology in Bytom [8]. The laboratory is equipped with the Vicon motion capture system based on 10 cameras operating in the near-infrared light and 4 reference cameras operating in the visible light range. In addition, we used a Kinect sensor. Actors walked on a straight line on the path of 5m length which allows performing about 8 steps by healthy people. According to the instructions on the manufacturer's website (http://support.xbox.com/), sensor was located at a height of 55 cm above the floor at the end of the path on which subjects walked. The study included six actors with not identified abnormalities in the musculoskeletal system. Actors performed four walking trials along the path.

Gait has a cyclical nature and is characterized by the repetition of events such as foot contact with the ground or forward leg movement [9], therefore it is assumed that every step should be similar to every other step. Limited area of sensor's visibility and restricted length of walking path used by actors to walk, caused that only the middle step was chosen to analyze as a gait representative. In addition, the middle step is not perturbed by the start phase or the stop phase. It should be determined separately for each side of the human body, because for the given side a step begins with the first contact of the opposite foot with the ground and ends with the first contact of the given foot with the ground.

3   Preprocessing

The data collected from the reference system (motion capture) are stored in binary .c3d files. In the preprocessing phase Java implementation based on the j3d.org library was used. The markers placed at the ankles (RHLL marker for the right ankle and LHLL marker for left ankle) have been selected for the analysis purposes. Events needed to determine the step length, that's events in which foot contact with the ground occurs, are defined as FootStrike in the Vicon system. On the basis of the coordinates of the position of the foot in 3D space the Euclidean distance was used to determine the step length.

The Kinect Studio application developed by Microsoft, which is designed for recording motion data, does not allow directly export of collected data. Therefore, based on Kinect for Windows Software Development Kit v. 1.7 an application was created, which allows the registration of movement and their export. Kinect Sensor captures image and audio data. An image contains: video, depth map and the simplified human body skeleton. Unlike video data and depth map, which record a scene, in order to record a human body skeleton the registered object must be seen for a longer time on the camera so that a sensor can detect the skeleton. As mentioned earlier, analysis of human gait should not include a start or stop phase. Therefore, the actors started and finished walk aside from a camera visibility. This distance was too large for a sensor to recognize a skeleton. For this reason, a step length was analyzed only on the basis of the depth map and video. Stride length was analyzed from the camera located across a path on which actors were walking. For this purpose, video data were synchronized with the depth map. Events representing foot contact with the ground were selected in the video data to get their coordinates. Based on collected coordinates a depth image value of pixel was read from the depth map. In the preprocessing phase the depth map has been scaled to 8-bit blue color component. Thus, distance between two points on the depth map is determined by the difference of their values for the blue channel. To determine the real distance between the points, the distance calculated from the depth map for six selected calibration marks 1 was compared with real distance between them. The change rate of the depth image value from the depth map to the real image is equal to 42.857 mm. This remark can be expressed by the following equation: $world\_depth\_difference = depth\_map\_difference * 42.857$.

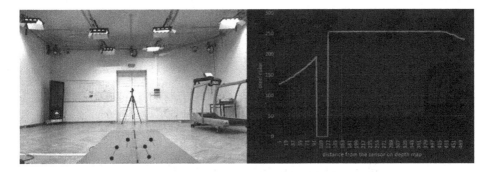

Fig. 1. Six calibration marks determining the real distance. Relationship between a depth image value and the distance.

The change of values ranges on the depth map which depends on the registered values of the calibration marks was also tested. The graph on the depth map in Fig. 1 shows linear increase of values on the depth map for a subject located farther away from the Kinect sensor. Black areas and a sharp drop in the graph illustrate the place for which the sensor has not registered the depth map.

4   Results

From recordings captured by the Kinect sensor three frames were selected. In each of them the first foot contact with the ground is visible, which is marked as a circle in Fig. 2.

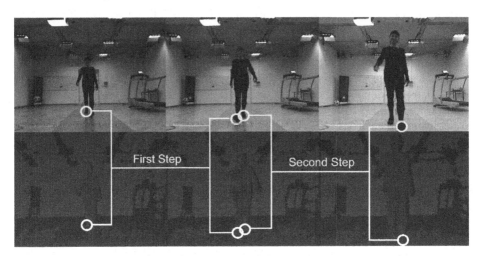

Fig. 2. Determination of steps

Table 1. Step length of the left foot

Motion capture	Depth Map				
	Most visible place	toes	ankles or heels	central part of an ankle	average
Actor1 1450,68	1532,14	1060,71	1285,71	1564,29	1342,86
Actor2 1290,61	1107,14	782,14	1060,71	1060,71	967,86
Actor3 1333,55	1333,55	964,29	1425,00	1542,86	1328,57
Actor4 1427,24	1617,86	1125,00	1596,43	1607,14	1442,86
Actor5 1330,79	1414,75	1157,14	1135,71	1467,86	1289,29
Actor6 1385,14	1574,66	1200,00	1510,71	1521,43	1410,71
correlation	0,87	0,58	0,61	0,74	0,74

Table 2. Step length of the right foot

Motion capture	Depth Map				
	Most visible place	toes	ankles or heels	central part of an ankle	average
Actor1 1397,53	1589,29	1467,86	1360,71	1639,29	1489,29
Actor2 1266,73	1178,57	1125,00	1125,00	1178,57	1142,86
Actor3 1374,53	1374,53	1617,86	1414,29	1746,43	1592,86
Actor4 1497,95	1628,57	1832,14	1553,57	1917,86	1767,86
Actor5 1364,17	1514,69	1660,71	1767,86	1778,57	1735,71
Actor6 1364,59	1634,62	1596,43	1028,57	1778,57	1467,86
correlation	0,77	0,87	0,48	0,84	0,81

Identification of a single step requires use of depth values from two images. It is worthwhile to mention that presented images are reflections of the real images. Hence, the first step constitutes a step made by the left foot and the second one – a step made by the right foot.

Since the analysis is based on images from one sensor with a resolution of 640x480, where depending on a foot orientation only its part is visible, the depth values were sampled from different locations. In the first experiment, depth map values were taken from the places chosen subjectively as locations where the foot is most visible on the image. Experiment was repeated three times for each actors walking.

In the second experiment values were collected from characteristic locations such as toes, ankles or heels (depending on the actor's gait) and the central part of an ankle. In the last experiment, the values taken from characteristic locations were averaged.

The step lengths (in millimeters) computed for each of the above described experiments are shown for a left and right foot in Table 1 and Table 2, respectively. Step length was averaged for each actor over his four walks.

The correlation between depth map data and data obtained using the motion capture system was examined for each variant of the depth map data collection method. For the left side the highest correlation 0.86 was obtained for data collected from the most visible places on human foot. In contrast, the lowest correlation 0.58 was related to the data collected from the toes.

For the right side the highest correlation 0.86 was obtained for data collected from the toes. The lowest correlation 0.47 was related to the data collected from both the ankle and the heel.

5 Conclusion

The Kinect sensor is designed to record movements in a restricted area of the scene. Thus, the research becomes difficult when the subject goes beyond this area. Nevertheless, analysis of the step length collected using the Kinect sensor revealed high correlation of 86% between recordings of different actors and recordings using motion capture technology. The relative difference of step length between two discussed approaches of movement recording averaged over all actors is equal to 13%. The method of retrieving data from three places of the feet obtained quite high correlation for both left and right side. Thus, the use of additional calibration marks could increase correlation, at the same time improving the ability to identify unnatural steps. Perhaps the more accurate results could be obtained using the newest version of the Kinect sensor because according to the producer the device should be charac-terized by higher fidelity depth map.

Acknowledgments. This project has been supported by the National Centre for Research and Development, Poland.(UOD-DEM-1-183/001 "Intelligent video analysis system for behavior and event recognition in surveillance networks").

References

1. Stawarz, M., Kwiek, S., Polański, A., Janik, Ł, Boczarska-Jedynak, M., Przybyszewski, A., Wojciechowski, K.: Algorithms for Computing Indexes of Neurological Gait Abnormalities in Patients after DBS Surgery for Parkinson Disease Based on Motion Capture Data. Machine Graphics and Vision 20(3), 299–317 (2011)
2. Świto'nski, A., Josi'nski, H., Michalczuk, A., Pruszowski, P., Wojciechowski, K.: Feature Selection of Motion Capture Data in Gait Identification Challenge Problem. ACIIDS, pp. 535–544 (2014)
3. Krzeszowski, T., Michalczuk, A., Kwolek, B., Świtoński, A., Josiński, H.: Gait recognition based on marker-less 3D motion capture. AVSS 2013, 232–237 (2013)
4. Arai, K., Asmara, R.A.: Human Gait Gender Classification using 3D Discrete Wavelet Transform Feature Extraction. International Journal of Advanced Re-search in Artificial Intelligence 3(2), 12–17 (2014)
5. Gabel, M., Gilad-Bachrach, R., Renshaw, E.,, S.: Full body gait analysis with Kinect Engineering in Medicine and Biology Society (EMBC). In: 2012 Annual International Conference of the IEEE, pp. 1964–1967 (2012)

6. Nebes, R.D., Pollock, B.G., Halligan, E.M., Kirshner, M.A., Houck, P.R.: Serum Anti-cholinergic Activity and Motor Performance in Elderly Persons. The Journals of Gerontology. Series A 62, 83–85 (2007)
7. Stone, E., Skubic, M.: Passive. In-Home Measurement of Stride-to-Stride Gait Variability Comparing Vision and Kinect Sensing. In: 33rd Annual International Conference of the IEEE EMBS, IEEE 2011, pp. 6491–6494 (2011)
8. Website of the Human Motion Laboratory of the Polish-Japanese Academy of Information Technology: http://hml.pjwstk.edu.pl/
9. Barlett, R.: Introduction to sports biomechanics, Analyzing Human Movement Patterns. Taylor & Francis Group (2007)

Computer User Verification
Based on Mouse Activity Analysis

Tomasz Wesolowski, Malgorzata Palys, and Przemyslaw Kudlacik

University of Silesia, Institute of Computer Science,
Bedzinska 39, 41-200 Sosnowiec, Poland
{tomasz.wesolowski,malgorzata.palys,
przemyslaw.kudlacik}@us.edu.pl

Abstract. The article concerns behavioral biometrics, where issues related to computer user verification based on analysis of the mouse activity in computer system are particularly studied. The work is devoted to the analysis of user activity in environments with a graphical user interface (GUI). The set of analyzed features is extended by introducing new features to characterize the mouse activity. A new method of aggregating the mouse activity basic events into a higher level events is presented. Additionally, an attempt to intrusion detection based on the k-NN classifier and various distance metrics is performed.

This article presents the preliminary research and conclusions.

Keywords: behavioral biometrics, mouse activity, user verification.

1 Introduction

The computers are present in almost every aspect of our life and the security of the computer systems and data became a crucial task. People use computers to perform calculations, manage finances, support health care [10,17] and rescue services or simply for entertainment. Institutions often process and store confidential information. Among them financial documents, projects documentations or personal and sensitive data of the users. Therefore, it is necessary to take appropriate security measures for the IT systems that will allow very high level of access control and only the authorized individuals should gain access to the specific resources. In order to allow such a security measures it is necessary to confirm the identity of the user. To increase the security level advanced methods are expected, e.g. biometric methods that identify users by their individual physical (face, fingerprint) or behavioral (signature [9], walking style) characteristics. Also using the peripherals of the computer (keyboard, mouse) is considered to be the human behavioral characteristics. Unfortunately the analysis of the keyboard use as a continuous process is very difficult due to the fear that the personal data (passwords, PIN codes) could leak or get stolen. This is why mouse activity, as more safe for the user, should be considered separately. Another important reason for analyzing mouse activity is that this approach is not involved with any additional costs (no sensors or devices are necessary to

© Springer International Publishing Switzerland 2015   61
D. Barbucha et al. (eds.), *New Trends in Intelligent Information and Database Systems*,
Studies in Computational Intelligence 598, DOI: 10.1007/978-3-319-16211-9_7

collect the data). The data acquisition methods used by the authors in these studies could be easily implemented in the security software like the host based intrusion detection system (HIDS).

When using a computer and an operating system (OS) with the graphical user interface (GUI) a user almost continuously works with computer mouse. This is how an individual style of interaction is developed. The assumption for using the mouse activity in biometrics is that this individual style could be unique for a user and could characterize this user in order to perform a user verification. Such an interaction could be very difficult to falsify because it is developed in a long period of time and depends on individual personal features of a user. At the time the mouse movements are already used in user authorization (image passwords). Unfortunately there is a major issue connected to these authorization methods - they are one-time operations usually performed at the beginning of work. The one-time authorization does not guarantee that an intruder will not overtake the access to the system where the authorized user is already logged in - this can happen when the authorized user leaves for some time the working place after logging in into the system. Such an intruder is called a masquerader and usually it is an insider. Within different kinds of cyber-attacks [12] masqueraders constitute the greatest threat [13]. To ensure a high level of security the activity of the user should be continuously monitored and the working person should be verified. Such a monitoring is performed by the IDS. Users' profiles are created and by means of various methods the reference profiles are compared with the users' activity data. If such a comparison is made on the fly (in real time) it is an on-line IDS (dynamic analysis methods are used) and if it's made after some time the IDS works off-line (static analysis methods are used).

The user profiling and intrusion detection method presented in this article were developed as dynamic to work with an on-line IDS to monitor the activity and verify the user in a real time. The presented approach represents the preliminary studies. The studies apply for the OS with GUI and the activity data is collected on the OS level so all the computer programs with GUI are covered (also web browsers, working with web pages and social media).

2   Related Works

The idea of user verification based on the activity in a computer system is well known in behavioral biometrics. At first, the interacting with the computer system was based on giving text commands. These commands were recorded and analyzed in order to create an individual user profile [14,15]. Another approach consisted of the direct use of the keyboard analysis [3]. Along with the development of computer systems due to the increasing computing power the way of human-computer interaction also has evolved. Nowadays, mostly OSs with GUI are used and some solutions combine the mouse and keyboard activity [1,16].

This paper is focused on using the information on computer mouse activity only. Some of the methods are developed to assist during a one-time logging in authorization. In [5] the mouse activity data set was acquired through a game placed on the internet website. The players were assigned individual id numbers and the game was collecting the information on mouse cursor position. As a classifier a Sequential Forward Selection algorithm was used. The methods for continuous analysis of the activity are more difficult to develop but they can be used in an on-line IDSs. The entertainment business has a big influence in this area because of the need to look for new methods to secure the computer games - especially the Massively Multiplayer Online games. These games are a part of a complex economical and financial mechanism. The personal and financial (credit card numbers) data of the users are stored. The analysis of mouse activity in games in order to identify the users was made in [7,8]. In [7] the trajectory of the mouse cursor was analyzed by using the signature verification methods, SVN and k-NN classification. Another solution is based on the separation of mouse events (move, click) and their subsequent analysis [11]. The distance, angle and speed of movement of the cursor was calculated and then the average values were determined. Gross deviations from the calculated average values consisted anomalies that were classified as the presence of an intruder. The analysis of continuous user activity was performed in [2] where the dynamics of mouse use based on the moves characteristics is presented. The activity data was collected during users' everyday activity. Basing on the data features were extracted and histograms characterizing individual users were depicted. Based on [2] further improvement was introduced in [4] where some additional features were analyzed and the Equal Error Rate (EER) was lowered in average of 1.01%. Other approach was based on calculating the vector of characteristic features [6]. The calculated vector includes information about the average number of the left and right mouse button clicks and a thorough analysis of the trajectory of the cursor. The effectiveness in verifying the identity of the person was around 80%.

In this paper the set of features is extended by proposing new features. Also a new way of dividing the mouse activity into a higher level events is presented.

3   Testing Data Set

The data set used in this studies was collected by the dedicated software. The software works in the background of the OS registering user activity at every moment of his work. Despite this, the application is unnoticeable to the user and does not affect the comfort of his work. The users were of various age, occupation and level of computer use experience. Each user was working on his/her own personal computer. The software records activity saving it in files compatible with the CSV format. The first line starts with a token RES followed by the resolution of the screen. Starting with the second line the activity data is stored and it consists of mouse and keyboard events. The software recorded even 60000 events per hour. Due to the security reasons the key codes and window titles are encoded. Each line starts with a prefix identifying the type of event. The prefix is always followed by the timestamp and optionally additional data related to the event. The prefixes for mouse activity events are: M - mouse move, L/l - left mouse button down/up, R/r - right mouse button down/up, S - mouse scroll.

Each mouse event consists of a vector [*prefix, timestamp, x, y*] where x and y are the coordinates of the mouse. The negative values of coordinates are possible when working with multiple screens. Despite the fact that the keyboard codes and window names are encoded the data acquisition process is very difficult as most of the users are not willing to voluntarily participate due to the fear of the personal data leaks. Initially the data from 10 users were collected but the software is still running to extend the testing data set and collect more data.

3.1   Data Pre-processing – Higher Level Event

As the acquired data set consists of activity data connected to mouse, keyboard use and activity in particular software/windows the pre-processing is necessary. The first step is to filter the data set in order to acquire only the mouse basic events (moves and button clicks). Next step is to extract such a data subsets that are connected to a mouse activity from which the features could be extracted.

Fig. 1. An example of mouse cursor path after an hour of work

After working for a longer period of time the mouse activity data have a very high number of basic events (Fig. 1). Extracting the features from such a volume of data is complex and time consuming. Therefore it is necessary to divide the mouse activity into sequences of basic events that constitute the *higher level event* (*HL-Event*) and then analyze the HL-Events separately. For the purpose of data analysis in this studies the HL-Event consists of all basic events that were recorded between two consecutive clicks of the mouse button (Fig. 2b). Because the activity in the close neighborhood of the click event and the click event itself could be important also these activities are analyzed separately.

There is some issues related to data pre-processing. The first problem is that the data recording software saves a double-click event as two single clicks, therefore it is necessary to extract double-click events during the pre-processing. Two single clicks are interpreted as a double-click when the time difference between them was less than 0.5 second and the distance less than 5 pixels. Next the double-click is interpreted as the border event for HL-Event. Another issue is related to Drag-and-Drop operations (DaD). The DaD consists of mouse button down event followed by one or more mouse move events and ended with mouse button up event. It is similar to the regular HL-Event where the border events are mouse clicks. This is why it is interpreted as HL-Event at the end of which instead of the mouse click the mouse button up event is registered. Next problem is the mouse inactivity - when the mouse cursor stops at the particular place for some time. The situation like this takes place for example when the user is: waiting for the software, reading or typing something on the keyboard. Such a situation has a negative influence on the features describing dynamics of the move. To eliminate the disadvantage the basic events till the mouse inactivity point are ignored and the basic events between inactivity and next mouse click constitute the HL-Event.

4   Mouse Activity Features

Every mouse move basic event is characterized by the vector of 3 values $[t, x, y]$ consisting of the horizontal (x) and vertical (y) coordinates of the mouse cursor in the time t. The HL-Event consists of $n + 1$ basic events and hence is a set of $n + 1$ such vectors. The value of n represents at the same time the number of points of the trajectory - without the initial point number 0. Therefore, the HL-Event is a set of $n + 1$ points and these points define n Euclidean vectors with a specified direction, initial point and length. The length of the HL-Event is a sum of n Euclidean vector lengths. Basing on these assumptions the set of mouse activity characteristics can be designated. The characteristics can be divided into following groups.

The first feature is velocity. The velocity at the first point of the HL-Event is zero $(v_0 = 0)$. The analyzed sequence of following n points allow to calculate the n successive values of local velocities $V = \{v_0, v_1, v_2, \ldots, v_n\}$. For each HL-Event *the maximal velocity*, the *arithmetic mean* and *median* of velocity were determined together with the *average horizontal* and *vertical velocities*. Additional features connected to velocity are: *jerk* - the mouse cursor acceleration at the start of the movement, *braking* - after determining the point where the velocity had a maximal value it is possible to calculate the braking as a ratio of braking distance to the entire distance traveled by the cursor.

Additionally the areas in the close neighborhood of the mouse click event (click neighborhood CN) were analyzed. CN is a square shaped area of side 100 pixels and the center at the mouse click coordinates. The example of activity in CN is presented in (Fig. 2a) - the big dot represents the mouse click event and the small dots coordinates of mouse move events. The left or right mouse button clicks are not distinguished, all basic mouse events in the CN are considered - both before and after the click. First the times were calculated: *waiting time* - the difference between the time when the cursor got to the click coordinates and the click event itself, *hold time* - time between button down and up events - to eliminate high values during DaD actions the value is zero, *reaction time* - time of mouse inactivity at click coordinates - in case of double-click event its counted from second click event. *Correction length* - describes how much too far did the mouse cursor move behind the click coordinates and then came back. It was assumed that the correction length is connected to velocity - higher the velocity of the cursor longer the expected correction length. This relation is described by *correction-velocity ratio* calculated as a ratio of the velocity before the click to the correction length. The last feature in this group is the *approaching* feature describing how does the velocity deceleration look like when approaching to the click coordinates.

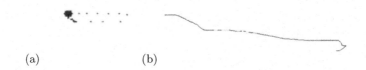

(a)   (b)

Fig. 2. Exemplary CN (a) and the trajectory of the HL-Event (b)

The basic feature is the *length* of the trajectory - the real distance traveled by the cursor between the border points. The cursor trajectory between the mouse button click events is an irregular line rarely covering the straight line passing through these points. The cursor trajectory features describe how the trajectory differs from the straight line: *filling* - the area between the trajectory and straight line in pixels, *filling ratio* - calculated by dividing the filling by the length, *filling-distance ratio* - the ratio of filling to the shortest distance between the border points, *deviation from the line* - standard deviation of the distances between the trajectory and the shortest distance, *maximum curvature* - the maximal change of the direction of the cursor.

The 3 features defined in [4] were used: trajectory center of mass *TCM*, scattering coefficient *SC* and velocity curvature *VCrv*. Additionally the *centre of mass coefficient* was defined as a difference between the TCM and the centre of mass of the straight line, together with the two features describing how much the trajectory differs from the straight line: *trajectory correctness* - calculated as ratio of length to the shortest distance between the border points and *trajectory excess* - the difference between the length and the shortest distance divided by the shortest distance between the border points.

5   User Profiling and Verification

In order to classify a user working with the computer system (as legitimate or non legitimate) it is necessary to designate user's profile. For classification method using k-NN classifier a profile should consist of reference samples that characterize the class. Having a profile of activity it is possible to compare data samples with the profile to determine class membership. In case of the presented approach, having the profile of an authorized user, a comparison of currently working user's activity samples with authorized user's reference profile is performed. To create a profile a set of training data consisting of 100 HL-Events of the authorized user is used - the number of HL-Events was determined experimentally. In this study the profile was defined as the vector of the average values of all the features extracted from the training set calculated as the arithmetic mean of the feature values.

5.1 Profile Verification

There are two types of classification: classification by determining the similarity of the sample to the class representative (hereinafter referred to as verification), and classification by defining classes based on the defining properties. The presented approach uses classification method to verify the user so the first type of classification is performed. The verification methods determine the similarity of the samples to previously designated reference profile - a representative of the class. A high rate of similarity proves the consistency of data. With this approach it is easy to analyze the data in order to detect anomalies, which are characterized by a smaller value of the similarity to the profile. The key issue is to establish the threshold determining whether the sample and profile belong to the same class. The result of the method may take the form of binary information: compliance or lack of compliance with the model.

The approach introduced in this paper is based on the k-NN classifier as a verification method. Since the studies relate to the process of verification and not classification, it is necessary to modify the k-NN algorithm - its role was limited to the determination of the similarity between the profile and the subsequent events of the tested set. The k-NN algorithm generally calculates the distance or similarity between the two objects. To determine the similarity between the k features of the tested set (A) and the profile (B) the following distance measures were used: *Euclidean Distance d_E* (1), *City Block d_{CB}* - also called Manhattan Distance (2), *Canberra Distance d_C* - a modified version of City Block (3) and *Bray Curtis Distance d_{BC}* - the result has a normalized form and therefore the result values are in the range of $< 0, 1 >$, where zero indicates that objects are identical (4).

$$d_E(A, B) = \sqrt{\sum_{i=1}^{k}(x_{Ai} - x_{Bi})^2} \tag{1}$$

$$d_{CB}(A, B) = \sum_{i=1}^{k}|x_{Ai} - x_{Bi}| \tag{2}$$

$$d_C(A, B) = \sum_{i=1}^{k}\frac{|x_{Ai} - x_{Bi}|}{x_{Ai} + x_{Bi}} \tag{3}$$

$$d_{BC}(A, B) = \frac{\sum_{i=1}^{k}|x_{Ai} - x_{Bi}|}{\sum_{i=1}^{k}(x_{Ai} + x_{Bi})} \tag{4}$$

5.2 Verification Improvement by Median Filter

Each method of biometric data acquisition is imperfect and is associated with the acquisition of relevant data as well as noise and distortion. The same situation takes place during the process of registering mouse movements. Each type of computer mouse is characterized by a limited sensitivity, which more or less affects the quality of the collected data. External factors such as dust accumulating on the lens of the mouse, or inequality and contamination of the surface on which the mouse is used affect the formation of interference in the data collected. The resulting noise can have a negative impact on the effectiveness of the analysis. Therefore, it is necessary to consider eliminating unwanted noise information, for example by performing the filtering process.

In this paper the use of median filtering is proposed, which is used primarily to reduce the distortion in digital images. For the purposes of performed work the filtering method has been adapted for the use with numbers. The median filter is composed of two operations: the sorting of the data and then extracting the middle values. In this way, the extreme values and outliers are discarded.

6   Experiments and Results

To evaluate the proposed solutions the designation of the EER was chosen. The lower the EER value of the more effective the method. The final EER value for the method is the arithmetic mean of the EER values obtained in all the experiments.

The number of samples was selected experimentally. The study showed EER of 36% for testing set consisted of 100 samples. The EER of the method for 200 samples raised to 42% to descend again to a level of 37% for 1000 samples. It should be noted that the aim is to obtain a method of the on-line type hence the number of samples for a single experiment should be low - for this reason it was decided to use 100 samples in the experiments. For each tested user the profile was created and then the similarity between the profile and a testing set was calculated. Testing set consisted of 200 random samples: the first 100 samples belong to the same user as the profile and further 100 samples to another user constituting the intruder.

6.1   Distances

The k-NN algorithm used in the experiments is based on a distance measure between two object (profile and tested sample). The smaller the calculated distance, the greater the compatibility of the two objects. The next stage of the research was to compare the proposed distance measures. To this end, users were put together in pairs in various combinations. One of the users was an authorized user while the second was an intruder. Then the method using proposed distance measures was tested. The results of the experiments presented in Table 1 indicate that the best method of calculating the distance of the proposed features is the City Block, for which the value of the EER is 30%. The EER value of 30% is still too high for user verification method. Therefore, it was necessary to modify the methods to improve its effectiveness.

Table 1. EER values for examined distance measures

Distance measure	EER
Euclidean Distance	30.88%
City Block	30%
Canberra Distance	36%
Bray Curtis Distance	36%

6.2   Filtering the Distances

When analyzing the distances between the profile and the samples it was observed that these distances are characterized by a high disharmony, which increases the EER. It is necessary to strive for the elimination of extreme values, for example by applying a median filter. Filtration has been designated to the distances and it was a success.

The calculated EER confirmed the achievement of objective pursued. In some of the experiments based on the Euclidean distances the EER was 12% (30% without filtering) and the average EER was decreased by 12.88% and after applying the filtering was 18%. However, the best results were obtained after applying the filter for City Block distances. The lowest value of EER for an experiment was approximately 11% and the average EER was 17.88%.

7   Conclusions

The main goal of the conducted research was to develop the biometric method of user verification based on the analysis of user's activity connected to mouse use. Additionally the method should allow the analysis of continuous activity of a user in an on-line mode. In order to achieve the goals the set of mouse activity features was defined basing on previous works and extended by introducing new features. A new method of aggregating the mouse activity basic events into a higher level event was also presented. Additionally, an attempt to intrusion detection based on the k-NN classifier and various distance metrics was performed. To improve the effectiveness of the method the calculated distances were filtered by using median filter what improved the EER of the method by more than 12%. The best results in both cases (with and without filtering) were observed for the City Block distance measure.

Despite the fact that the article presents the preliminary research end conclusions the results are promising. In future works the further improvements are necessary. The arithmetic mean used to create the profile is very sensitive and a large impact on its value has the noise in the activity data. To reduce the influence of outliers a different model for creating a profile using e.g. the median or dominant could be proposed. Another improvement could be achieved by analyzing the feature set in order to extract the most distinctive features or by using different classifiers.

References

1. Agrawal, A.: User Profiling in GUI Based Windows Systems for Intrusion Detection, Master's Projects, Paper 303 (2013)
2. Ahmed, A.A.E., Traore, I.: A New Biometric Technology Based on Mouse Dynamics. IEEE Transactions on Dependable and Secure Computing 4(3), 165–179 (2007)
3. Banerjee, S.P., Woodard, D.L.: Biometric Authentication and Identification Using Keystroke Dynamics: A survey. J. of Pattern Recognition Research 7, 116–139 (2012)
4. Feher, C., Elovici, Y., Moskovitch, R., Rokach, L., Schclar, A.: User Identity Verification via Mouse Dynamics. Information Sciences 201, 19–36 (2012)
5. Gamboa, H., Fred, A.: A Behavioral Biometric System Based on Human Computer Interaction. Biometric Technology for Human Identification, 381–392 (2004)
6. Garg, A., Rahalkar, R., Upadhyaya, S., Kwiat, K.: Profiling Users in GUI Based Systems for Masquerade Detection. In: Proc. of the 7th IEEE Workshop on Information Assurance, pp. 48–54 (2006)
7. Hsing-Kuo, P., Junaidillah, F., Hong-Yi, L., Kuan-Ta, C.: Trajectory analysis for user verification and recognition. Knowledge-Based Systems 34, 81–90 (2012)
8. Kaminsky, R., Enev, M., Andersen, E.: Identifying Game Players with Mouse Biometrics, University of Washington, Technical Report (2008)
9. Palys, M., Doroz, R., Porwik, P.: On-line signature recognition based on an analysis of dynamic feature. In: IEEE Int. Conf. on Biometrics and Kansei Engineering, pp. 103–107. Metropolitan University Akihabara, Tokyo (2013)
10. Porwik, P., Sosnowski, M., Wesolowski, T., Wrobel, K.: A computational assessment of a blood vessel's compliance: A procedure based on computed tomography coronary angiography. In: Corchado, E., Kurzyński, M., Woźniak, M. (eds.) HAIS 2011, Part I. LNCS, vol. 6678, pp. 428–435. Springer, Heidelberg (2011)
11. Pusara, M., Brodley, C.E.: User re-authentication via mouse movements. In: Proc. of the 2004 ACM Workshop on Visualization and Data Mining for Computer Security, pp. 1–8 (2004)
12. Raiyn, J.: A survey of cyber attack detection strategies. International Journal of Security and Its Applications 8(1), 247–256 (2014)
13. Salem, M.B., Hershkop, S., Stolfo, S.J.: A survey of insider attack detection research. Advances in Information Security, vol. 39, pp. 69–90. Springer, US (2008)
14. Schonlau, M.: Masquerading user data, http://www.schonlau.net
15. Wesolowski, T., Kudlacik, P.: Data clustering for the block profile method of intruder detection. J. of MIT 22, 209–216 (2013)
16. Wesolowski, T., Kudlacik, P.: User profiling based on multiple aspects of activity in a computer system. J. of MIT 23, 121–129 (2014)
17. Wesolowski, T., Wrobel, K.: A computational assessment of a blood vessel's roughness. In: Burduk, R., Jackowski, K., Kurzynski, M., Wozniak, M., Zolnierek, A. (eds.) CORES 2013. AISC, vol. 226, pp. 231–240. Springer, Heidelberg (2013)

Part II

Intelligent Computational Methods
in Information Systems

Finger Knuckle Print Identification with Hierarchical Model of Local Gradient Features

Rafał Kozik and Michał Choraś

Institute of Telecommunications and Computer Science,
UTP University of Science and Technology, Poland
{rafal.kozik,michal.choras}@utp.edu.pl

Abstract. In this paper we present new biologically inspired method for biometric human identification based on the knuckle finger print (FKP). Knuckle is a part of hand, and therefore, is easily accessible, invariant to emotions and other behavioral aspects (e.g. tiredness) and most importantly is rich in texture features which usually are very distinctive. The proposed method is based on the hierarchical feature extraction model. We also showed the results obtained for PolyU knuckle image database.

Keywords: Finger Knuckle Print, FKP, Biometry, Local Gradient Features.

1 Introduction

In this paper the new biologically inspired feature extraction method for finger-knuckle-print (FKP) is presented. Knuckle is a part of hand, and therefore, is easily accessible, invariant to emotions and other behavioral aspects (e.g. tiredness) and most importantly is rich in texture features which usually are very distinctive. Requirements of users and stakeholders prove that biometric systems will soon have to be user-centric meaning requirements-free, friendly, accepted and mobile [1,2]. One of such emerging modalities satisfying those requirements is knuckle also termed as FKP (finger-knuckle-print). It has several advantages, such as: invariance to emotions and other behavioural aspects, high-textured region (experiments prove it to be that very distinctive), up to 5 distinctive biometrics samples per one hand, and it is easily accessible. Knuckle biometrics methods can be used in biometric systems for user-centric, contactless and un-restricted access control e.g. for medium-security access control or verification systems dedicated for mobile devices (e.g. smartphones and mobile telecommunication services) [3]. The samples of knuckles images (index and middle finger for left hand and middle finger for right hand respectively) from PolyU Database are presented in Fig. 1.

Most of the applied methods originate from known signal processing transformations or image processing methodologies. In general those can be categorized as approaches based on: Gabor-based approach (including e.g. 1D Log-Gabor),

© Springer International Publishing Switzerland 2015   73
D. Barbucha et al. (eds.), *New Trends in Intelligent Information and Database Systems,*
Studies in Computational Intelligence 598, DOI: 10.1007/978-3-319-16211-9_8

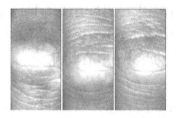

Fig. 1. Sample knuckle images from PolyU Database for one person (index and middle finger for left hand and middle finger for right hand)

Ridgelets and transforms (Radon, Riesz), (Probabilistic) Hough transform (PHT), SIFT and SURF as well as Phase correlation functions.

For example, Shariatmadar et al. [4] apply Gabor feature extraction followed by PCA and LDA for four fingers. Then feature level fusion is made before the matching process. Yang et al. [5] use Gabor filtering features and OLDA (Orthogonal LDA). Similarly, 8 Gabor filtering orientations and 5 scales are also used in FKP recognition based on Local Gabor Binary Patterns [6]. Meraoumia et al. [7] propose to use result of 1D Log-Gabor filtering in a palmprint and knuckle multimodal system.

Later Zhang et al. [8] proposed Gabor filtering to create improved competitive coding (ImCompCode) and magnitude coding (MagCode) representations of knuckles. The same authors also proposed to use known Gabor characteristics to obtain ensemble of local and global features/information. They proposed an LGIC (local-global information combination) scheme basing on CompCode as well as Phase Only Correlation (POC) and Band Limited Phase Only Correlation [9].

Global subspace methods such as PCA, LDA and ICA can be used not only for dimensionality reduction (of. e.g. Gabor filtration based vectors) but also as global appearance feature extractors as shown in [10]. In [11] image is analyzed by means of Probabilistic Hough Transform (PHT), which is used both for determining the dominant orientation and also for building the intermediate feature vector. SIFT (scale invariant feature transform) was proposed as knuckle feature extractor in [12] and [13]. SURF stands for Speeded Up Robust Features and it was used in [11] find the closest matching (if any) between querying image and the knuckle templates. Authors of [14] fused both extractors: SIFT and SURF, which allows to describe local patterns (of texture) around key characteristic points. Of course, the images are to be enhanced before feature extraction step.

Another approach to knuckle feature extraction is to use ridgelets or known signal processing transforms such as Radon Transform and Riesz Transform. Such approach was used in several papers by Goh et al., e.g. in both [15] and [16] in their proposition of bi-modal knuckle-palm biometric system (however, they proposed other feature extractors to palmprints).

Another approach to represent local knuckle features is based on Phase Correlation Function (PCF) also termed as Phase Only Correlation (POC) [17,18].

In most realizations, in order to eliminate meaningless high frequency components (in classic PCF/POC all frequencies are involved), the Band-Limited Phase Only Correlation (BLPOC) is used [17].

This paper is structured as follows. First, the proposed method overview is presented. Particularly it is explained how the original images are preliminary processed and how the information about relevant features are extracted and encoded. In the following section, the experiments and results are described. The conclusions are given thereafter.

2   Proposed Approach

The proposed approach follows the idea of HMAX model (see Fig.2) proposed by Riesenhuber and Poggio [19]. It exploits a hierarchical structure for the image processing and coding. The model is arranged in several layers that process information in a bottom-up manner. The lowest layer is fed with a grayscale image. The higher layers of the model are either called "S" or "C". These names correspond to simple the (S) and complex (C) cells discovered by Hubel and Wiesel [20].

In contrast to original model an additional layer that mimics the "retina codding" mechanism is added. Our previous experiments [24] showed that this step increases the robustness of the proposed method. This process is described in section 2.1.

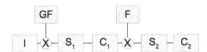

Fig. 2. The structure of a hierarchical model HMAX. (used symbols: I - image, S - simple cells, C - complex cells, GF - Gabor Filters, F - prototype features vectors, X - convolution operation).

The second modification includes a different method for calculating the S_1 layer response. The original model adapts the 2D Gabor filters computed for four orientations (horizontal, vertical, and two diagonal) at each possible position and scale. The Gabor filters are 11x11 in size, and are described by:

$$G(x,y) = e^{-(X^2+\gamma Y^2)/(2\sigma^2)} cos(\frac{2\pi}{\lambda})$$ (1)

where $X = x\cos\phi - y\sin\phi$ and $Y = x\sin\phi + y\cos\phi$; x, y $\in< -5;5 >$, and $\phi \in< 0; \pi >$. The aspect ratio (γ), effective width (σ), and wavelength (λ) are set to 0.3, 4.5 and 5.6 respectively. In this approach, the responses of 4 Gabor filters (two diagonals, horizontal and vertical) responses are approximated using Prewitt filters and voting algorithm described in section 2.2.The original HMAX model is intended to solve object detection problem with ability to preserve (to some extend) scale and transposition invariance. Therefore, two additional layers (S_2 and C_2) are introduced. In the case of PolyU dataset, the scale and transposition effects of biometric samples are so small, that can be suppressed by S_1. This is further explained in section 2.2. As a result, we replace these layers with classifier and use the C_1 layer responses as an input information.

2.1   Preprocessing Stage

In this work Difference of Gaussians (DoG) filter is used in preprocessing stage. It allows for feature enhancement and it involves the subtraction of two images blurred with different Gaussians filters (different standard deviation). It can be expressed with equation 2, where "$*$" represents convolution operation and σ_1 and σ_2 mentioned above standard deviations.

$$DoG_{\sigma_1 \sigma_2}(x, y) = I * \frac{1}{\sigma_1 \sqrt{2\pi}} e^{-\frac{x^2+y^2}{2\sigma_1^2}} - I * \frac{1}{\sigma_2 \sqrt{2\pi}} e^{-\frac{x^2+y^2}{2\sigma_2^2}} \qquad (2)$$

This preprocessing stage is intended to mimic the "retina coding" mechanism. Human retina shows remarkable and interesting properties of image enhancement. From a general point of view, the the retina serves as the first step of visual information processing. In the literature, there are several models explaining the basic mechanism the retina uses to encode visual information before it reaches visual cortex [22]. Basically, this model works as a filter that whitens the image spectrum and corrects luminance thanks to local adaptation. It has also the ability to filter out spatio-temporal noise and enhance the image details.

More simplistic approach to retina-based image enhancement was proposed in [23]. Authors adapted a local method that is based on a contrast equalisation. Within the sliding window authors normalised the luminance in the way, that the mean value is set to zero while the Euclidean norm is set to 1. This allowed to enhance image details and reduce the noise.

2.2   Gradient Information Coding

In our approach the S_1 layer covers whole knuckle sample image of WxH size. In this layer there are NxMx4 simple cells arranged in a grid of rectangular blocks. In each block there are 4 cells. Each cell is assigned a receptive field (pixels inside the block). Each cell is activated according depending on a stimuli. In this case there are four possible stimulus, namely vertical, horizontal, left diagonal, and right diagonal edges. As a result the S_1 simple cells layer output has dimensionality of a size 4 (x,y,scale and 4 cells). In order to compute responses of all four cells inside a given block (receptive field), an algorithm 1 is applied.

The algorithm computes the responses of all cells using only one iteration over the whole input image I. For each pixel at position (x, y) a vertical and horizontal gradients are computed (G_x and G_y). Given the pixel position (x, y) and gradient vector $[G_x, G_y]$ the algorithm indicates the block position (n, m) and type of cell (*active*) that response has to be incremented by magnitude $|G| = \sqrt{G_x^2 + G_y^2}$. In order to classify given gradient vector $[G_x, G_y]$ as horizontal, L,R-diagonal (left or right) or vertical the *get_cell_type*$(\cdot, \cdot)$ uses the wheel shown in Fig.3.

If a point $(|G_x|, |G_y|)$ is located between line $y = 0.3 \cdot x$ and $y = 3.3 \cdot x$ it is classified as a diagonal. If G_y is positive then the vector is classified as a right diagonal (otherwise it is a left diagonal). In case the point $(|G_x|, |G_y|)$ is located above line $y = 3.3 \cdot x$ the gradient vector is classified as vertical and as horizontal when it lies below line $y = 0.3 \cdot x$.

Data: Grayscaled image I of WxH size
Result: S_1 layer of size NxMx4
Assign G_{min} the low-threshold for gradient magnitude.
for *each pixel (x, y) in image I* **do**
  Compute horizontal G_x and vertical G_y gradients using Prewitt operator;
  Compute gradient magnitude $|G|$ in point (x, y);
  if $|G| < G_{min}$ **then**
    go to next pixel;

  else
    // Get cell type (horizontal, L,R-diagonal or vertical)
    $active \leftarrow get\_cell\_type(G_x, G_y)$;
    // Calculate block indices
    n $\leftarrow \lfloor x \cdot \frac{N}{W} \rfloor$;
    m $\leftarrow \lfloor y \cdot \frac{M}{W} \rfloor$;
    // Increment response
    $S_1[n, m, active] \leftarrow S_1[n, m, active] + |G_{x,y}|$;
  end
end

Algorithm 1. Algorithm for calculating S_1 response

The C_1 complex layer response is computed using max-pooling filter, that is applied to S_1 layer.

3 Experiments and Results

The proposed approach was tested using PolyU Knuckle Database [25]. The knuckle images were obtained from 165 individuals. Each individual contributed 10-12 image samples per left index, left middle, right middle and right index finger.

For evaluation purposes we have adapted the stratified 10-fold cross-validation technique. For that approach the data obtained for learning and evaluation purposes is divided randomly into 10 parts (sets). For each part it is intended to preserve the proportions of labels in the full dataset. One part (10% of full dataset) is used for evaluation while the remaining 90% is used for training.

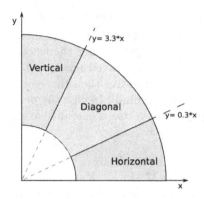

Fig. 3. A part of wheel that is used by $get\_cell\_type(\cdot, \cdot)$ to recognise a given gradient vector $[G_x, G_y]$ as horizontal, diagonal or vertical

When the classifier is learnt the evaluation data set is used to calculate the error rates. The whole procedure is repeated 10 times, so each time different part is used for evaluation and different part of data set is used for training. The result for all 10 runs (10-folds) are averaged to yield an overall error estimate.

For evaluation purposes we have used WEKA toolkit. In these experiments we evaluated different classifiers but kNN (where k=1), NaiveBayes, and RandomTrees gave comparable and promising results. The numbers of correctly and incorrectly classified instances are presented in Table 1.

Table 1. Effectiveness for PolyU Dataset

	Correctly Classified	Incorrectly Classified
kNN	**98.94%**	**1.06%**
NaiveBayes	95.81%	4.19%
50-RandomTrees	91.92%	8.08%
150-RandomTrees	96.87%	3.13%

In the Fig.4 the ROC curve for the kNN classifier is presented. The experiments showed that it is possible to achieve 0.9% of ERR (Equal Error Rate). However, the process of decreasing the False Acceptance Ratio causes rapid growth of False Rejection Ratio.

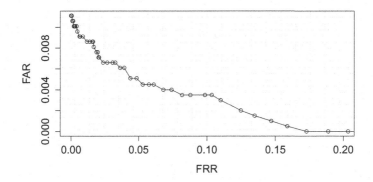

Fig. 4. FAR vs. FRR

4 Conclusions

In this paper new developments in human identification based on knuckle texture features are presented. The major contribution of the paper is the proposal of new biologically inspired knuckle feature extraction methodology based on hierarchical HMAX model. The method was evaluated using the benchmark PolyU database and we report the promising results.

In our opinion, the effectiveness of FKP based identification, should shortly allow for the application of knuckles in multi-modal biometric systems, mainly with palmprint, hand features or hand veins [7,15], as well as for application of knuckles in contactless (touchless) scenarios, especially with images acquired by mobile phones [15].

References

1. Choraś, M.: Novel techniques applied to biometric human identification. Electronics 3, 35–39 (2009)
2. Choraś, M.: Emerging Methods of Biometrics Human Identification. In: Proc. of 2nd International Conference on Innovative Computing, Information and Control (ICICIC 2007), pp. 365–373. IEEE CS Press, Kumamoto (2007)
3. Kozik, R., Choraś, M.: Combined Shape and Texture Information for Palmprint Biometrics. Journal of Information Assurance and Security 5(1), 58–63 (2010)
4. Shiariatmadar, Z.S., Faez, K.: A Novel Approach for Finger-Knuckle Print Recognition Based on Gabor Feature Extraction. In: Proc. of 4th International Congress on Image and Signal Processing, pp. 1480–1484 (2011)
5. Yang, W., Sun, C., Sun, Z.: Finger-Knuckle Print Recognition Using Gabor Feature and OLDA. In: Proc. of 30th Chinese Control Conference, Yantai, China, pp. 2975–2978 (2011)
6. Xiong, M., Yang, W., Sun, C.: Finger-Knuckle-Print Recognition Using LGBP. In: Liu, D., Zhang, H., Polycarpou, M., Alippi, C., He, H. (eds.) ISNN 2011, Part II. LNCS, vol. 6676, pp. 270–277. Springer, Heidelberg (2011)
7. Meraoumia, A., Chitroub, S., Bouridane, A.: Palmprint and Finger Knuckle Print for efficient person recognition based on Log-Gabor filter response. Analog Integr Circ Sig Process 69, 17–27 (2011)

8. Zhang, L., Zhang, L., Zhang, D., Zhu, H.: Online finger-knuckle-print verification for personal authentication. Pattern Recognition 43, 2560–2571 (2010)
9. Zhang, L., Zhang, L., Zhang, D., Zhu, H.: Ensemble of local and global information for finger-knuckle print-recognition. Pattern Recognition 44, 1990–1998 (2011)
10. Kumar, A., Ravikanth, C.: Personal authentication using finger knuckle surface. IEEE Trans. Information Forensics and Security 4(1), 98–110 (2009)
11. Choraś, M., Kozik, R.: Knuckle Biometrics Based on Texture Features. In: Proc. of International Workshop on Emerging Techniques and Challenges for Hand-based Biometrics (ETCH B2010), pp. 1–5. IEEE CS Press, Stambul (2010)
12. Morales, A., Travieso, C.M., Ferrer, M.A., Alonso, J.B.: Improved finger-knuckle-print authentication based on orientation enhancement. Electronics Letters 47(6) (2011)
13. Hemery, B., Giot, R., Rosenberger, C.: Sift Based Recognition of Finger Knuckle Print. In: Proc. of Norwegian Information Security Conference, pp. 45–56 (2010)
14. Badrinath, G.S., Nigam, A., Gupta, P.: An Efficient Finger-Knuckle-Print Based Recognition System Fusing SIFT and SURF Matching Scores. In: Qing, S., Susilo, W., Wang, G., Liu, D. (eds.) ICICS 2011. LNCS, vol. 7043, pp. 374–387. Springer, Heidelberg (2011)
15. Goh, K.O.M., Tee, C., Teoh, B.J.A.: An innovative contactless palm print and knuckle print recognition system. Pattern Recognition Letters 31, 1708–1719 (2010)
16. Goh, K.O.M., Tee, C., Teoh, B.J.A.: Bi-modal palm print and knuckle print recognition system. Journal of IT in Asia 3 (2010)
17. Zhang, L., Zhang, L., Zhang, D.: Finger-knuckle-print verification based on band-limited phase-only correlation. In: Jiang, X., Petkov, N. (eds.) CAIP 2009. LNCS, vol. 5702, pp. 141–148. Springer, Heidelberg (2009)
18. Aoyama, S., Ito, K., Aoki, T.: Finger-Knuckle-Print Recognition Using BLPOC-Based Local block Matching, pp. 525–529 (2011)
19. Riesenhuber, M., Poggio, T.: Hierarchical models of object recognition in cortex. Nature Neuroscience 2, 1019–1025 (1999)
20. Hubel, D.H., Wiesel, T.N.: Receptive fields, binocular interaction and functional architecture in the cat's visual cortex. The Journal of physiology 160, 106–154 (1962)
21. Mutch, J., Object, D.G.: class recognition and localization using sparse features with limited receptive fields. International Journal of Computer Vision (IJCV) 80(1), 45–57 (2008)
22. Benoit, A., Caplier, A., Durette, B., Herault, J.: Using Human Visual System Modeling For Bio-Inspired Low Level Image Processing. Computer Vision and Image Understanding 114, 758–773 (2010)
23. Brumby, S.P., Galbraith, A.E., Michael Ham, G., Kenyon, J.S.: George: Visual Cortex on a Chip: Large-scale, real-time functional models of mammalian visual cortex on a GPGPU. In: GPU Technology Conference (GTC), pp. 20–23 (2010)
24. Kozik, R.: A proposal of biologically inspired hierarchical approach to object recognition. Journal of Medical Informatics & Technologies 22, 169–176 (2013)
25. The Hong Kong Polytechnic University (PolyU) Finger-Knuckle-Print Database, http://www4.comp.polyu.edu.hk/~biometrics/FKP.htm

Type-Reduction for Concave Type-2 Fuzzy Sets

Bing-Kun Xie and Shie-Jue Lee

Department of Electrical Engineering,
National Sun Yat-Sen University,
Kaohsiung 80424, Taiwan
bkshie@water.ee.nsysu.edu.tw,
leesj@mail.ee.nsysu.edu.tw

Abstract. An efficient centroid type-reduction strategy for general type-2 fuzzy set is proposed by Liu. In Liu's method, a type-2 fuzzy set will be decomposed into several interval type-2 fuzzy sets. However, if the membership function of the type-2 fuzzy set is concave, the primary membership of these interval type-2 fuzzy sets on some points may not have only one continuous interval. Existing type-reduction algorithms, such as Karnik-Mendel algorithm and Enhanced Karnik-Mendel algorithm, can not deal with this problem. We propose a method to decompose this problem into several subproblems which can then be solved by existing type-reduction algorithms. The union of the solutions to the subproblems is the final solution to the original problem.

Keywords: Karnik-Mendel (KM) algorithm, type-reduction, type-2 fuzzy set, interval type-2 fuzzy set.

1 Introduction

The type-2 fuzzy system has evolved from the type-1 fuzzy system [1], [4], [5], [7]. One of the major differences between these two systems lies on the defuzzification involved in the inference process. The defuzzification of a type-2 fuzzy set is composed of type-reduction and type-1 defuzzification. Therefore, type-reduction plays an important role in type-2 fuzzy systems [6].

Two kinds of type-reduction have been introduced. Liu proposed a method [2] to perform type-reduction for general type-2 fuzzy sets. A type-2 fuzzy set is decomposed into interval type-2 fuzzy sets, and one only needs to perform type-reduction for each resulting interval type-2 fuzzy set [2], [9], [10]. There are many type-reduction algorithms for interval type-2 fuzzy sets, such as Karnik-Mendel algorithm (KM) [3] and Enhanced Karnik-Mendel algorithm (EKM) [8]. EKM is an iterative algorithm which is a faster version of KM. The Enhanced Centroid-Flow algorithm [12] is a more accurate version of the Centroid-Flow algorithm [11]. This algorithm utilizes KM or EKM only at the central α-level, and then lets its result flow upward to the maximum α-level and downward to the minimum α-level.

However, if the membership function of the type-2 fuzzy set is concave, the primary membership of these interval type-2 fuzzy sets on some points may

© Springer International Publishing Switzerland 2015
D. Barbucha et al. (eds.), *New Trends in Intelligent Information and Database Systems*,
Studies in Computational Intelligence 598, DOI: 10.1007/978-3-319-16211-9_9

not have only one continuous interval. The type-reduction methods mentioned above can not deal with this problem. We propose a method to decompose this problem into several subproblems which can then be solved by any of the above mentioned algorithm. The union of the solutions to the subproblems is the final solution to the original problem.

The rest of this paper is organized as follows. Section 2 provides a brief introduction to fuzzy sets. Our proposed method is described in Section 3. An illustrating example is shown in Section 4. Finally, a conclusion is given in Section 5.

2   Background

Some background about fuzzy sets is given here. For more details, please refer to [1] and the other cited literature.

2.1   Type-2 Fuzzy Set

A type-2 fuzzy set $\widetilde{A}$ can be expressed as

$$\widetilde{A} = \{(x, u), \mu_{\widetilde{A}}(x, u) | \forall x \in X, \forall u \in J_x\} \tag{1}$$

where X is the universe for primary variable x, and J_x is the primary membership of $\widetilde{A}$ at x. $\widetilde{A}$ can also be expressed as

$$\widetilde{A} = \int_{x \in X} \int_{u \in J_x} \mu_{\widetilde{A}}(x, u)/(x, u) \tag{2}$$

where $\int$ denotes the union of all admissible x and u.

2.2   Interval Type-2 Fuzzy Set

An interval type-2 fuzzy set is a special case of type-2 fuzzy set. It's membership function degree is 1. An interval type-2 fuzzy set $\widetilde{A}$ can be expressed as

$$\widetilde{A} = \{(x, u), \mu_{\widetilde{A}}(x, u) = 1 | \forall x \in X, \forall u \in J_x\} \tag{3}$$

or

$$\widetilde{A} = \int_{x \in X} \int_{u \in J_x} 1/(x, u) \tag{4}$$

The footprint of uncertainty of $\widetilde{A}$, denoted by $FOU(\widetilde{A})$, is defined by

$$FOU(\widetilde{A}) = \bigcup_{x \in X} J_x = \bigcup_{x \in X} [\underline{I}(x), \bar{I}(x)] \tag{5}$$

where $[\underline{I}(x), \bar{I}(x)]$ is an interval set, and $[\underline{I}(x), \bar{I}(x)] \subseteq [0, 1]$.

2.3   Liu's Method

An efficient centroid type-reduction strategy for general type-2 fuzzy set is proposed by Liu [2]. The key idea of Liu's method is to decompose a type-2 fuzzy set into several interval type-2 fuzzy sets, called α-plane representation, as illustrated in Fig. 1. The α-plane for a type-2 fuzzy set $\widetilde{A}$, denoted by $\widetilde{A}_\alpha$, is defined

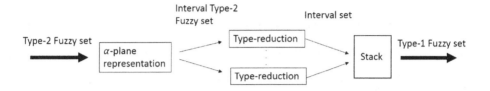

Fig. 1. Liu's method

to be

$$\widetilde{A}_\alpha = \bigcup_{x \in X} \{(x, u) | \mu_{\widetilde{A}}(x, u) \geq \alpha\} \tag{6}$$

The $\widetilde{A}_\alpha$ is an interval type-2 fuzzy set. For each interval type-2 fuzzy set, an interval set can be obtained by applying the centroid type-reduction. Finally, The union of the resulting interval sets form a type-1 fuzzy set. This type-1 fuzzy set is the centroid of the type-2 fuzzy set $\widetilde{A}$.

2.4   Enhanced Karnik-Mendel Algorithm

The Enhanced Karnik-Mendel (EKM) algorithm [8] is a type-reduction algorithm for interval type-2 fuzzy sets. It is an enhanced version of Karnik-Mendel algorithm (KM) [3]. An interval type-2 fuzzy set $\widetilde{A}$ can be regarded as composed of many type-1 fuzzy sets A_i such as

$$\widetilde{A} = A_1 \cup A_2 ... \cup A_i \cup ... \tag{7}$$

where

$$A_i = \{(x, u) | \forall x \in X, u \in J_x\} \tag{8}$$

The centroid type-reduction of the interval type-2 fuzzy set $\widetilde{A}$ is composed of all the centers of the type-1 fuzzy sets A_i. As a result, one can get an interval set $[c_l, c_r]$. This interval set is the centroid of the interval type-2 fuzzy set $\widetilde{A}$. EKM is a fast algorithm to find such c_l and c_r, and can be summarized in Table 1.

3   Proposed Method

The problem of Liu's method is the membership function of type-2 fuzzy set must be a convex function. If the membership function is concave function, J_x

Table 1. EKM to calculate the centroid of an interval type-2 fuzzy set

Step	EKM for c_l
1	Set $k = [N/2.4]$ (the nearest integer to $N/2.4$) for $N = |X|$.
2	Compute

$$c = \frac{\sum_{i=1}^{k} x_i \bar{\mu}_{\tilde{A}}(x_i) + \sum_{i=k+1}^{N} x_i \underline{\mu}_{\tilde{A}}(x_i)}{\sum_{i=1}^{k} \bar{\mu}_{\tilde{A}}(x_i) + \sum_{i=k+1}^{N} \underline{\mu}_{\tilde{A}}(x_i)}.$$

3	Find $k' \in [1, N-1]$ such that $x_{k'} \leq c \leq x_{k'+1}$.
4	Check $k = k'$:
	if yes, stop and set $c_l = c$, $L = k$;
	if no, set $k = k'$ and go to step 2.

Step	EKM for c_r
1	Set $k = [N/2.4]$ (the nearest integer to $N/1.7$) for $N = |X|$.
2	Compute

$$c = \frac{\sum_{i=1}^{k} x_i \underline{\mu}_{\tilde{A}}(x_i) + \sum_{i=k+1}^{N} x_i \bar{\mu}_{\tilde{A}}(x_i)}{\sum_{i=1}^{k} \underline{\mu}_{\tilde{A}}(x_i) + \sum_{i=k+1}^{N} \bar{\mu}_{\tilde{A}}(x_i)}.$$

3	Find $k' \in [1, N-1]$ such that $x_{k'} \leq c \leq x_{k'+1}$.
4	Check if $k = k'$:
	if yes, stop and set $c_r = c$, $R = k$;
	if no, set $k = k'$ and go to step 2.

may be a union which is composed of two or more intervals for the α-plane representation $\tilde{A}_\alpha$. The $FOU(\tilde{A}_\alpha)$ may be

$$FOU(\tilde{A}_\alpha) = \bigcup_{x \in X} J_x \qquad (9)$$

where

$$J_x = [\underline{I_1}(x), \bar{I}_1(x)] \cup [\underline{I_2}(x), \bar{I}_2(x)]... \cup [\underline{I_j}(x), \bar{I}_j(x)] \cup ...$$

If J_x is composed of more than one interval, we can't use EKM algorithm to calculate the centroid.

We propose a method to perform the centroid type-reduction of a concave type-2 fuzzy set $\tilde{A}$. The first step is to decompose $\tilde{A}$ into several interval type-2 fuzzy sets $\tilde{A}_{\alpha_i}$ for $\alpha_i \in [0,1]$ by the α-plane representation as shown in Fig. 2. Since $\tilde{A}$ is a concave type-2 fuzzy set, the primary membership of $\tilde{A}_{\alpha_i}$ may be the union of several interval sets, denoted by $J_x^{\alpha_i}$. The footprint of uncertainty of $\tilde{A}_{\alpha_i}$ is defined by

$$FOU(\tilde{A}_{\alpha_i}) = \bigcup_{x \in X} J_x^{\alpha_i}$$

and

$$J_x^{\alpha_i} = I_{x1}^{\alpha_i} \cup I_{x2}^{\alpha_i}... \cup I_{xj}^{\alpha_i} \cup ... \qquad (10)$$

where $I_{xj}^{\alpha_i}$ is an interval set, and $I_{xj}^{\alpha_i} \subseteq [0,1]$.

For each interval type-2 fuzzy set $\tilde{A}_{\alpha_i}$, we decompose it into several interval type-2 fuzzy sets $\tilde{A}_{\alpha_i}^l$ as shown in Fig. 3. The footprint of uncertainty of $\tilde{A}_{\alpha_i}^l$ is

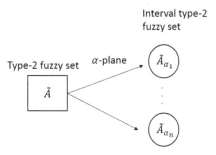

Fig. 2. The α-plane representation

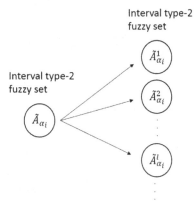

Fig. 3. The decomposition of an interval type-2 fuzzy set

defined by

$$FOU(\widetilde{A}_{\alpha_i}^l) = \bigcup_{x \in X} I_x^{\alpha_i} \tag{11}$$

where $I_x^{\alpha_i}$ is an interval set, and $I_x^{\alpha_i} \subseteq J_x^{\alpha_i}$. Then we can have S interval type-2 fuzzy sets from $\widetilde{A}_{\alpha_i}^l$, i.e.,

$$S = \prod_{x \in X} s_x \tag{12}$$

where s_x is the number of intervals contained in $J_x^{\alpha_i}$. If $\widetilde{A}$ is a convex type-2 fuzzy set, s_x is 1 for all $x \in X$.

The primary membership of each interval type-2 fuzzy set $\widetilde{A}_{\alpha_i}^l$ is an interval set for all $x \in X$, so we can perform the centroid type-reduction by EKM for each interval type-2 fuzzy set $\widetilde{A}_{\alpha_i}^l$. Then, we collect all the results from the application of EKM to each interval type-2 fuzzy set $\widetilde{A}_{\alpha_i}^l$ and get a set W_{α_i}

$$W_{\alpha_i} = \bigcup_{l=1}^{S} w_{\alpha_i}^l \tag{13}$$

where $w_{\alpha_i}^l$ is the result of applying EKM on $\tilde{A}_{\alpha_i}^l$. Note that $w_{\alpha_i}^l$ is an interval set, but W_{α_i} is not necessarily an interval set. The W_{α_i} we obtain is the centroid of the interval type-2 fuzzy set $\tilde{A}_{\alpha_i}$.

Finally, we can obtain a type-1 fuzzy set by doing a union of W_{α_i} for all i. This type-1 fuzzy set is then the centroid of the concave type-2 fuzzy set $\tilde{A}$. The whole process of our proposed method for computing the centroid of a concave type-2 fuzzy set can be summarized in Fig. 4.

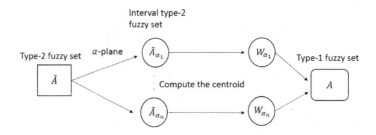

Fig. 4. Our method for computing the centroid of a concave type-2 fuzzy set

4   Numerical Results

In this section, we show the result of applying our method on a type-2 fuzzy set. Liu's method is not applicable for this case. The universal set X involved is $[0.0, 0.2, 0.4, ..., 4]$ and the α set adopted is $[0.000, 0.025, 0.050, 0.075, ..., 1]$ for this example.

This example is a concave type-2 fuzzy set $\tilde{A}$ which is shown in Fig. 5. The primary membership function $f_{\tilde{A}}(x)$ and secondary membership function $\mu_{\tilde{A}}(x, u)$ of $\tilde{A}$ are defined by

$$f_{\tilde{A}}(x) = exp\left[-\frac{(x-2)^2}{2*(0.3)^2}\right] \tag{14}$$

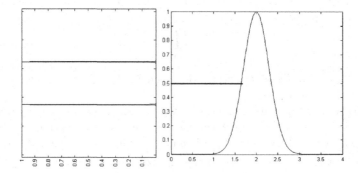

Fig. 5. $\tilde{A}$ for illustration

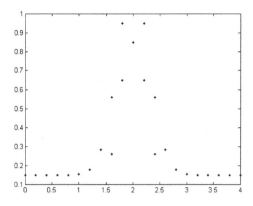

Fig. 6. The α-plane with $\alpha = 0.5$

and

$$\mu_{\widetilde{A}}(x, u) = \begin{cases} exp\left[-\frac{(u-(f_{\widetilde{A}}(x)-0.15))^2}{2*(0.001)^2}\right] , u \leq f_{\widetilde{A}}(x) \\ exp\left[-\frac{(u-(f_{\widetilde{A}}(x)+0.15))^2}{2*(0.001)^2}\right] , u > f_{\widetilde{A}}(x) \end{cases} \qquad (15)$$

Firstly, we get α-planes from $\widetilde{A}$. For example, the α-plane with $\alpha = 0.5$ is shown in Fig. 6. Note that each point in this figure signifies an interval. Secondly, each α-plane is further decomposed into several interval type-2 fuzzy sets. For the α-plane with $\alpha = 0.5$, it is decomposed into 16 interval type-2 fuzzy sets. One of the decomposed interval type-2 fuzzy set is shown in Fig. 7, where each point signifies an interval. Finally, we do type-reduction for each decomposed interval type-2 fuzzy set using EKM. Since the secondary membership function of $\widetilde{A}$ is a concave function, W_{α_i} may contain several intervals. It means that the result of $\widetilde{A}$ can be a concave type-1 fuzzy set. The centroid obtained for $\widetilde{A}$ is shown in Fig. 8.

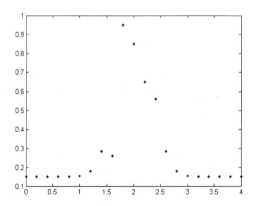

Fig. 7. A decomposed interval type-2 fuzzy set of Fig. 6

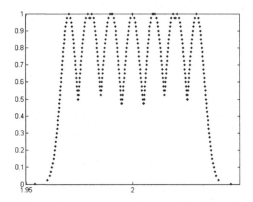

Fig. 8. The type-reduction result of $\widetilde{A}$

5   Conclusion

We have presented a method to perform type-reduction for type-2 fuzzy sets. Our method is based on Liu's method which can only handle the type-2 fuzzy sets with convex membership functions. By our method, an underlying type-2 fuzzy set is decomposed into several interval type-2 fuzzy sets by the α-plane representation. Then, we decompose each interval type-2 fuzzy set into several new interval type-2 fuzzy sets. These new interval type-2 fuzzy sets are then handled by existing type-reduction algorithms, e.g., EKM, and we collect the type-reduction results from the new interval type-2 fuzzy sets to form the type-reduction result of each interval type-2 fuzzy set of the α-plane representation. The union of the type-reduction results of all the α planes is the centroid of the original type-2 fuzzy set. In this way, type-reduction of both convex and concave type-2 fuzzy sets can be done properly.

References

1. Klir, G.J., Yuan, B.: Fuzzy Sets and Fuzzy Logic. Pearson Education Taiwan (2005)
2. Liu, F.: An Efficient Centroid Type-Reduction Strategy for General Type-2 Fuzzy Logic System. Information Sciences 178(9), 2224–2236 (2008)
3. Karnik, N.N., Mendel, J.M.: Centroid of a Type-2 Fuzzy Set. Information Sciences 132(1–4), 195–220 (2001)
4. Karnik, N.N., Mendel, J.M., Liang, Q.: Type-2 Fuzzy Logic Systems. IEEE Transactions on Fuzzy Systems 7(6), 643–658 (1999)
5. Mendel, J.M.: Type-2 Fuzzy Sets and Systems: an Overview. IEEE Transactions on Computational Intelligence Magazine 2(1), 20–29 (2007)
6. Mendel, J.M.: On KM Algorithms for Solving Type-2 Fuzzy Set Problems. IEEE Transactions on Fuzzy Systems 21(3), 426–445 (2013)
7. Mendel, J.M., Liang, Q.: Type-2 Fuzzy Sets Made Simple. IEEE Transactions on Fuzzy Systems 10(2), 117–127 (2002)

8. Wu, D., Mendel, J.M.: Enhanced Karnik-Mendel Algorithms. IEEE Transactions on Fuzzy Systems 17(4), 923–934 (2009)

9. Wu, H.-J., Su, Y.-K., Lee, S.-J.: A Fast Method for Computing the Centroid of a Type-2 Set. IEEE Transactions on Systems, Man, And Cybernetics-Part B 42(3), 764–777 (2012)

10. Yeh, C.-Y., Jeng, W.-H.R., Lee, S.-J.: An Enhanced Type-Reduction Algorithm for Type-2 Fuzzy Sets. IEEE Transactions on Fuzzy Systems 19(2), 227–240 (2011)

11. Zhai, D., Mendel, J.M.: Computing the Centroid of a General Type-2 Fuzzy Set by Means of the Centroid-Flow Algorithm. IEEE Transactions on Fuzzy Systems 19(2), 401–422 (2011)

12. Zhai, D., Mendel, J.M.: Enhanced Centroid-Flow Algorithm for Computing the Centroid of General Type-2 Fuzzy Sets. IEEE Transactions on Fuzzy Systems 20(5), 939–956 (2012)

Evaluating Customer Satisfaction: Linguistic Reasoning by Fuzzy Artificial Neural Networks

Reza Mashinchi[1], Ali Selamat[1,3], Suhaimi Ibrahim[2], Ondrej Krejcar[3], and Marek Penhaker[4]

[1] Universiti Teknologi Malaysia, Faculty of Computing, 81310 Johor Baharu, Johor, Malaysia
r_mashinchi@yahoo.com, aselamat@utm.my
[2] Universiti Teknologi Malaysia, Advanced Informatics School,
54100 Kuala Lumpur, Malaysia
suhaimiibrahim@utm.my
[3] University of Hradec Kralove, Faculty of Informatics and Management,
Center for Basic and Applied Research,
Rokitanskeho 62, Hradec Kralove, 500 03, Czech Republic
ondrej@krejcar.org
[4] Department of Cybernetics and Biomedical Engineering,
Faculty of Electrical Engineering and Computer Science,
VSB - Technical University of Ostrava, 17. Listopadu 15, Ostrava Poruba
70833, Czech Republic
Marek.Penhaker@vsb.cz

Abstract. Customer satisfaction is a measure of how a company meets or surpasses customers' expectations. It is seen as a key element in business strategy; and therefore, enhancing the methods to evaluate the satisfactory level is worth studying. Collecting rich data to know the customers' opinion is often encapsulated in verbal forms, or linguistic terms, which requires proper approaches to process them. This paper proposes and investigates the application of fuzzy artificial neural networks (FANNs) to evaluate the level of customer satisfaction. Genetic algorithm (GA) and back-propagation algorithm (BP) adjust the fuzzy variables of FANN. To investigate the performances of GA- and BP-based FANNs, we compare the results of each algorithm in terms of obtained error on each alpha-cut of fuzzy values.

Keywords: Prediction, customer satisfactory index (CSI), rich data, computing with words, genetic algorithms, fuzzy artificial neural networks.

1 Introduction

Customers' opinion has a key role in business success. Earlier studies have shown that success chance greatly depends on meeting customers' satisfaction. The impact is such that increasing the satisfactory level of customers has been placed into business plan strategies; and consequently, evaluating satisfactory level is become noticed. There are many related studies from different departments of a company to varieties of business sectors; such as marketing [1,2], sales [3,4], finance [5,6], human resource

© Springer International Publishing Switzerland 2015
D. Barbucha et al. (eds.), *New Trends in Intelligent Information and Database Systems,*
Studies in Computational Intelligence 598, DOI: 10.1007/978-3-319-16211-9_10

[7], product design [8,9], airline [10,11], auto-motive [8], [12], E-commerce [13,14], retailing [15] and so on. Few studies have been even deepen to gender-based analysis [16], where few others have been conducted through lean manufacturing [17,18].

All mentioned studies are to support achieving the targets of business outcome. Though, from processing aspect, two complexities hamper a reliable evaluation. They include: (i) the input data that steams the results, and (ii) the evaluator method (modeling) that processes and build a model base on the input data. The former is related to nature of data, and the latter is to select a proper method compatible with nature of input data – whereby it is also a matter that a method provides the outcome in the required nature. In fact, data can be collected in various methods, e.g., verbal-based or numeric-based, which then a pre-processing method prepares the data to feed the evaluator model. Though, keeping the originality of data is always a matter of reliability of results in the outcome. Undoubtedly, verbal expression of a customer on satisfactory level is the foremost, which is original and rich in content. Many studies have been conducted using verbal data [3], [19,20]. However, the problem is that rich piece of information comes with complexities, and then, it causes the uncertainties that require proper methods to cope with it.

Fuzzy model widely exposes to confront with complexities concealed in linguistic terms. It was emerged in 1996 to deal with complexities in social sciences, and, it suites to compute with words [21]. Fuzzy model keeps the richness of original value; however, it requires certain circumstances for a method to process a set of data of this kind. A fuzzy value is represented by levels of cuts on a membership function (MF). Therefore, α-cuts divide an MF to represent degree of belief in a value. Consequently, a method should be able to deal with α-cuts to process a fuzzy value. Among the reasoning methods that can cooperate with fuzzy values, fuzzy artificial neural networks (FANNs) play the foremost to evaluate the level of customer satisfactory. FANNs were proposed [22] as a generalized form of artificial neural networks (ANNs) through the concept of fuzzy logic (FL), and then, rapidly enhanced for applications [23]. Two directions of enhancing FANNs have been based on genetic algorithms (GAs) [23,24,25] and back-propagation algorithms (BP) [26,27,28]. GAs is employed for its strength to seek for a globally optimum solution [29,30]; in contrast, local optimizers such as BP may trap into local minima [31,32]. Undoubtedly, performances of either GA- or BP-based FANNs are worth studying to evaluate the level of customer satisfaction, as it will transparent the effect of global or local optimization for this application. Trade off between the performances of either method can differentiate them to find whether or not: an absolute prediction is worth taking the complexities of a global optimization – this will then furnish the decision being made.

The rest of this paper is organized in four sections as follows. Section 2 describes the mechanism that this paper uses to evaluate the level of customer satisfaction. Section 3 describes fuzzy evaluator neural networks as the body of idea to construct the customer satisfactory system; subsequently, section 4 explains GA- and BP-based FANNs. Section 5 presents the results and analyses of the proposed methods using fuzzy value data set; where, generated errors on each α-cut of fuzzy variables conduct the comparisons. Section 6 concludes the results and addresses the future works.

2   Satisfactory Evaluation Mechanism

Evaluation mechanism involves with modeling the input and the output, and the structure of evaluator system. It must deal with rich data to furnish decision makers of a company with reliable information in the outcome. This paper deals with verbal words expressed by a customer, as the richest data, and then keeps the original richness for the outcome – therefore, both input and outcome sides of an evaluator are in verbal words. Though, the verbal words are revealed through fuzzy representation to mathematically express the original data. To this end, an MF defines an expressed word – as a linguistic variable – and then α-cuts sharpen the MF of each fuzzy value. This paper uses two levels of α-cuts – support and core – to represent a triangular fuzzy number. Equations below defines the support and core of a triangular fuzzy number $\widetilde{A}$.

$$\widetilde{A} = \{(x, \mu_{\widetilde{A}}(x)) | x \in R, \mu_{\widetilde{A}} : R \mapsto [0,1]\} \tag{1}$$

$$\widetilde{A_1} = Core(\widetilde{A}) = \{x \in R | \mu_{\widetilde{A}}(x) = 1\} \tag{2}$$

$$\widetilde{A_0} = Support(\widetilde{A}) = \{x \in R | \mu_{\widetilde{A}}(x) = 0\} \tag{3}$$

$$\widetilde{A_\alpha} = \{x \in R | \mu_{\widetilde{A}}(x) \geq \alpha, \alpha \in (0,1)\} \tag{4}$$

where, $\mu_{\widetilde{A}}$ is a continuous MF, and, R is the set of all real numbers. We use the equations above to provide the evaluator system with fuzzy represented input and outcome data. An overall schema is shown in Fig. 1.

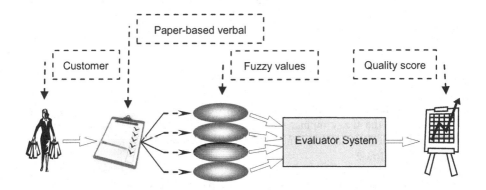

Fig. 1. Mechanism of customers' satisfactory evaluation

The mechanism above associates with Customer Satisfaction Index (CSI). It defines the indicators to measure the satisfactory level of customers. The indicators and its parameters are varies upon each case study, though this paper uses the CSI employed in [19] – which has nine fuzzy instances to predict the gap based on following variables: product, service, network system, and, payment.

Once the linguistic values are provided, evaluating the performance depends on evaluator system. This paper uses FANNs [24,27], as an evaluator, based on GA and BP optimization algorithms. These networks deliver the outcome in fuzzy numbers – as a representation of linguistic variables. It keeps the originality of customer expressions. Eventually, non-expert decision makers in business panel can refer to this outcome.

3   Fuzzy Evaluator Neural Networks

Evaluating the level of customer satisfaction is based on expressed preferences of customers by his/her current expectation. This paper performs the evaluation process by learning the collected data. To perform the learning, we use FANNs as the body of evaluator system to predict the satisfactory level. The FANNs were firstly proposed in 1993 [22]; a mathematically revised version of FANNs, which was enhanced through genetic algorithms, has been investigated in [23,24], [27]. An FANNs version that utilized back-propagation algorithms are studied in [27,33]. This paper uses type three of FANNs; which, all variables of network are based on fuzzy numbers and shown by FANN-3 [24]. The reason of using FANNs-3 is to keep the originality of input data. The variables are input, output, weights, and, biases as shown and defined in Fig. 2.

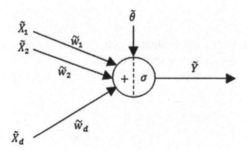

Fig. 2. A fuzzy neuron that constructs a FANN-3

d number of input variables and connection weights are represented by $\tilde{X}_i$ and $\tilde{W}_i$, respectively. All the fuzzy inputs, weights, and biases are fuzzy numbers in $F_0(R)$, where $F_0(R)$ is defined in [34]. The input-output relationship of the fuzzy neuron is determined as follows:

$$\tilde{Y} \triangleq F(\tilde{X}_1, \dots, \tilde{X}_d) = \sigma\left(\sum_{i=1}^{d} \tilde{x}_i \cdot \tilde{w}_i + \tilde{\theta}\right) = \sigma(\langle \tilde{X}, \tilde{W} \rangle + \tilde{\theta}) \tag{5}$$

where, $\tilde{\theta}$ means a fuzzy bias (or a fuzzy threshold); $\tilde{X}_1, \dots, \tilde{X}_d$, $\tilde{W}_1, \dots, \tilde{W}_d \in F_0(R)^d$ is fuzzy vectors; and, $\sigma: R \to R$ is a transfer function.

4   GA- and BP-Based Networks

GAs and BP approaches have been hybridized previously to enhance the performance of FANNs [23,24]. The advantage of GA-based FANNs is that they find a global optimum solution; in contrast, BP-based FANNs perform the local search to find a solution. However, the results are unexpected in different applications [31]. Besides that, taking the complexities to find a globally optimum solution is not always required for every application.

GAs perform based on principles of natural evolution. They gradually find the solution in generation based process, where each generation operates reproduction functions. The selection, crossover, and mutation functions operate on simulated genomes to replace the genes. Crossover function is known to be more effective; though, the result of optimization tightly depends on initial population. Thus, in practice, it hinders GAs to promise achieving global solution. In contrast, BP is based on strong mathematical fundamentals to backward the propagation of error when training a network. It calculates the gradient of loss function in respect to variables of network. BP is a local optimizer that is more likely to trap in local minima; however, theoretically, it finds the optimal value faster than GA [31]. Algorithm 1 proposes the evaluator systems constructed by BP and GA to probe the performance of each in hybrid with FANNs.

```
Begin
Initialize
(x1_ante, x2_ante) ← create random values(x)
While ≠ termination Criterion( ) do
    x1_post ← update GA(x1_ante)
    x2_post ← update BP(x2_ante)
    If f(x1_post) < f(x1_ante) then (x1_ante ← x1_post)
    If f(x2_post) < f(x2_ante) then (x2_ante ← x2_post)
Return (x1_ante, x2_ante)
End
```

Fig. 3. The steps of two evaluator systems

In Algorithm 1, the network is first initialized; and then, frequently, set the better solutions in terms of obtained errors by updating the outcome of each network. The updating procedure is continued until a method meets the stopping criterion (or criteria). Eq. (5) is a stopping condition used in this paper; where, f is fitness function, $\|.\|$ is the distance measure, and ε is an imperceptible positive error.

$$\left\|f(x_{post}) - f(x_{ante})\right\| < \varepsilon \qquad (6)$$

5   Results and Analysis

This section provides the results of applying two algorithms to construct the evaluator systems, i.e., GA- and BP-based FANNs algorithms. The results are based on data set used by [35,36,37]; which consists of nine fuzzy instances, four predictor variables, and one response variable. We applied one-leave-out cross-validation on each fold to give the results by average in 100 runs. To build each FANN method, we designed two networks of fully-connected with four input and one output units in three layers. The outcome results are analyzed to find the effect of using GA- and BP-based FANNs. The analyses are carried out in terms of obtained error from the following aspects: (i) α-cuts of fuzzy numbers, (ii) predictor variables, and (iii) response variables. At the end, the results manifest the superiority of BP-based method to decrease the error rate of response variable.

Fig. 4 shows the obtained error on each α-cut of actual and target values. We applied one-leave-out validation to test each FANN method for predicting each variable of data set. The prediction abilities for each method are given in percentage. The error rate is computed by difference between actual and desired values of support and core boundaries of each fuzzy number.

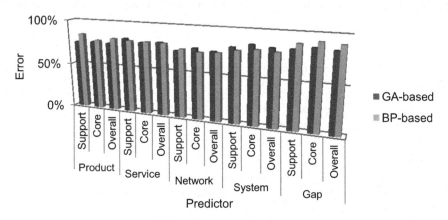

Fig. 4. Rate-Analysis for support and core α-cuts

Fig. 5 shows the analysis for each α-cut, which is based on portion of errors that support and core generates in compared to each other. We can find the superiority of a method by tracing the errors on each α-cut as they lead the overall error. In fact, the smaller size of an error we observe; the more ability of a method reveals to achieve the desired value.

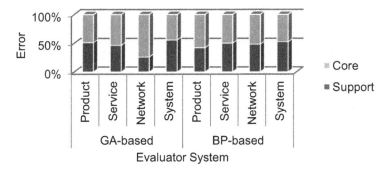

Fig. 5. Portion-Analysis for support and core α-cuts

Consequently, Fig. 6 gives the overall errors resulted by errors on each α-cut. The performance of a method depends on overall error on each predictor; obviously, the higher overall error we have on each predictor, the less performance we achieve for a method.

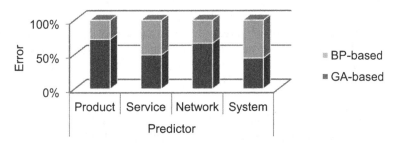

Fig. 6. Overall Input Errors

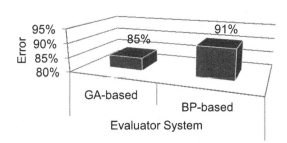

Fig. 7. Overall Output Error

Fig. 7 compares the final performance of two methods. We observe that the error rates on each α-cut of predictor affects on the response value; and thus, it reflects on performances of the applied methods. The obtained results show for FANNs that local optimizer BP is superior to global optimization GA. In other words, BP can better tune α-cuts of fuzzy variables for FANN by better finding the interconnections between input and target data. This achieved is possible only by avoiding the local minima. However, increasing the training iterations may change the performances of each method.

6   Conclusion

This paper proposed an evaluator system to predict the level of customer satisfaction. Keeping the originality of verbal opinions, expressed by customers, requires an evaluation system to process rich data; and therefore, linguistic terms demand a true representation. We used fuzzy artificial neural networks (FANNs) to process fuzzy-represented opinions, and consequently, FANNs carried the outcome in fuzzy-represented values. We applied genetic algorithm (GA) and back-propagation algorithm (BP) to find the best performance of FANNs. GA and PB were selected to find the effect of global and local optimization in performance of FANN evaluator. The results showed BP-based FANNs superior by 6% – though, analyzes on α–cuts showed that GA-based FANNs performs better in some parts. The reason is avoidance from traps of local minima, which directs the performances of each method to a rich or poor level. Future works can study whether BP-based FANNs will stand superior to GA-based FANNs with higher complex data. In addition, one can perform the attribute relevance analysis to increase the efficiency of evaluator system.

Acknowledgements. The Universiti Teknologi Malaysia (UTM) and Ministry of Education Malaysia under research university grants 00M19, 01G72 and 4F550 are hereby acknowledged for some of the facilities that were utilized during the course of this research work. It, also, has been supported by project "Smart Solutions for Ubiquitous Computing Environments" FIM, University of Hradec Kralove, Czech Republic, and by research and development in the Moravian-Silesian Region 2013 DT 1 - International research teams" (RRC/05/2013), financed from the budget of the Moravian-Silesian Region.

References

1. Kim, K.-P., Kim, Y.-O., Lee, M.-K., Youn, M.-K.: The Effects of Co-brand Marketing Mix Strategies on Customer Satisfaction, Trust, and Loyalty for Medium and Small Traders and Manufacturers (2014)
2. Walters, D.: Market Centricity And Producibility: An Opportunity For Marketing And Operations Management To Enhance Customer Satisfaction. Journal of Manufacturing Technology Management 25, 9–9 (2014)
3. Johansson, U., Anselmsson, J.: What's the Buzz about the Store? A Comparative Study of the Sources of Word of Mouth and Customer Satisfaction and their Relationships with Sales Growth. European Retail Research 26, 97–128 (2013)

4. Gómez, M.I., Shapiro, M.: Customer Satisfaction and Sales Performance in Wine Tasting Rooms. International Journal of Wine Business Research 26, 3–3 (2014)
5. Swaminathan, V., Groening, C., Mittal, V., Thomaz, F.: How Achieving the Dual Goal of Customer Satisfaction and Efficiency in Mergers Affects a Firm's Long-Term Financial Performance. Journal of Service Research (2013)
6. Luo, X., Zhang, R., Zhang, W., Aspara, J.: Do Institutional Investors Pay Attention To Customer Satisfaction And Why? Journal of the Academy of Marketing Science 1-18 (2013)
7. Rogg, K.L., Schmidt, D.B., Shull, C., Schmitt, N.: Human Resource Practices, Organizational Climate, And Customer Satisfaction. Journal of Management 27, 431–449 (2001)
8. Chandramouli, S., Krishnan, S.A.: An Empirical Study on Customer Satisfaction in Indus Motors Pvt. Journal of Business Management & Social Sciences Research 44, 38–44 (2013)
9. Nahm, Y.-E.: New Competitive Priority Rating Method Of Customer Requirements For Customer-Oriented Product Design. International Journal of Precision Engineering and Manufacturing 14, 1377–1385 (2013)
10. Pezak, L., Sebastianelli, R.: Service Quality In The US Airlines Industry: Factors Affecting Customer Satisfaction. Pennsylvania Economic Association 132 (2013)
11. Baker, D.M.A.: Service Quality and Customer Satisfaction in the Airline Industry: A Comparison between Legacy Airlines and Low-Cost Airlines. American Journal of Tourism Research 2, 67–77 (2013)
12. Chougule, R., Khare, V.R., Pattada, K.: A Fuzzy Logic Based Approach For Modeling Quality And Reliability Related Customer Satisfaction In The Automotive Domain. Expert Systems with Applications 810, 800–810 (2013)
13. Goyanes, M., Sylvie, G.: Customer Orientation On Online Newspaper Business Models With Paid Content Strategies: An Empirical Study. First Monday 19 (2014)
14. Wu, I.-L., Huang, C.-Y.: Analysing Complaint Intentions In Online Shopping: The Antecedents of Justice And Technology Use And The Mediator Of Customer Satisfaction. Behaviour & Information Technology, 1–12 (2014)
15. Subramanian, N., Gunasekaran, A., Yu, J., Cheng, J., Ning, K.: Customer Satisfaction And Competitiveness In The Chinese E-Retailing: Structural Equation Modeling (SEM) Approach To Identify The Role Of Quality Factors. Expert Systems with Applications 41, 69–80 (2014)
16. Kuo, Y.-F., Hu, T.-L., Yang, S.-C.: Effects Of Inertia And Satisfaction In Female Online Shoppers On Repeat-Purchase Intention: The Moderating Roles Of Word-Of-Mouth And Alternative Attraction. Managing Service Quality 23,168 23, 168–187 (2013)
17. Zhao, D.-Y., Ye, W.-M., Gao, C.-J., Zhang, M.-F.: Customer Requirements Analysis Method In Lean Six Sigma Project Selection Based On RAHP. In: 2013 International Conference on Quality, Reliability, Risk, Maintenance, and Safety Engineering (QR2MSE), pp. 1224–1228. IEEE (2013)
18. Zhao, X.-J., Zhao, Y.: Optimizing Six Sigma Processes to Satisfy Customers by TRIZ Innovation Methodology. In: Proceedings of 2013 4th International Asia Conference on Industrial Engineering and Management Innovation (IEMI 2013), pp. 753–759. Springer (2014)
19. Liu, P.: Evaluation Model of Customer Satisfaction of B2C E_Commerce Based on Combination of Linguistic Variables and Fuzzy Triangular Numbers. In: Eighth ACIS International Conference on Software Engineering, Artificial Intelligence, Networking, and Parallel/Distributed Computing, vol, vol. 3, pp. 450–454. IEEE (2007)

20. de Araújo Batista, D., de Medeiros, D.D.: Assessment Of Quality Services Through Linguistic Variables. Benchmarking: An International Journal 21, 29–46 (2014)
21. Zadeh, L.A.: Fuzzy logic= computing with words. IEEE Transactions on Fuzzy Systems 4, 103–111 (1996)
22. Hayashi, Y., Buckley, J.J., Czogala, E.: Fuzzy Neural Network With Fuzzy Signals And Weights. International Journal of Intelligent Systems 8, 527–537 (1993)
23. Reza Mashinchi, M., Selamat, A.: An Improvement On Genetic-Based Learning Method For Fuzzy Artificial Neural Networks. Applied Soft Computing 9, 1208–1216 (2009)
24. Aliev, R.A., Fazlollahi, B., Vahidov, R.M.: Genetic Algorithm-Based Learning Of Fuzzy Neural Networks. Fuzzy Sets and Systems 118, 351–358 (2001)
25. Krishnamraju, P., Buckley, J., Reilly, K., Hayashi, Y.: Genetic Learning Algorithms For Fuzzy Neural Nets. In: Proceedings of the Third IEEE Conference on Fuzzy Systems, pp. 1969–1974. IEEE Press (1994)
26. Zhang, X., Hang, C.-C., Tan, S., Wang, P.-Z.: The Min-Max Function Differentiation And Training Of Fuzzy Neural Networks. IEEE Transactions on Neural Networks 7, 1139–1150 (1996)
27. Liu, P., Li, H.: Approximation Analysis Of Feedforward Regular Fuzzy Neural Network With Two Hidden Layers. Fuzzy Sets and Systems 150, 373–396 (2005)
28. Ishibuchi, H., Nii, M.: Numerical Analysis Of The Learning Of Fuzzified Neural Networks From Fuzzy If–Then Rules. Fuzzy Sets and Systems 120, 281–307 (2001)
29. Mashinchi, M.H., Mashinchi, M.R., Shamsuddin, S.M.H.: A Genetic Algorithm Approach for Solving Fuzzy Linear and Quadratic Equations. World Academy of Science, Engineering and Technology 28 (2007)
30. Mashinchi, M.R., Mashinchi, M.H., Selamat, A.: New Approach for Language Identification Based on DNA Computing. In: BIOCOMP, pp. 748–752 (2007)
31. Sexton, R.S., Dorsey, R.E., Johnson, J.D.: Toward Global Optimization Of Neural Networks: A Comparison Of The Genetic Algorithm And Backpropagation. Decision Support Systems 22, 171–185 (1998)
32. Gupta, J.N., Sexton, R.S.: Comparing Backpropagation With A Genetic Algorithm For Neural Network Training. Omega 27, 679–684 (1999)
33. Horikawa, S.-I., Furuhashi, T., Uchikawa, Y.: On Fuzzy Modeling Using Fuzzy Neural Networks With The Back-Propagation Algorithm. IEEE Transactions on Neural Networks 3, 801–806 (1992)
34. Liu, P., Li, H.-X.: Fuzzy Neural Network Theory And Application, vol. 59. World Scientific (2004)
35. Reza Mashinchi, M.: Ali Selamat: Measuring Customer Service Satisfactions Using Fuzzy Artificial Neural Network with Two-phase Genetic Algorithm. InTech (2010)
36. Fasanghari, M., Roudsari, F.H.: The Fuzzy Evaluation Of E-Commerce Customer Satisfaction. World Appl. Sci. J. 4, 164–168 (2008)
37. Mashinchi, M.R., Selamat, A.: Constructing A Customer's Satisfactory Evaluator System Using GA-Based Fuzzy Artificial Neural Networks. World Appl. Sci. J. 5, 432–440 (2008)

Hybrid Particle Swarm Optimization Feature Selection for Crime Classification

Syahid Anuar, Ali Selamat, and Roselina Sallehuddin

Faculty of Computing, Universiti Teknologi Malaysia
syah2105@yahoo.com, {aselamat,roselina}@utm.my

Abstract. In this study, we propose a hybrid crime classification model by combining artificial neural network (ANN), particle swarm optimization (PSO) and grey relation analysis (GRA). The objective of this study is to identify the significant features of the specific crimes and to classify the crimes into three different categories. The PSO as the feature selection method, reduce the dimension of datasets by selecting the most significant features. The reduction of the datasets' dimension may reduce the complexity thus shorten the running time of ANN to classify the crime datasets. The GRA is used to rank the selected features of the specific crimes thus visualize the importance of the selected crime's attribute. The experiment is carried out on the Communities and Crime dataset. The result of PSO feature selection will then compare with the other feature selection methods such as evolutionary algorithm (EA) and genetic algorithm (GA). The classification performance for each feature selection method will be evaluated. From our experiments, we found that PSO select less features compare with EA and GA. The classification performance results show that the combination of PSO with ANN produce less error and shorten the running time compare with the combination of EA with ANN and GA with ANN.

Keywords: Crime Classification, Particle Swarm Optimization, Grey Relation Analysis, Artificial Neural Network, Hybrid Artificial Neural Network, Crime Prevention.

1 Introduction

The increasing volume of the crimes had brought serious problems to the community in any country. However, the increase in realization of information technology has open a new door for government to include crime prevention component as a strategy to reduce crime. Classification is one of the data mining technique which can be use to analyze the crime patterns. The data mining approach can help to detect the crime patterns and speed up the process of solving crime [17]. Due to this reasons, research on crime classification has increase because of the potential and effectiveness of the classification in crime prevention programs. Several study on crime classification have been done by several researchers [10, 11, 16]. The crime can be divided into several types and the most common findings at the city level are crimes against property such as

© Springer International Publishing Switzerland 2015   101
D. Barbucha et al. (eds.), *New Trends in Intelligent Information and Database Systems*,
Studies in Computational Intelligence 598, DOI: 10.1007/978-3-319-16211-9_11

burglary, robbery and theft and crime of aggression such as assault, homicides and rape [9]. Each crime's types can be classified into several categories such as low, medium and high. Thus, the objective of this study is to propose a new hybrid classifier to classify the crimes into given categories.

The artificial neural network (ANN) is a model inspired by the connected neurons in the brain. ANN with back-propagation learning algorithm is usually use as a benchmark model for any classifier. However, ANN is a black-box learning approach where it cannot determine automatically the significant input features. To overcome the limitation of ANN in choosing relevant features as input, thus the feature selection is needed. Inspired by bird flocking or fish schooling, particle swarm optimization (PSO) is one of the popular optimization techniques that have the capability to perform the feature selection task [19]. Grey relation analysis (GRA) is a method of analysis proposed in Grey system theory [8]. The GRA also can be used to rank the data according to its importance [3]. Therefore, in this paper, we intend to combine PSO and ANN as a hybrid classifier and using GRA to rank the crime attributes to its importance.

2   Related Works

A general framework for crime data mining has been introduced by Chen et al. [7] as the initial idea of using data mining technique in crime's domain. After that, the data mining techniques in crime was applied by Adderley [2] where in this study, self organising map (SOM) was applied to recognize burglary offences. Nath [17] has proposed data mining technique such as k-means to identify crime pattern using real crime data from a sheriff's office. Ozgul et al. [20] have investigated the sources of crime data mining to forms knowledge discovery that is suitable for which methodologies. Nissan [18] has studied the application of data mining in two forensic fields such as intelligence and investigative tasks.

Classification is one of the data mining techniques. Abu Hana et al. [1] have proposed crime scenes classification using neural network and support vector machine to classify violent crime into two classes which are attack from inside or outside of the scene. Futhermore, Yang et al. [22] have used classification methods such as decision tree, random forest, support vector machine and neural network to predict the relationships between murder victims and the offenders. Kim et al. [13] examine the relationship between weather and crime using decision tree algorithm. Kamiran et al. [12] have proposed discrimination aware classification to predict whether an individual is a crime suspect or not.

Iqbal et al. have proposed a classification methods to applied on crime dataset to predict crime categories. The crime dataset can be downloaded from UCI machine learning repository website [5]. The crime categories which are low, medium and high were added based on attribute name 'Violent Crime Per Pop'. The crime categories need to be added as goals or classes to enable classification algorithms to perform the prediction [10]. However, only statistical techniques such as Naive Bayes and decision tree were considered to classify the crime dataset. Later, the feature selection method and machine learning method have been applied by Anuar et al. [4] as a continuation of previous work and experimental results show improvement in terms of classification accuracy. To further improve both experiments, we proposed a crime classification model. The classification model consist of feature selection method and classification method. PSO has been selected as feature selection method based on the encouraging experiment result. The GRA will be used to rank the crime's attributes in order of importance and ANN will be used as a classification method. The next section will discuss in detail the proposed crime classification model.

3   Proposed Crime Classification Model

3.1   Particle Swarm Optimization

A population in PSO is called a swarm. The candidate solutions are encoded as particles in the search space. The PSO starts by randomly initialize the population of particles. The whole swarm move in the search space to search for the best solution by updating the position of each particle according to Eqn 1.

$$x_{id}^{t+1} = x_{id}^t + v_{id}^{t+1} \tag{1}$$

The movement of the current position of the particle i is represented by a vector $x_i = (x_{i1}, x_{i2}, ..., x_{iD})$, where D is the dimension of the search space. The velocity of the particle i is represented as $v_i = (v_{i1}, v_{i2}, ..., v_{iD})$ which is limited by a predefined maximum velocity, $v_{id}^t \in [-v_{max}, v_{max}]$. The equation for velocity is given by Eqn 2

$$v_{id}^{t+1} = w * v_{id}^t + c_1 * r_{1i} * (P_{id} - x_{id}^t) + c_2 * r_{2i} * (P_{gd} - x_{id}^t) \tag{2}$$

where t denotes the tth iteration, $d \in D$ denotes the dth dimension in the search space, w is the inertia weight, $c1$ and $c2$ are acceleration constants or sometimes called the learning rate, r_{1i} and r_{2i} are the random numbers uniformly distributed within the range of $[0,1]$, p_{id} represent the element of the solution for the particle's individual best *pbest*, and p_{gd} is the element of the solution for the particle's global best, *gbest* in the dth dimension. Generally, the PSO algorithm consists of three steps which are repeated until some stopping condition is met [21]:

1. Evaluate the fitness of each particle.
2. Update the individual and the global best fitnesses and positions.
3. Update the velocity and the position of each particle.

For the purpose of the feature selection, the required particle number and the initial coding alphabetic string for each particle is randomly produced. Each particle encoded to imitate a chromosome, like in the genetic algorithm. Each particle was coded to a binary alphabetic string $S = F_1, F_2, ..., F_n, n = 1, 2, ..., m$ where m are the number for features. The bit value 1 represents a selected feature and the bit value 0 represents a non-selected feature. The *gbest* value consider as the fitness value.

3.2   Grey Relation Analysis(GRA)

The grey relation analysis (GRA) is a method of analysis, proposed in the Grey system theory. GRA is a distinct similarity measurement that use data series to obtain grey relational order to describe the relationship between the related series. Therefore, it is significant to use GRA to analyze the relationship between features and the type of crimes. GRA consists of two parts which are reference sequence and compared sequence. The following are the basic steps in the GRA [14]:

1. The type of crimes is set as the reference sequence, $X_0(k), k = 1, 2, 3, ..., M$ where M = number of instances in datasets and compared sequence $X_i(k)$, $i = 1, 2, ..., N$ where N = number of features, is the features after feature selection process. Fig. 1 shows the example of decision matrix for crime dataset.

$$(x_0, x_1, ..., x_N) = \begin{bmatrix} x_0(1) & x_1(1) & ... & x_N(1) \\ x_0(2) & x_1(2) & ... & x_N(2) \\ \cdot & \cdot & \cdot & \cdot \\ \cdot & \cdot & \cdot & \cdot \\ \cdot & \cdot & \cdot & \cdot \\ x_0(M) & x_1(M) & ... & x_N(M) \end{bmatrix}$$

Fig. 1. M × N decision matrix [15]

2. Normalization of the sequence to dimensionless form, is calculated according to Eqn 3:

$$x_i(k) = \frac{X_i(k) - X_{min}(k)}{X_{max}(k) - X_{min}(k)} \tag{3}$$

3. Calculate the grey relational coefficient, ξ, between X_0 and X_i is calculated according to Eqn 4:

$$\xi_i(k) = \frac{\min_i \min_k \triangle_i(k) + \rho \max_i \max_k \triangle_i(k)}{\triangle_i(k) + \rho \max_i \max_k \triangle_i(k)} \tag{4}$$

where $\triangle_i(k) = |x_0(k) - x_i(k)|$ is the absolute difference of index k. ρ is the distinguishing coefficient and in most situations takes the value of 0.5 because this value usually offers moderate distinguishing effects and good stability [6].

4. Calculate the grey relational grade (GRG) according to Eqn 5:

$$r_i = \frac{1}{n} \sum_{i=1}^{n} \xi_i(k) \tag{5}$$

4   Experiment Setup

4.1   Dataset Preparation

The Communities and Crime dataset is obtained from the UCI Machine Learning Repository. This dataset focus on the communities in United States of America (USA). The data comprises of socio-economic data from the 90 Census, law enforcement data from the 1990 Law Enforcement Management and Admin Stats survey and crime data from the 1995 FBI UCR. The dataset consists of 147 attributes and 2215 instances including missing values [5]. Data preparation is essential for successful data classification. Poor quality data typically results in incorrect and unreliable classification results. Thus, the following data preparation mechanisms were carried out:

- The dataset is divided to three type of crimes which are assault, burglaries and murders.
- All the attributes with large number of missing values are remove
- Following the work done by Iqbal et al. [10], the newly added nominal attribute named 'Categories'is created from each type of crime for classification purpose - if the value is less than 25% than the Categories is 'Low'. If the value is equal to or greater than 25% than the Categories is 'Medium'. If the value is equal to or greater than 40% than the Categories is 'High'.
- All the data will be normalize to [0, 1] using the min-max method represented by Eqn 3.

4.2   Parameters Setting

Table 1 shows the parameter setting for GA, EA, and PSO. The parameters setting for all algorithms is based on the best result obtained from several self-experiments.

Table 1. Parameters Setting

PSO	EA	GA
C1:2.0	Crossover rate: 0.6	Crossover rate: 0.6
C2:2.0	Mutation rate: 0.01	Mutation rate: 0.033
Number of particles: 200	Population size: 200	Population size: 200

5   Result and Discussion

5.1   Feature Selection Results

Fig. 2 shows the result of feature selection for each method. The result shows
that PSO selects lowest features for assault, burglaries and murders. For assault,
PSO selects 4 significance features compare with EA and GA which are 7 and
15. For burglaries, PSO selects only 3 significance features compare with EA
which is 7 and GA which is 12. The PSO selects 7 significance features while EA
selected 12 and GA selected 8 significance features for the murders.

For all types of crimes, PSO has successfully selected the lowest number of
features. Hence, after the feature selection process, GRA will rank the selected
features thus visualize the importance of each feature.

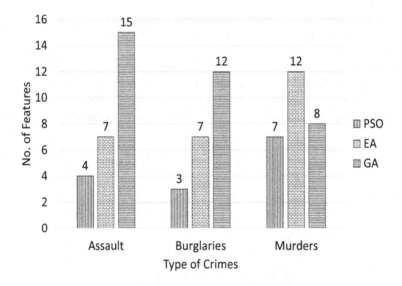

Fig. 2. Feature selection results

Table 2 shows the result of GRA after the feature selection process by PSO.
Each selected features are ranked to its importance based on GRG. Higher GRG
meaning that the features are the most important. This ranked features may
reflect the importance of the attributes of crime and can make it easier for the
police to prioritize the work force.

Table 2. The rank of crime features

Type of crimes	Features Number	Features Name	Features Description	GRG	Rank
Assault	4	pctAsian	percentage of population that is of Asian heritage	0.8349	1
	9	pct65up	percentage of population that is 65 and over in age	0.7235	2
	24	pctEmployProfServ	percentage of people 16 and over who are employed in professional services	0.6184	3
	64	pctUsePubTrans	percent of people using public transit for commuting	0.3997	4
Burglaries	49	pctWfarm	percentage of households with farm or self employment	0.8391	1
	63	pctSpeakOnlyEng	percent of people who speak only English	0.3902	2
	64	pctSameCounty-5	percent of people living in the same state	0.3328	3
Murders	4	pctAsian	percentage of population that is of Asian heritage	0.8427	1
	29	pctFemDivorc	percentage of households with farm or self-employment income	0.7713	2
	31	pct2Par	percentage of families (with kids) that are headed by two parents	0.6217	3
	35	pctWorkMom-6	percentage of moms of kids 6 and under in labour force	0.4849	4
	37	pctKidsBornNevrMarr	percentage of kids born to never married	0.3682	7
	41	pctFgnImmig-10	percentage of immigrants who immigrated within last 10 years	0.5399	6
	63	pctSameHouse-5	percent of people living in the same house	0.4208	5

5.2 Classification Results

The classification performances are evaluated using several criteria. Root Mean Square Error (RMSE) and Mean Absolute Error (MAE) will be used to measure the performance of ANN. Accuracy, precision, recall and f-measure will be used to measure the classification's performance. Table 3 shows the results of RMSE and MAE for each crime. The results for assault indicate that PSO-ANN gives lowest RMSE compare with EA-ANN and GA-ANN. For Burglaries, the PSO-ANN generates lowest RMSE but ANN generate lowest MAE. For murders, the ANN give lowest MAE value.

Table 3. Artificial neural network performance results

Type of crimes	ANN		PSO-NN		EA-NN		GA-NN	
	RMSE	MAE	RMSE	MAE	RMSE	MAE	RMSE	MAE
Assault	0.2155	**0.0641**	**0.1792**	0.0647	0.1797	0.0651	0.1830	0.0674
Burglaries	0.1296	**0.0203**	**0.1212**	0.0290	0.1222	0.0283	0.1238	0.0287
Murders	0.2025	**0.0566**	0.1827	0.0671	0.1852	0.0688	0.1827	0.0673

Note: The bold value indicates the lowest error.

Table 4 shows the results of classification performance and processing time for various classification models. These results aim to demonstrate the effectiveness of feature selection methods on various crime datasets.

Table 4. Comparison of classification result and processing time

Type of crimes	Models	Accuracy (%)	Precision (%)	Recall (%)	F-measure	Time(s)
Assault	ANN	92.27	90.60	92.30	91.40	1480
	PSO-ANN	94.99	90.20	95.00	92.50	26
	EA-ANN	94.98	90.20	95.00	92.50	43
	GA-ANN	94.89	90.20	94.90	92.50	108
Burglaries	ANN	97.33	95.50	97.30	96.40	1661
	PSO-ANN	97.74	95.50	97.70	96.60	24
	EA-ANN	97.69	95.50	97.70	96.60	43
	GA-ANN	97.65	95.50	97.70	96.60	74
Murders	ANN	93.45	90.00	93.50	91.70	1504
	PSO-ANN	94.76	89.90	94.80	92.30	44
	EA-ANN	94.76	89.90	94.80	92.30	74
	GA-ANN	94.80	89.90	94.80	92.30	47

The results from Table 4 indicate that PSO-ANN produce lowest running time for each type of crimes. For assault, PSO-ANN give 26 seconds compared with

EA-ANN which give 43 seconds. The GA-ANN give the highest running times which are 108 seconds. The PSO-ANN give 24 seconds for burglaries compare with EA-ANN which give 43 seconds, and the highest 74 seconds for GA-ANN. For murders, the PSO-ANN give 44 seconds compared to EA-ANN which give 74 seconds and GA-ANN which give 47 seconds. The results prove that PSO has produced the lowest number of features thus reduce the running time of ANN. Although the results for accuracy, precision, recall and f-measure are so close, the processing time for each model shows the effectiveness of feature selection. This suggests that by using a small number of features can also produce high accuracy in a short time.

5.3   Conclusions

This paper presents a hybrid application of PSO and GRA in crime classification modelling. The PSO as a feature selection method to identify and obtain the significant features for different type of crimes such as assault, burglaries and murders. The GRA is used to rank the significant features of each type of crimes. The ANN is used to classify each crime into three different categories which are low, medium and high. The experiment result indicates that the PSO select lowest number of features thus shortens the running time of ANN. Moreover, the GRA has ranked the selected features of it important thus visualize the crimes dataset to be more specific. This would help the police to prioritize the important attributes for the specific type of crimes. Finally, the experimental results have proved that PSO-ANN is an acceptable model to analyze crime data.

Acknowledgement. The Universiti Teknologi Malaysia (UTM) and Ministry of Education Malaysia under research Grant 01G72 and Ministry of Science, Technology & Innovations Malaysia, under research Grant 4S062 are hereby acknowledged for some of the facilities that were utilized during the course of this research work.

References

1. Abu Hana, R.O., Freitas, C.O., Oliveira, L.S., Bortolozzi, F.: Crime scene classification. In: Proceedings of the 2008 ACM Symposium on Applied computing, pp. 419–423. ACM Press (2008)
2. Adderley, R.: The use of data mining techniques in operational crime fighting. In: Chen, H., Moore, R., Zeng, D.D., Leavitt, J. (eds.) ISI 2004. LNCS, vol. 3073, pp. 418–425. Springer, Heidelberg (2004)
3. Alwee, R., Shamsuddin, S.M.H., Sallehuddin, R.S.: Economic indicators selection for crime rates forecasting using cooperative feature selection. In: Proceeding of the 20th National Symposium on Mathematical Sciences Research in Mathematical Sciences: A Catalyst for Creativity and Innovation, vol. 1522, pp. 1221–1231. AIP Publishing (2013)
4. Anuar, M.S., Selamat, A., Sallehuddin, R.: Particle swarm optimization feature selection for violent crime classification. In: Advanced Approaches to Intelligent Information and Database Systems, pp. 97–105. Springer (2014)

5. Bache, K., Lichman, M.: UCI machine learning repository (2013),
   http://archive.ics.uci.edu/ml
6. Chang, T.C., Lin, S.J.: Grey relation analysis of carbon dioxide emissions from
   industrial production and energy uses in taiwan. Journal of Environmental Management 56(4), 247–257 (1999)
7. Chen, H., Chung, W., Xu, J.J., Wang, G., Qin, Y., Chau, M.: Crime data mining:
   a general framework and some examples. Computer 37(4), 50–56 (2004)
8. Deng, J.-L.: Introduction to grey system theory. The Journal of Grey System 1(1),
   1–24 (1989)
9. Gorr, W., Olligschlaeger, A., Thompson, Y.: Assessment of crime forecasting accuracy for deployment of police. International Journal of Forecasting (2000)
10. Iqbal, R., Murad, M.A.A., Mustapha, A., Panahy, S., Hassany, P., Khanahmadli-
    ravi, N.: An experimental study of classification algorithms for crime prediction.
    Indian Journal of Science & Technology 6(3) (2013)
11. Theresa, M.J., Raj, V.J.: Fuzzy based genetic neural networks for the classification
    of murder cases using trapezoidal and lagrange interpolation membership functions.
    Applied Soft Computing 13(1), 743–754 (2013)
12. Kamiran, F., Karim, A., Verwer, S., Goudriaan, H.: Classifying socially sensitive
    data without discrimination: an analysis of a crime suspect dataset. In: 2012 IEEE
    12th International Conference on Data Mining Workshops (ICDMW), pp. 370–377.
    IEEE (2012)
13. Kim, J.-M., Ahn, H.-K., Lee, D.-H.: A study on the occurrence of crimes due
    to climate changes using decision tree. In: IT Convergence and Security 2012,
    pp. 1027–1036. Springer (2013)
14. Kung, C.-Y., Yan, T.-M., Chuang, S.-C., Wang, J.-R.: Applying grey relational
    analysis to assess the relationship among service quality customer satisfaction and
    customer loyalty. In: 2006 IEEE Conference on Cybernetics and Intelligent Sys-
    tems, pp. 1–5. IEEE (2006)
15. Lu, J., Chen, P., Shen, J., Liang, Z., Yang, H.: Study on the prediction of gas
    content based on grey relational analysis and bp neural network. In: Proceedings
    of The Eighth International Conference on Bio-Inspired Computing: Theories and
    Applications (BIC-TA), 2013, pp. 677–685. Springer (2013)
16. Nasridinov, A., Ihm, S.-Y., Park, Y.-H.: A decision tree-based classification
    model for crime prediction. In: Information Technology Convergence, pp. 531–538.
    Springer (2013)
17. Nath, S.V.: Web Intelligence and Intelligent Agent Technology Workshops. In: 2006
    IEEE/WIC/ACM International Conference on. WI-IAT 2006 Workshops, pp. 41–44.
    IEEE (2006)
18. Nissan, E.: An overview of data mining for combating crime. Applied Artificial
    Intelligence 26(8), 760–786 (2012)
19. Omar, N., Shahizan, M., Othman, o.b.: Particle swarm optimization feature se-
    lection for classification of survival analysis in cancer. International Journal of
    Innovative Computing 2(1) (2013)
20. Ozgul, F., Atzenbeck, C., Çelik, A., Erdem, Z.: Incorporating data sources and
    methodologies for crime data mining. In: 2011 IEEE International Conference on
    Intelligence and Security Informatics (ISI), pp. 176–180. IEEE (2011)
21. Shi, Y., Eberhart, R.: A modified particle swarm optimizer. In: The 1998 IEEE
    International Conference on Evolutionary Computation Proceedings, IEEE World
    Congress on Computational Intelligence, pp. 69–73. IEEE (1998)
22. Yang, R., Olafsson, S.: Classification for predicting offender affiliation with murder
    victims. Expert Systems with Applications 38(11), 13518–13526 (2011)

Identification of Shill Bidding for Online Auctions Using Anomaly Detection

Grzegorz Kołaczek and Sławomir Balcerzak

Wroclaw University of Technology, Wybrzeze Wyspianskiego 27 str.
50-370 Wroclaw, Poland
Grzegorz.Kolaczek@pwr.edu.pl

Abstract. The paper presents a novel method of shill biding frauds detection in online auctions. The main idea behind the method is a reputation system using anomaly detection techniques. The system focuses on cases where the final price can be inflated by interference of persons who are colluding with the seller. The main aim of the work was to support users of online auctions systems by mechanisms which would be able to detect this type of frauds. The proposed method of shill bidding identification has been implemented using statistical analysis software and data derived from the test bed provided by one of the leading online auction houses. The other aim of the research was to assess whether the proposed solution is better than previous approaches described in the literature and how well the systems are able to detect real frauds. The presented system has been validated using some experimental data obtained from real world auction systems and specially generated with application of domain specific tools. Study confirmed that proposed system was able to detect most frauds related to the artificial price inflation.

Keywords: anomaly detection, reputation systems, security of online auctions, bid shilling, frauds.

1 Introduction

Internet auctions are largely a reflection of traditional auctions. The vendor sells the items while the participant of the auction should give their offers. Each new offer should be higher than the previous one. The winner of the auction is the person who offered the highest bid. However, auction websites offer something more, which makes using them much more than the traditional ones. This is of course thanks to the wide availability through the Internet, but also by viewing many different offers from different, often very distant places of the world and giving the feeling that the goods are purchased at a bargain price. The main idea about online auction systems, was to create a place for virtual meetings customer and the seller, where they can transact as is the case with traditional trade. It turned out that the new form of contact between the customer and the seller, in addition to many of the benefits, carries a lot of risk. No direct contact makes the seller more than in the traditional trade is susceptible to act unfairly in relation to their customers [1,2]. Despite the higher risk of fraud,

© Springer International Publishing Switzerland 2015
D. Barbucha et al. (eds.), *New Trends in Intelligent Information and Database Systems*,
Studies in Computational Intelligence 598, DOI: 10.1007/978-3-319-16211-9_12

shopping via the internet and also Internet auctions, are very popular. This can be seen in the increase in the number of users in the world's leading auction sites.

There are some typical types of frauds associated with online auctions. One of the most frequent is fraud related to the delivery of goods inconsistent with the description. Another type of dishonest behavior is difficult to qualify as a crime and it is called "shill bidding" A person who wants to sell an item at a profit greater than the market indicates, hires some people, who give higher bids, so the final price reached is the highest possible value. "Shilling" is possible in real world auction but in the virtual world, this thing is pretty simple. Putting more accounts on the most popular auction sites is no longer tightened such restrictions as it was a few years ago. Auction sites allow to create new accounts without thorough verification of the identity of users. Thus, the organization of a sufficiently large group of "Shills" is not a very difficult task [5,6].

The effective discovery of frauds in online auction systems requires the examination of the vast amount of data using appropriate methods in the field of data mining [7]. However there are some specific elements which can make more effective the detection of acts of shilling. Accounts used by dishonest seller usually have little feedback comments. This is because cheaters use alternative accounts because they do not want to risk their main account to be banned. The main account is necessary to build a reputation of seller and to attract new buyers [8]. Another important symptom of cheating may be when the same bidder frequently appears at various auctions of the seller. Typically, the scammer has a limited number of accounts that can make bids on their auctions [9]. Accounts, which are used by scammers usually are relatively new. If the account have been created recently, has a small number of comments and takes part in several auctions of the same seller, it is likely that it is used to shilling [5]. Additionally some online auction platforms (eg. eBay) allows to change final results of betting. This is when even if you win, you can withdraw your offer, thus making the winner the second bidder. Accounts that are involved in the shilling process sometimes have a history with such a withdrawal. They are obviously a bad sign in the history of the user, so for normal accounts are very rare, and at the hand side are often in the case of rogue users accounts.

The paper presents a reputation system which is an extension of the proposition presented by Rubin at.al. in [4]. The system defines new methods taking into account the specialization of the seller, the number of rejected bids, and the number of bidders strongly associated with the dealer. New methods are designed to increase the efficiency of detection of incidents.

2   Reputation Systems

The main objective of the reputation system is grading a set of objects. Grading is the result of appropriate algorithms which perform data analysis on data sets describing the history related to a particular object [10]. The objects are usually defined as users of auction systems or web shops, web pages, the borrower (credit rating), or even countries. The assessment may involve a wide range of aspects depending on the field of application. In the case of online auction platforms, may be taken into account opinions issued by counterparties or additionally the prices between competing entities could be compared. When it comes to shopping online, this is the most important customer rating or comparing prices between shops competing [11]. The reputation is therefore an essential tool in decision making process that require a good level of security.

2.1   Online Auctions and Frauds

The relation between the final price of the item and the shill bidding has been analyzed in the work [14]. The main thesis of this work was "the final price of the item that is higher than it was expected is a sign of the shill bidding".   This thesis was verified using several datasets containing the records related to real auctions. The proposed method was based on artificial neural network called Large Memory Storage and Retrieval (LAMSTAR). The network was used in prediction of the final price of the item. The shortcoming of the method is the network can only be trained to predict correctly the prices of one item. Additionally, the experiment has been performed using 'Nintendo Wii' prices. During the experiment this was a new product and so the spread of the prices was very limited. Probably, this also could have some impact on the final result of the experiment. Additionally, there are many product which are the subjects of online auctions for which the final price depends from very many different elements. In such cases it would be very difficult and inefficient to train the neural network to produce the correct predictions. The examples of such items could be PCs or laptops which have very many attributes which induces the final value of the product.

The work [12] presents the very important element for accurate anomaly detection in online auctions. This element which strongly influence the final price is the context of the auction. In this research the context was defined by time of the auction, more strictly it was the time remaining to the Christmas. The results obtained in this work show not only how difficult is to predict correct auction prices but also it presents the complexity of the reasoning process which must be done in any attempt to analysis online auction systems.

The research results presented in [15] demonstrates the way how it is possible to extend and improve the native reputation systems. The authors aim was to show how it is possible to fix the main problems of the simple reputation systems which use buyers and sellers comments. The main elements of the prosed method include: the maximum value of the reputation is equal to 1, the new comments have greater impact on the reputation value, each comment is connected to value within the range from 0 till 1, the value of the transaction influences the reputation value, the comments of the buyer have greater impact on the final reputation than the comments of the seller. Additionally the buyer can get the chart illustrating the last changes of seller reputation. The proposed in [15] method of reputation value calculation is a good example that it is possible to improve the functionality of reputation system by adding only a few modifications to existing naïve method.

3   Models of User Activity in Online Auction System

The proposition presented in this paper extends the approach proposed in the "An Auctioning Reputation System Based on Anomaly Detection" [4]. It verifies the relation between the average number of bids per auction (Bid's Per Auction – BPA) and the total number of all auctions of the seller. The observations of the online auctions resulted in the statement that the average number bids per auction BPA is inversely proportional to the number of all auctions governed by the selected subject. It means that sellers with the high number of their auctions have low values of BPA. Contrary, the sellers with only single auction have a high values of BPA. Let call this approach to buyer/seller behavior modeling - Model I.

The next model (Model II) uses the relation between the average value of BPA and the minimal price. This minimum price is just starting price. To compare the minimum prices of different types of items it can be used a measured called "relative minimum starting bid (RMB)" introduced in [4]. For each auction this coefficient can be calculated in the following way

$$RMB = (Ns-Cm)/Ns \tag{1}$$

where: RMB – coefficient of the minimum price, Ns – the highest bid, Cm – the minimum price.

This model can explain why some points plotted as an illustration of the previously described relation, lie far from the majority of other points. The model has been derived from the observation that the items with the high minimum price are much less attractive for bidders, so they get less bids.

The third model (Model III) described in [4] investigates the population of buyers related to the particular seller. In this model the aim of the analysis is to find out if there is some specific group of users which takes part in the auctions of the seller. Model III can be used to detect anomalous behavior during the selected seller auctions. This anomalous behavior can be identified by drawing 'cumulative bidder wins curve'. This curve can used to identify the situations where the majority of bids is performed by minority of users and additionally, the users more active during the auctions do not win.

3.1   Extension of the Reputation Models

The proposed in this paper a new Model IV focuses on the problem of identification of the group of buyers which take part in the auctions of the selected seller constantly. This information is interesting because such group may be a part of group of associated accounts which enable to perform shill bidding. To calculate the appropriate for this model coefficients the information provided by eBay has been used. For example eBay provides up to date statistics called 'bid activity (%) with this seller' and 'items bid on'. The first one corresponds to the number of auctions of particular seller in which the buyer was active. The second informs about the number of auctions of in which the selected buyer was active. From the analysis performed where the data collected from eBay has been used it can be stated that there is a strong correlation between the value 'Bid Activity with this seller' and 'Items bid on'. The type of the relation is reversely proportional it means that the more often buyer takes part in different auctions it is less probably that he/she takes part in auction of some   particular seller (Fig. 1).

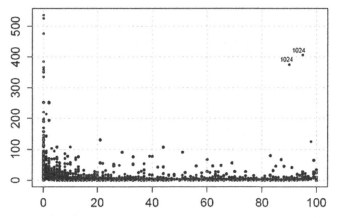

Fig. 1. Number of bids vs number of auctions with the same seller

The user who made a separate bids on multiple auctions of the seller, could do it because he/she was especially interested in the assortment of this seller. However it could be recognized as the anomalous behavior. Such situation can be noticed in the Fig. 1 as two points 1024. This two points denotes the buyers who were very active in the period of observation, but theirs offers were related only to auctions led by the seller described as 1024.

Model IV take into account the number 'Bid retractions' value. The option of bid retraction is available in some auction platforms and for example eBay platform allows to retract the bid under several conditions described in the platform regulations. Despite that 'retractions' are legal, it is important to investigate more closely this element because also it can be used in bid shilling. The exemplary plot describing the relation between the 'retraction' value and number of bids is presented in the Fig. 2 This plot illustrates the regularity that the users which are taking part in auctions more frequently, less frequently retract their bids. There is also another important information that can be derived from this plot, that sellers no. 39, 36 and 13 have buyers who very frequently retracted their bids (from 40 till 140 bids were retracted by them).

Next proposed model (Model V) adds another feature to the process of shill bidding detection. This feature is the specialization of a seller. The assumption is that the seller with a greater specialization has more constant buyers than the seller with low specialization. The specialization coefficient will be denoted by $SPEC_s$ and its value will be calculated using the following formula:

$$SPECs = (LMs - LKs)/LMs \tag{2}$$

where:

$SPEC_s$ – specialization of the seller s, s – number of seller, $LK_s = \max(K_s)$ – number of auctions in the category where the seller sells the greatest number of items, $LM_s = \sum_{i=1}^{n} k_i$, where $\{k \in K\}, n = |K|$ - the total number of auctions for the seller s.

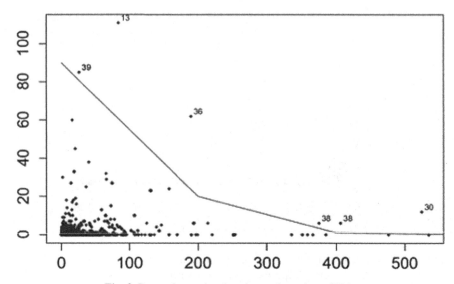

Fig. 2. Retractions related to the total number of bids

The values of $SPEC_s$ are closer to 0 when the seller is highly specialized and the value is closer to 1 when seller has no specialization.

Intuition in the context of seller specialization is as follows. The seller specialized in a particular category can have a group of trusted buyers who often follow the auctions of the seller. This assumption is confirmed by linear regression analysis calculated for the variables: degree of specialization, and the percentage of auctions in repeating group of bidders. The result is shown in Fig. 3. For high specialization (specialization ratio close to 0) more frequently buyers come from the 'trusted' group of bidders. The area outside the upper limit of the confidence interval is regarded as abnormal. In turn, little specialized sellers do not have such a trusted group of bidders, hence the results oscillate between 40% participation in the auction.

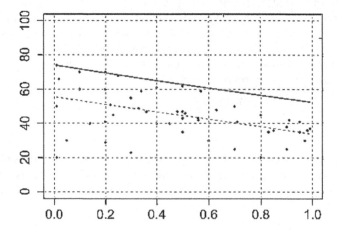

Fig. 3. Relation between the percentage of trusted buyers and the specialization of the seller

4   Reputation Model and Its Verification

The reputation system must allow for comparison of activity of different users within online auction platform. The value returned by reputation system must be strictly related to user behavior and the values must reflect the trust level and be based on the historical records. The proposed in this work reputation system intends to be specialized in detection of specific form of frauds undergoing in online auction systems, which are called 'shill bidding'. The presented solution benefits from all five previously described models of buyer/seller activity. It means that the final reputation level is calculated using the results of analysis with all above presented models. The final value of reputation will be calculated using the formula:

$$R = \frac{\sum_{i=1}^{5} p_i}{5} \tag{3}$$

Where: R – reputation $R \in [0,1]$, p - the partial evaluation [0,1] (1 - normal, 0 - abnormal).

The reputation system uses an eBay API and the local database which collects all needed data for analysis. The aim of the performed experiments was a verification of the effectiveness of the reputation system. The problem is so complex that using only data from the real eBay auctions, it would be impossible to verify whether the anomaly is definitely related to fraud [3],[13]. This is also because the auction platforms do not provide data containing the list of frauds and list of blocked accounts and history of their last activity. Verification of the proposed models will be also carried out by methods using crafted datasets.

4.1   Verification with the Original Dataset from eBay

The data used in the experiment were collected from eBay.com during the observation period from 01.10.2013 till 01.11.2013. For each model were prepared corresponding charts that facilitated the understanding of the relations among buyers and sellers, but due to lack of space, in this work only numerical results will be presented. The reputation level observed at the output of the system gives the degree of abnormality of the observation in comparison to the behavior of the rest of the population.

The results of reputation system considering all five proposed models have been presented in Table 1. As it can be noticed, both approaches have produced different results. The first of them qualified for the user with id: 8 as a potential shill bidder (two models indicated an anomaly), and the user with id: 19 as a shill bidder with strong evidence of fraud (three models indicated an anomaly). The second approach defined user with ID 8 as normal, and with ID: 19 as a potential shill bidder, but with a smaller certainty level than the first system. In addition, a second system indicated as a potential shill bidder user with id: 11, and a user with id: 61 with strong premises. Both systems give different results, and using only this type of experiment and data available it is impossible to find out which one is better. The next experiments described in next sections will give more detailed view of the similarities and differences of both approaches.

Table 1. Summary of reputation value calculated using Model I-V

Seller	Model I	Model II	Model III	Model IV	Model V	Rep.	Classification
8	A	N	N	N	N	N(4/5)	Normal
11	A	N	N	A	A	A(2/5)	Potential shill bidder
13	A	N	N	N	A	N(3/5)	Normal
21	A	N	N	N	N	N(4/5)	Normal
19	A	A	A	N	A	A(2/5)	Potential shill bidder
61	A	N	A	A	A	A(1/5)	Shill bidder

4.2   Verification with the Crafted Datasets

To evaluate the proposed reputation system to the original data set there were added some specially crafted elements. These elements represent the history of the entity which nature is typical when performing bid shilling. Particularly it means that were added records with: high average number of bids per transaction, while a large number of listed items; high average number of bids per transaction, while relatively high minimum price; large average percentage of the auction, while a large number of auction. In this experiment a population of 150 sellers carrying its business on eBay has been used. The data includes 4721 auctions issued by these sellers, which made 28326 offers. To this dataset were introduced artificially crafted records in such a way as to remind sellers committing bid shilling. The number of such records was 20 and the sum of all data records (observations taken from eBay + crafted observations) was 170. Data were entered into both reputation systems so as to allow comparison of results. Both have set a reputation with its individual accuracy, but for the purposes of the test have been simplified to binary: A-abnormal, and N -Normal. The result of the systems was published in Table 2.

Table 2. Verification of reputation system with crafted datasets

No	Mod.	Models I-III Class	P	R	Z	Mod.	Models I-V Class	P	R	Z
001	0/3	st	A	A	1	0/5	st	A	A	1
002	0/3	st	A	A	1	0/5	st	A	A	1
...	...	...	...	...	...	...	...	...	...	...
015	2/3	u	N	A	0	2/5	p	A	A	1
016	2/3	u	N	A	0	2/5	p	A	A	1
017	2/3	u	N	A	0	3/5	u	N	A	0
018	3/3	unknown	N	A	0	3/5	u	N	A	0
019	3/3	unknown	N	A	0	4/5	u	N	A	0
020	3/3	unknown	N	A	0	5/5	unknown	N	A	0
...	...	...	...	...	...	...	...	...	...	...
089	2/3	u	N	N	1	2/5	p	A	N	0

The meaning of the column headers: No - the number of the seller; Mod – number of models which returned the abnormal result; Class - a class of behavior returned by reputation system; P - Planned reputation [A-anomaly, N-Normal]; R - the actual membership; Z - the compatibility of the system as a result of actual membership (1-consistent, 0 incompatible).

The new reputation system proposed in this paper and the other one acts as a classifier which assigns tested observations to one of the classes (unknown, unlikely (u), very probable, probable (p), strong class (st)).

Verification of effectiveness of both systems has been done using the classical method of evaluation of classifiers. For each system have been calculated classification error rates (CER), and measures of effectiveness accuracy (ACC). Classification error is expressed by the ratio of the number of incorrectly classified objects to all test items.

In the case of original reputation system (Models I-III), the CER value was 0.035, and 0.029 for the system proposed in the paper (Models I-V). The obtained values are small, and therefore the efficiency of both the classifiers is high, however the new reputation system gives better results. The next comparison has been done using a measure of the effectiveness of ACC, which is more accurate and allows to distinguish between two types of errors. FN - type I error (Falsle Negative) - consists in classifying the observation which is in fact an anomaly, as normal, and the problem of the second FP (False Positive) - occurs when the observation is actually a normal system qualifies as an anomaly. This way one can assign appropriate weights errors.

The native reputation system (Model I-III) reaches level of accuracy value 0.96 while the proposed system reaches accuracy value 0.97. Also this measure shows that the introduction of new models to evaluate the final value of reputation makes the system to give more precise results.

5   Conclusions

The results of the several test validating the behavior of the proposed reputation system show that the system is characterized by high precision and most of the observations which are anomalies were correctly classified. Analysis of the results presented in this paper suggests that the reputation systems using Models I-V is slightly better than the original one. This result is a consequence that the reputation system using Models I-V examines more variables describing user/seller relationship during the online auctions and so, it can produce recommendations, which enables to find out the shill bidding cases more correctly. The further work related to the proposed reputation model will include more validation scenarios with tests conducted in eBay testbed environment with the real test users simulating bid shilling.

References

1. Gregg, D.G., Scott, J.E.: The Role of Reputation Systems in Reducing On-Line Auction Fraud. 10, 95–120 (2006)
2. Chirita, P.-A., Nejdl, W., Zamfir, C.: Preventing shilling attacks in online recommender systems. In: Proceedings of the Seventh ACM International Workshop on Web Information and Data Management WIDM 2005, vol. 55, p. 67 (2005)

3. Kołaczek, G.: Multi-agent platform for security level evaluation of information and communication services. Studies in Computational Intelligence 457, 107–116 (2013)
4. Rubin, S., et al.: An auctioning reputation system based on anomaly (2005)
5. Dong, F., Shatz, S.M., Xu, H.: Combating online in-auction fraud: Clues, techniques and challenges. Computer Science Review 3, 245–258 (2009)
6. Dong, F., Shatz, S.M., Xu, H., Majumdar, D.: Price comparison: A reliable approach to identifying shill bidding in online auctions? Electronic Commerce Research and Applications 11, 171–179 (2012)
7. Myerson, R.B.: Optimal Auction Design, pp. 58–73 (1981)
8. Berkhin, P.: A survey of clustering data mining techniques. Grouping Multidimensional Data, 25–71 (2006)
9. Ott, R.L., Longnecker, M.T.: An Introduction to Statistical Methods and Data Analysis. Cengage Learning (2008)
10. Kolaczek, G.: Trust modeling in virtual communities using social network metrics. In: Intelligent System and Knowledge Engineering, ISKE 2008, pp. 1421–1426 (2008)
11. eBay API, https://www.x.com/developers/ebay/documentation-tools
12. Chakraborty, I., Kosmopoulou, G.: Auctions with shill bidding. Economic Theory 24, 271–287 (2004)
13. Andrews, T., Benzing, C., Fehnel, M.: The price decline anomaly in Christmas season internet auctions of PS3s. Journal of the Northeastern Association of Business, 1–12 (2011)
14. Juszczyszyn, K., Kolaczek, G.: Motif-Based Attack Detection in Network Communication Graphs. Communications and Multimedia Security, 206–213 (2011)
15. Resnick, P., Zeckhauser, R., Friedman, E., Kuwabara, K.: Reputation Systems: Facilitating Trust in Internet Interactions. Communications of the ACM 43, 45–48 (2000)

Investigation of Time Interval Size Effect on SVM Model in Emotion Norm Database

Chih-Hung Wu[1], Bor-Chen Kuo[1], and Gwo-Hshiung Tzeng[2]

[1] National Taichung University of Education, Taichung, Taiwan
{chwu,kbc}@mail.ntcu.edu.tw
[2] National Taipei University, Taipei, Taiwan
ghtzeng@mail.ntpu.edu.tw

Abstract. Few studies have focused on investigating the time interval effects on support vector machine (SVM) model in emotion recognition. This study tested original averaged value models and difference value models with three different time-intervals (5 seconds, 10 seconds, and 30 seconds) for SVM emotion recognition in our developed emotion norm database. Forty one elementary school students were recruited as participants to see some emotion pictures in international affective picture system (IAPS), and to collect their affective information—attention, meditation, electroencephalography (EEG), electrocardiogram (ECG), and SpO2 for developing the affective norm recognition system. This study selected 5480 IPAS photos physiological data as the tested dataset from our emotion norm database. The bio-physiology signals were averaged by seconds or difference the value by second and then to serve as the input variables value for C-SVM with RBF kernel function. The results showed that the original averaged models have better performance than the difference models. In addition, the original averaged value model with 30 seconds time interval is the optimal classification model. This study suggested that future research can adopt 30 seconds as the time interval for determining the size of time interval for their training dataset in emotion recognition problem.

Keywords: EEG (Electroencephalography); ECG (Electrocardiogram), Affective Computing, Support Vector Machine, Emotion and Attention Recognition System, Eye Tracker, Time interval.

1 Introduction

Since Affective Computing was proposed, there has been a burst of research that focuses on creating technologies that can monitor and appropriately respond to the affective states of the user (Picard, 1997). Because of this new Artificial Intelligence area, computers are able to recognize human emotions in different ways. Why is human emotion an important research area? The latest scientific findings indicate that emotions play an essential role in decision-making, perception, learning, and more [1]. However, how to develop an efficient emotion recognition system? Suited time-interval for setting the optimal training data for emotion recognition system and classifiers are unknown.

© Springer International Publishing Switzerland 2015
D. Barbucha et al. (eds.), *New Trends in Intelligent Information and Database Systems,*
Studies in Computational Intelligence 598, DOI: 10.1007/978-3-319-16211-9_13

Therefore, this study compared two types of models--original averaged value models and difference value models with three different time-intervals (5 seconds, 10 seconds, and 30 seconds) for SVM emotion recognition in our developed emotion norm database [2] to determine the optimal size of time-interval for SVM emotion recognition system.

2   The Physiological Input Signals for Emotion Recognition

The physiological input signals of eye movement, EEG and ECG will be selected to input our learning state recognition system. Attention and emotion is a complex phenomenon. A single physiological parameter can't evaluate the state of attention and emotion. For including more objective physiological indices when recognize learning state, several techniques need to be combined for classifying the state of attention and emotion based on the past studies. .With the emergence of Electroencephalography (EEG) technology, learner's brain characteristics could be accessed directly and the outcome may well hand-in-hand supported the conventional test, recognize a learner's Learning Style [3]. The arousal state of the brain [4], alertness, cognition, and memory [5, 6] also can be measure. Heart rate variability (HRV) from ECG, has gained widespread acceptance as a sensitive indicator of mental workload [7]. The positive emotions may change the HF components of HRV [8].

3   Method

3.1   Participants and Procedure

Forty-one elementary school students in the fifth grade participated in this study. Participants were tested individually. On arrival and after attaching physiological sensors, Participants will be asked to rest quietly. Participants will be told that pictures differing in emotional content would be displayed for 5 seconds on a screen in front of them, and that each picture should be viewed during this moment. Each rating will be preceded by a 5 seconds preparatory slide showing the number (1–24) of the next photograph to be presented (baseline period). The photograph will be rated and then be screened for 5 seconds (stimulus period), while the physiological responses will be recorded. The inter-trial interval was set at 30 seconds permitting recovery from the previous slide on all physiological measures. Slide presentation will be randomized for all subjects. Prior to the onset of the experimental trials, three pictures (highly pleasant-arousing, neutral on both valence/arousal, and highly unpleasant-arousing) will serve as practice stimuli. We collected participant's physiology signals (EEG, ECG, heart beating, SpO2) during the experiment every one second.

3.2   Materials

The aim of emotion norm construction is to determine (calculate) the relations between physiological changes and subjective ratings of International Affective Picture System (IAPS) photographs. Providing a standardized pool of affective and attention stimuli, the International Affective Picture System (IAPS) has become a highly useful methodological tool used in numerous investigations [9] IAPS includes normative ratings of each photograph with respect to valence (pleasure), arousal, and dominance and consists of over 700 standardized color photographs evoking a range of affective responses.

A total number of 18 IAPS were used in the experiment. The pictures were divided into three different groups; 6 pleasant, 6 neutral and 6 unpleasant pictures [10]. Twenty-four International Affective Picture System (IAPS) photographs are grouped into 3 sets of 8 photographs: highly pleasant-arousing, neutral on both valence/arousal, and highly unpleasant-arousing ones. IAPS picture sequence: 1750(pleasant), 5480(pleasant), 1740(neutral), 1050(unpleasant), 9584(unpleasant), 6260(unpleasant), 2210(neutral), 7004(neutral), 7330(pleasant), 1300(unpleasant), 7050(neutral), 1670(neutral), 1463(pleasant), 7460(pleasant), 6370(unpleasant), 7150(neutral), 9592(unpleasant), 1710(pleasant) [11]. We selected 5480 picture as our testing database in our study.

4   Database

We used the emotion norm database in our previous studies [2] via NeuroSky. The database includes 10 brainwave variables (attention score, meditation score, delta1, theta, beta1, beta2, alpha1, alpha2, gamma1, gamma2), heart beating, SpO2, and emwave emotion variable.

Based on the distribution of frequency zones, when a learner's emotional state is negative, peaceful or positive, the Coherence score will be calculated as 0, 1 and 2, respectively .The value of Heart Rate Artifacts (HRAs) is zero when the human emotion detected is in normal situations, whereas the value of HRAs is one when the human emotion is detected in abnormal situations. The emWave system identifies learner emotional states every 5 s. In this study, identifying the percentages spent in positive or negative emotions was applied to assess the effects of two learning methods on learning emotions. In computing the percentages of positive and negative emotions, the Accumulated Coherence Score (ACS) has the key role, whereas the method for computing ACS based on different coherence states and HRAs is as follows:

$$CV(t) = \begin{cases} -1, \text{if Coherence}(t) = 0 \text{ and HRA}(t) = 0 \text{ (negative emotion)} \\ +1, \text{if Coherence}(t) = 1 \text{ and HRA}(t) = 0 \text{ (peaceful emotion)}, CV(0) = 0, t = 0, 1, 2, ..., m \\ +2, \text{if Coherence}(t) = 2 \text{ and HRA}(t) = 0 \text{ (positive emotion)} \end{cases} \quad (1)$$

The entire emotion norm database that collected data from 42 elementary school students. Each student was asked to see 20 IAPS pictures. The database included brain wave data, hear beating data, SpO2, and emotion data. There are totally 14972 records in this database. We selected photo number 5480 and average all the participants' physiological data by second as our tested dataset. Our tested Dataset that selected from our emotion norm database [2] is shown in Table 1. There are totally 59 records in this tested dataset.

Table 1. Our tested dataset that selected from our emotion norm database

Photo Id	Attention	Meditation	Delta1	Theta	Beta1	Beta2	Alpha1	Alpha2	Gamma1	Gamma2	Emwave	HeartBeat	SPO2	Output
5480	53	61	44.9	2.1	0.6	1.4	1.1	0.2	0.2	0.1	0	87	98	2
5480	51	61	44.9	2.1	0.6	1.4	1.1	0.2	0.2	0.1	0	87	98	2
5480	51	50	44.9	2.1	0.6	1.4	1.1	0.2	0.2	0.1	0	87	98	2
5480	51	50	86.8	11.1	1.4	0.5	1.7	0.8	0.2	0.4	0	87	98	2
5480	51	50	86.8	11.1	1.4	0.5	1.7	0.8	0.2	0.4	0	85	98	2
5480	51	50	86.8	11.1	1.4	0.5	1.7	0.8	0.2	0.4	0	85	98	2
5480	51	57	86.8	11.1	1.4	0.5	1.7	0.8	0.2	0.4	0	85	98	2
5480	51	57	67.9	14.9	2.7	9.4	5.8	1.3	0.8	0.4	0	85	98	2
5480	51	57	67.9	14.9	2.7	9.4	5.8	1.3	0.8	0.4	0	82	98	2

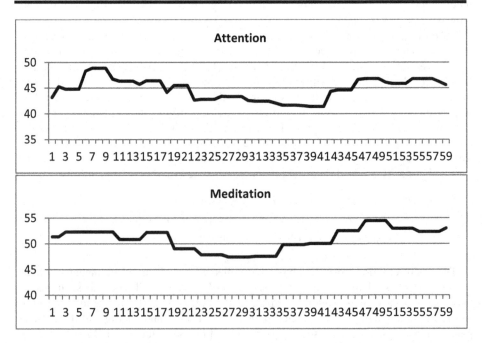

Fig. 1. Trend of Physiological Signals in Photo 5480 (Average by 41 students)

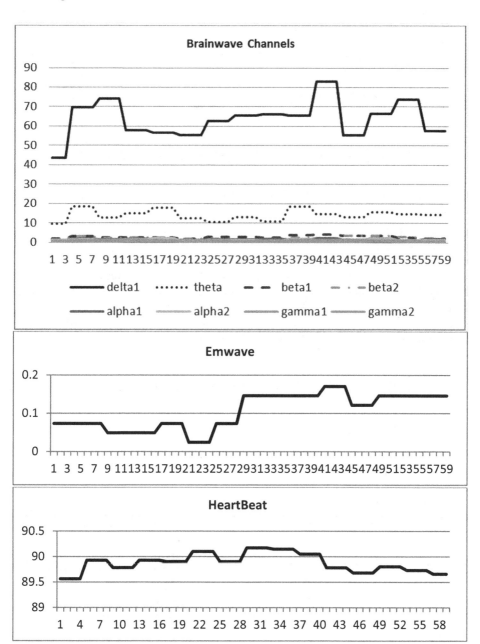

Fig. 1. (*continued*)

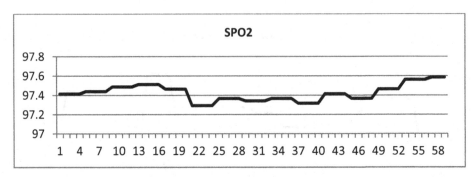

Fig. 1. (*continued*)

4.1 Classification Accuracy

This study uses C-support vector machine that cost set to 1 and degree set to 3 with radial basis function kernel via 10-folds validation method. The results are summarized as shown in Table 2. Among six types of SVM models, Model 2 has the best SVM classification accuracy (98.31%). The model 1 to model 3 used the averaged value of original physiological signal values. The model 4 to model 6 (difference model) used the difference value of original physiological values via equation 2. Generally speaking, the averaged value models have better performance than the difference models.

The model 4-6 used difference model that used the formula as below:

$$\Delta f\left(x_t\right) = f\left(x_{t+1}\right) - f\left(x_t\right) \tag{2}$$

where X denotes the physiological signal variables, and t denotes the time in second.

Table 2. Classification results of SVM (C-SVM) among six models (10 folds)

| | Original Value Model | | | Difference Value Model | | |
	Model1 5s	Model2 10s	Model3 30s	Model4 5s	Model5 10s	Model6 30s
Accuracy	96.61%	**98.31%**	94.92%	91.53%	83.05%	42.37%
TP Rate	0.966	0.983	0.949	0.915	0.831	0.424
FP Rate	0.366	0.083	0.050	0.915	0.831	0.583
Precision	0.967	0.983	0.950	0.838	0.690	0.404
Recall	0.966	0.983	0.949	0.915	0.831	0.424
F-Measure	0.962	0.983	0.949	0.875	0.754	0.397

5   Conclusion and Suggestion

This study investigated the window size effect on SVM in emotion recognition via our developed emotion norm database in our previous study [2]. We compared two models—one is original value model (model 1 to model 3) that averaged the original physiological signal values, and the other one is difference value models (model 4 to model 6) that used difference values of physiological signal values. In order to understand the optimal time-interval for emotion recognition, three different time-interval sizes (5 seconds, 10 seconds, and 30 seconds) are selected in this study for SVM model. The results showed that the original value models have better performance than difference value models. Among all models, the model 3 that used 30 seconds as emotion time-interval has the best performance. The implications for researchers reveal that the participant's emotion was triggered in the first 30 seconds. The limitation of this study is the sample size. We only selected IAPS 5480 as the tested dataset. This study suggested that future research can use more different IAPS photo data to test the external validity. .

Acknowledgements. Authors thank the National Science Council of Taiwan for support (grants NSC 101-2410-H-142-003-MY2 and MOST 103-2410-H-142-006).

References

1. Ben Ammar, M., Neji, M., Alimi, A.M., Gouardères, G.: The Affective Tutoring System. Expert Systems with Applications 37, 3013–3023 (2010)
2. Wu, C.-H., Tzeng, Y.-L., Kuo, B.-C., Tzeng, G.-H.: Integration of affective computing techniques and soft computing for developing a human affective recognition system for U-learning systems. International Journal of Mobile Learning and Organisation 8, 50–66 (2014)
3. Rashid, N.A., Taib, M.N., Lias, S., Sulaiman, N., Murat, Z.H., Kadir, R.S.S.A.: Learners' Learning Style Classification related to IQ and Stress based on EEG. Procedia - Social and Behavioral Sciences 29, 1061–1070 (2011)
4. Zhang, Q., Lee, M.: Emotion development system by interacting with human EEG and natural scene understanding. Cognitive Systems Research 14, 37–49 (2012)
5. Berka, C., Levendowski, D.J., Cvetinovic, M.M., Petrovic, M.M., Davis, G., Lumicao, M.N., Zivkovic, V.T., Popovic, M.V., Olmstead, R.: Real-Time Analysis of EEG Indexes of Alertness, Cognition, and Memory Acquired With a Wireless EEG Headset. International Journal of Human-Computer Interaction 17, 151–170 (2004)
6. Berka, C., Levendowski, D.J., Lumicao, M.N., Yau, A., Davis, G., Zivkovic, V.T., Olmstead, R.E., Tremoulet, P.D., Craven, P.L.: EEG correlates of task engagement and mental workload in vigilance, learning, and memory tasks. Aviat Space Environ Med 78, 231–244 (2007)
7. Lin, T., Imamiya, A., Mao, X.: Using multiple data sources to get closer insights into user cost and task performance. Interacting with Computers 20, 364–374 (2008)
8. von Borell, E., Langbein, J., Després, G., Hansen, S., Leterrier, C., Marchant-Forde, J., Marchant-Forde, R., Minero, M., Mohr, E., Prunier, A., Valance, D., Veissier, I.: Heart rate variability as a measure of autonomic regulation of cardiac activity for assessing stress and welfare in farm animals — A review. Physiology & Behavior 92, 293–316 (2007)

9. Tok, S., Koyuncu, M., Dural, S., Catikkas, F.: Evaluation of International Affective Picture System (IAPS) ratings in an athlete population and its relations to personality. Personality and Individual Differences 49, 461–466 (2010)

10. Waters, A.M., Lipp, O.V., Spence, S.H.: The effects of affective picture stimuli on blink modulation in adults and children. Biological Psychology 68, 257–281 (2005)

11. Rantanen, A., Laukka, S.J., Lehtihalmes, M., Seppänen, T.: Heart Rate Variability (HRV) reflecting from oral reports of negative experience. Procedia - Social and Behavioral Sciences 5, 483–487 (2010)

Part III

Semantic Web, Social Networks and Recommendation Systems

Twitter Ontology-Driven Sentiment Analysis

Liviu-Adrian Cotfas[1,2], Camelia Delcea[2], Ioan Roxin[1], and Ramona Paun[3]

[1] Franche-Comté University, Montbéliard, France
{liviu-adrian.cotfas,ioan.roxin}@univ-fcomte.fr
[2] Bucharest University of Economic Studies, Bucharest, Romania
camelia.delcea@csie.ase.ro
[3] Webster University, Bangkok, Thailand
paunrm@webster.ac.th

Abstract. As the usage of micro-blogging services has rapidly increased in the last few years, services such as Twitter have become a rich source of opinion information, highly useful for better understanding peoples' feelings and emotions. Making sense of this huge amount of data, would provide invaluable benefits to companies, organizations and governments alike, by better understanding what the public thinks about their services and products. However, almost all existing approaches used for social networks sentiment analysis are only able to determine whether the message has a positive, negative or neutral connotation, without any information regarding the actual emotions. Besides, critical information is lost, as the determined perception is only associated with the entire tweet and not with the distinct notions present in the message. For this reason, the present paper proposes a novel semantic social media analysis approach, TweetOntoSense, which uses ontologies to model complex feeling such as happiness, affection, surprise, anger or sadness. By storing the results as structured data, the possibilities offered by the semantic web technologies can be fully exploited.

Keywords: twitter, opinion mining, ontology, big data, sentiment analysis.

1 Introduction

Twitter has gradually become the most commonly used micro-blogging service, which allows users to broadcast 140 character status messages, also known as tweets. With over 240 million monthly active users, who post more than 500 million tweets every day, as reported in April 2014, Twitter contains opinions on virtually everything, which can be exploited through sentiment analysis [1].

Sentiment analysis, also known as opinion mining, is a growing area of Natural Language Processing, commonly used in social media analysis to determine whether a text expresses a positive, negative or neutral perception [2, 3]. Besides simply determining the perception, some papers also investigate how the strength of the perception should be evaluated [4].

An ontology based approach was proposed in [5], that while still evaluates emotions in terms of negative, positive and neutral perceptions, similar to previous

© Springer International Publishing Switzerland 2015
D. Barbucha et al. (eds.), *New Trends in Intelligent Information and Database Systems,*
Studies in Computational Intelligence 598, DOI: 10.1007/978-3-319-16211-9_14

attempts, also takes into consideration the fact that users express opinions about the various characteristics of the analyzed subject and not only about the subject as a whole. Moving away from simple perception scores, [6] proposes an emotion ontology for annotation tweets, which however only includes seven basic emotions. Besides the limited set of emotions, the paper also doesn't deal with modelling the analyzed products or service and also does not propose any approach for further analyzing the extracted information.

In this paper, an end-to-end semantic approach – TweetOntoSense - is proposed, that uses ontologies to model the various emotions expressed in social media messages, the analyzed concepts and their characteristics, as well as the relationships between the Twitter users. It has been considered that while knowing the perception of the user is definitely important, analyzing the categories of emotions contained in twitter messages provides far more information. Moreover, by storing the extracted information as triples, advanced analysis can be performed using the technologies associated with the semantic web. Thus, the proposed approach expands upon both traditional approaches and the semantic ones presented in [5] and [6].

The paper is organized as follows. In the second section, the Emotions, Twitter and TweetOntoSense ontologies, which form the bases of the proposed approach, are presented. The third section of the paper includes the steps taken to perform emotion analysis. The fourth section shows how the extracted information can be further exploited using semantic web inference and SPARQL queries, to create the bases for developing an advanced social media analysis platform. The last section summarizes the paper and introduces some of the future research directions.

2   Emotion Analysis Ontology

According to [7], ontologies are defined as a "formal, explicit specification of a shared conceptualization". They formally represent knowledge as a hierarchy of concepts, using a shared vocabulary to denote the types, properties and interrelationships of those concepts. Currently, ontologies have become the means of choice for representing knowledge, by both providing a common understanding for concepts and being machine processable.

The concepts needed in order to perform sentiment analysis on Twitter messages can be grouped in three main categories:

- concepts that express human emotions;
- concepts that describe Twitter specific knowledge;
- concepts that provide a connection between the twitter message, the expressed emotion and the analyzed subject and it's properties.

While for the first category of concepts, the ones describing emotions, several existing ontologies were found in the scientific literature, for the last two categories no appropriate ontology was identified. Therefore, as an initial step, a Twitter ontology modelling the relations between users, tweets and their associated properties had to be created. Afterwards, a sentiment analysis ontology, named TweetOntoSense, which connects the expressed sentiments, the twitter messages and the analyzed concepts and their properties was defined.

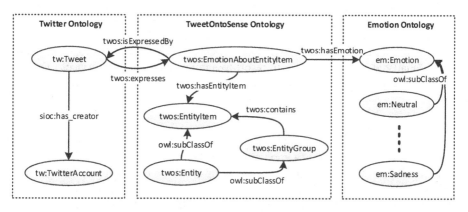

Fig. 1. Ontology-based sentiment analysis

The main concepts from the three ontologies are shown in Fig. 1, together with the object properties that connect them.

The following subsections describe in further details the proposed ontologies, used to enable social media analysis using semantic web technologies.

2.1 Emotion Ontology

Several emotion ontologies such as the ones proposed in [6] and [8] currently exist. From them, it has been chosen the emotional categories ontology presented in [8], as besides being inspired by recognized psychological models, it also structures the different human emotions in a taxonomy.

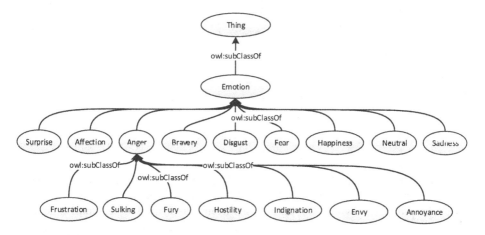

Fig. 2. Ontology of emotions

As shown in Fig. 2, the ontology contains for each class a number of individuals, representing words associated with the particular type of emotion. In order to obtain a better coverage of the words used to express emotions, we have chosen to enrich the ontology using some of the values in the corresponding WordNet synsets [9].

Even though the ontology currently supports only English and Spanish, it can easily be extended with other languages as shown in [10], where the ontology was extended to include concepts in Italian. Thus, tweets in other languages can be more precisely analyzed, without having to resort to automatic translation services. This is highly important, as almost 49% of all the Twitter messages are written in other languages than English.

2.2   Twitter Ontology

When producing semantic data, a good practice is to reuse classes and properties from existing ontologies [11], as it facilitates mappings with other ontologies such as the ones in the Linked Open Data[1] project.

Therefore, given the fact that the existing Twitter REST API ontology presented in [12] does not provide any mappings to well-known ontologies, a new Twitter ontology, for which the main classes and properties are shown in Fig. 3, is proposed, that both reuses well-known vocabularies such as Dublin Core[2], FOAF[3] and SIOC[4] and also facilitates social media network analysis using SPARQL queries.

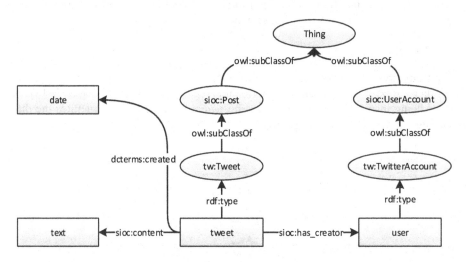

Fig. 3. Twitter ontology

As shown in [13], several generic widely used vocabularies for annotating the data extracted from social media networks currently exist. One of the best well-known is the Friend of a friend – FOAF ontology, used to represent people and their relationships. The proposed Twitter ontology reuses from FOAF the foaf:accountName and the foaf:homepage properties. Another widely used ontology is The Semantically-Interlinked Online Communities – SIOC ontology, dedicated to the description of

---

[1] http://linkeddata.org
[2] http://dublincore.org/
[3] http://xmlns.com/foaf/spec/
[4] http://sioc-project.org/

information exchanges in online communities such as blogs and forums, from which the proposed ontology reuses several properties, including sioc:has_topic, sioc:content and sioc:links_to. Moreover, the tw:Tweet and tw:TwitterAccount classes are derived from the sioc:Post and sioc:UserAccount classes, defined in the SIOC ontology.

The Dublin Core ontology provides terms to declare a large variety of document's metadata, from which the dcterms:created property was reused, in order to specify the date when the tweet was published.

2.3   TweetOntoSense Ontology

The application specific ontology describes the analyzed entities, like products, services or events, together with their characteristics, using the twos:Entity, twos:EntityItem and twos:EntityGroup classes. The Entity class serves as a base class for twos:EntityItem and twos:EntityGroup and defines the twos:hasQueryTerm data property, containing the keywords or hashtags that will be used to retrieve the analyzed tweets. The analyzed entity is modeled by the twos:EntityGroup class, representing the particular product, service or event for which social media analysis is performed. Given the fact that people usually express opinions not only about the concept, but also about its characteristics [5], the twos:EntityItem class models the relevant characteristics. Finally, the twos:EmotionAboutEntityItem class provides the necessary link with the Emotion and Twitter ontologies, previously described.

3   Ontology Based Emotion Analysis

Extracting emotions from tweets is known to be a challenging task for several reasons. Among the difficulties that were encountered while performing sentiment analysis, it can be mentioned the huge variety of topics covered, the informality of the language, as well as the extensive usage of abbreviations and emoticons. Also, the concise nature of the twitter messages can be considered both an advantage and a drawback. Further reasons are explained in [14, 15].

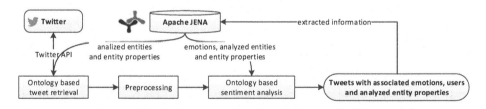

Fig. 4. Sentiment analysis steps

The steps used for sentiment analysis are shown in Fig. 4 and further described in the subsections bellow.

3.1   Tweet Retrieval

First, the tweets are retrieved using the Twitter Search API, using as search terms all the combinations between the keywords associated with the individuals belonging to the EntityGroup and the corresponding individuals from the EntityItem class.

3.2   Preprocessing

The second step represents the preprocessing phase in which tokenization, normalization and stemming are applied. Given the fact that many users write messages using a casual language, the normalization process includes:

- Removing URLs, as they don't convey any emotional meaning. Removing duplicated letters, which frequently occur in twitter messages and emphasize a particular word, in order not to interfere with the stemmer. For example the first tweet in the Sentiment140 corpus, presented in [15] is:

> "I loooooooovvvvvveee my Kindle2. Not that the DX is cool, but the 2 is fantastic in its own right."

- Converting all-caps words to lower case. While it can be argued that further information regarding the intensity of a sentiment could be extracted from the use of all-caps [7], in this paper it has been chosen to only focus on extracting the associated emotions. An example of a tweet that uses All-caps is:

> "My Kindle2 came and I LOVE it! :)".

- Replacing hashtags with the corresponding words.
- Replacing abbreviations with the corresponding regular words taken from the Internet Lingo Dictionary.

> AHH YES LOL IMA TELL MY HUBBY TO GO GET ME SUM MCDONALDS

- Replacing emoticons with the corresponding emotions from the ontology of emotional categories. Table 1 includes the emoticons that are mapped to the word happiness during the preprocessing phase.

Table 1. Emotions mapped to happiness

:)	:)	:-)	:-))	:-)))	;)	;-)	^_^	:-D	:D
=D	C:	=)							

The last operation of the preprocessing phase consists in applying the Porter stemmer on the resulting sequence of words.

3.3   Emotion Identification

In the last step, emotions are determined by comparing the processed tweet with the individuals in the enriched ontology of emotion categories. As the novelty of the proposed approach lies in the ontology-based analysis of tweets preceding and following the sentiment analysis phase, we have chosen a simple opinion mining approach, which only focuses on extracting explicit emotions. The resulting knowledge is saved in the triple store for further analysis using SPARQL queries, as it is shown in the fourth section of this paper.

The proposed approach was tested on a publically available corpus[5], which contains tweets collected for the search terms "Microsoft", "Apple", "Twitter" and "Google" that were annotated with the following sentiment labels: positive, negative, neutral and irrelevant. From the above mentioned corpus we only analyze the positive and negative tweets as they are the ones that could express emotions. From 973 analyzed tweets, 79 tweets were found to contain words from the emotion ontology, the most frequent emotions being like (34), love (25), hate (10) and hope (4).

4   Social Media Analysis

The proposed semantic sentiment analysis approach can be used to develop an end-to-end ontology-based social media analysis platform, as shown in Fig. 5. By using semantic web inference, new relations between the collected information are discovered automatically. Thus, if the "offending" emotion is associated to a tweet, during the emotion identification phase, the inference engine also associates the more general "indignation" emotion.

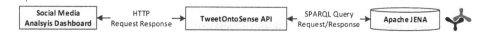

Fig. 5. Social media analysis platform

SPARQL queries provide the necessary mean for performing advanced analysis, while their structured result can easily be processed for creating meaningful charts and data tables in the user interface. For example, the following query retrieves from the triple store all the studied entities together with the detected emotions.

```
SELECT ?group ?emotion WHERE
{ ?eaei rdf:type twos:EmotionAboutEntityItem;
    twos:hasEmotion ?emotion;
    twos:hasEntityItem ?entityItem.
  ?group twos:contains ?entityItem.
} ORDER BY ASC(?group)
```

Using also inference, the query bellow returns the users that have written tweets which express emotions derived from happiness, ordered by the influence of each user, measured as the number of followers. Influencers can thus be easily determined for each type of emotion.

---

[5] http://www.sananalytics.com/lab/twitter-sentiment/

```
SELECT ?user ?tweetContent ?followerCount WHERE
{?eaei rdf:type twos:EmotionAboutEntityItem;
    twos:hasEmotion ?emotion;
    twos:isExpressedBy ?tweet.
  ?tweet sioc:content ?tweetContent;
    sioc:hasCreator ?user.
  ?user tw:hasFollowerCount ?followerCount.
  ?emotion rdf:type em:Happiness.
} ORDER BY DESC(?followerCount)
```

5   Concluding Remarks

The present paper proposes a novel ontology-based social media analysis approach that better captures the wide array of emotions expressed in the millions of tweets published every day. While existing approaches only associate simple positive, negative or neutral perceptions, TweetOntoSense paves the way towards fine-grained analysis using semantic web technologies, thus unlocking a vast amount of emotional information that has previously been unavailable to companies and public authorities, trying to better understand their customers' opinions through social media analysis.

Among the further research directions, we consider both extending the proposed approach to other online social media networks, such as Facebook, LinkedIn and Google+ and also analyzing how the expressed emotions change over time as a result of the changes in user perception. After adapting, improving, and testing the opinion mining algorithm on different social media networks, a complete semantic social analysis platform will be developed. The proposed ontologies are available for download at https://github.com/lcotfas/TweetOntoSense.

Acknowledgments. This paper was co-financed from the European Social Fund, through the Sectorial Operational Programmee Human Resources Development 2007-2013, project number POSDRU/159/1.5/S/138907 "Excellence in scientific interdisciplinary research, doctoral and postdoctoral, in the economic, social and medical fields -EXCELIS", coordinator The Bucharest University of Economic Studies. Also, the authors gratefully acknowledge partial support of this research by Webster University Thailand.

References

1. Pak, A., Paroubek, P.: Twitter as a Corpus for Sentiment Analysis and Opinion Mining. In: Proceedings of the Seventh International Conference on Language Resources and Evaluation, Valletta, pp. 1320–1326 (2010)
2. Ghiassi, M., Skinner, J., Zimbra, D.: Twitter brand sentiment analysis: A hybrid system us-ing n-gram analysis and dynamic artificial neural network. Expert Syst. Appl. 40, 6266–6282 (2013)

3. Mostafa, M.M.: More than words: Social networks' text mining for consumer brand sentiments. Expert Syst. Appl. 40, 4241–4251 (2013)
4. Thelwall, M., Kevan, B., Georgios, P.: Sentiment strength detection for the social web. J. Am. Soc. Inf. Sci. Technol. 63, 163–173 (2012)
5. Kontopoulos, E., Berberidis, C., Dergiades, T., Bassiliades, N.: Ontology-based sentiment analysis of twitter posts. Expert Syst. Appl. 40, 4065–4074 (2013)
6. Roberts, K., Roach, M., Johnson, J.: EmpaTweet: Annotating and Detecting Emotions on Twitter. In: Proceedings of the Eighth International Conference on Language Resources and Evaluation, Istanbul, pp. 3806–3813 (2012)
7. Studer, R., Benjamins, V.R., Fensel, D.: Knowledge Engineering: Principles and methods. Data Knowl. Eng. 25, 161–197 (1998)
8. Francisco, V., Hervás, R., Peinado, F., Gervás, P.: EmoTales: creating a corpus of folk tales with emotional annotations. Language Resources and Evaluation 46, 341–381 (2012)
9. Montejo-Ráez, A., Martínez-Cámara, E., Martín-Valdivia, M.T., Ureña-López, L.A.: Ranked WordNet graph for Sentiment Polarity Classification in Twitter. Comput. Speech Lang. 28, 93–107 (2014)
10. Baldoni, M., Baroglio, C., Patti, V., Rena, P.: From tags to emotions: Ontology-driven sentiment analysis in the social semantic web. Intelligenza Artif 6, 41–54 (2012)
11. Shadbolt, N., Berners-Lee, T., Hall, W.: The Semantic Web Revisited. IEEE Intell. Syst. 21, 96–101 (2006)
12. Togias, K., Kameas, A.: An Ontology-Based Representation of the Twitter REST API. In: 2012 IEEE 24th International Conference on Tools with Artificial Intelligence, pp. 998–1003 (2012)
13. Breslin, J.G., Passant, A., Decker, S.: The Social Semantic Web. Springer (2009)
14. Maynard, D., Bontcheva, K., Rout, D.: Challenges in developing opinion mining tools for social media. In: Proceedings of the Eighth International Conference on Language Resources and Evaluation, Istanbul, pp. 15–22 (2012)
15. Kouloumpis, E., Wilson, T., Moore, J.: Twitter sentiment analysis: The good the bad and the omg! In: Proceedings of the Fifth International Conference on Weblogs and Social Media, Barcelona, pp. 538–541 (2011)

A Virtual Reality Based Recommender System for Interior Design Prototype Drawing Retrieval

Kuo-Sui Lin and Ming-Chang Ke

Department of Information Management, Aletheia University, Taiwan, R.O.C.
{au4234,au1130}@mail.au.edu.tw

Abstract. There is a lack of interactive rapid and visual recommender systems for recommending interior design prototype drawing to the consigner. Therefore, the purpose of this study is to propose a virtual reality based recommender system as a platform to retrieve a design drawing from a historical interior design drawings database, and to recommend the retrieved drawing to the consigner as a prototype drawing. The as yet untapped recommender system consists of a virtual reality based query system, a database system and a pattern matching engine. A preliminary case study of the recommender system was made, including a front end data collecting and preprocessing phrase and a back end pattern matching and recommendation phase. The proposed recommender system has been shown to greatly help the designer in storing historical drawing items, extracting relevant design features, as well as to direct the consigner from the query system to the historical database so as to access the most matching design drawing that best suites the consigner's interests and requirements.

Keywords: Interior design prototype drawing, Pattern matching, Recommender system, Similarity measure function.

1   Introduction

During requirements analysis stage of the interior design drawing development life cycle, ambiguity, inaccuracy, incompleteness, inconsistency and errors at this beginning stage propagates through the development life cycle and are harder and more expensive to correct later. An interior design prototype drawing is better to be created earlier at this requirements analysis stage; thereafter, any adjustment, correction, or addition to the design prototype drawing can be implemented, bridging the design requirements gap between the consigner and the designer.

In the current industry practice, while proposing a prototype drawing, the designer can either select a design drawing randomly from some existing historical drawings, or prepare a new design drawing as a prototype drawing for the consigner. By using this prototype drawing, the consigner can get a reality feeling of the design concept of the designer, enabling the consigner to better communicate and interact with the designer. Recently, accompanying with the development of data mining techniques, processing and selecting among large amounts of complex and unstructured objects by a recommender system has become one of the major domains of knowledge in the

© Springer International Publishing Switzerland 2015   141
D. Barbucha et al. (eds.), *New Trends in Intelligent Information and Database Systems,*
Studies in Computational Intelligence 598, DOI: 10.1007/978-3-319-16211-9_15

next decade. However, there is a lack of interactive rapid and visual recommender system for recommending an interior design drawing to the consigner. Therefore, the purpose of this study is to propose a virtue reality based recommender system to retrieve an interior design drawing from a historical drawings database. The retrieved design drawing can be recommended and displayed to the consigner as a prototype drawing through a virtue reality platform. Thereafter, necessary modification to the prototype drawing can be made. Thus, bridging the communication gap between the designer's design concept and consigner's design requirements can be easier.

The rest of this study is organized as follows. Section 2 briefly reviews the research background and related works. Section 3 provides the target pattern recommendation problem and the similarity measure function for pattern matching of this study. Section 4 describes a preliminary case study on interior design prototype drawing recommender system and discusses the experimental results. Finally in Section 5, conclusions of this work are drawn and future areas of research are discussed.

2   Related Work

2.1   Historical Database for Interior Design Prototype Drawings

An interior design prototype drawing can be created using various materials and methods. Among them, the virtual reality based application system has its most advantages. The virtual reality based interior design prototype drawing can create quick design, facilitate and foster discussions by offering participants a visual representation of what they are talking about. Thus, in the interior design industry, the most promising trend for interior design drawing retrieval is to integrate the virtual reality based recommender system with the data mining techniques.

While recommending a new interior design drawing form a historical drawing database for the consigner, the designer has to search from some historical drawings to find one drawing which may suits consigner requirements. In order to recommend an interior design prototype drawing for the designer and the consigner to proceed to the requirement analysis stage, the designer can either develop new design drawings or store old design drawings to form a historical drawing database. From the historical drawing database, the selected prototype drawing can be regarded as a first version interior design drawing and provide an effective and efficient means of design communication for the designer and the consigner.

2.2   Recommender Systems

Recommender systems or recommendation engines are active information filtering systems that seek to predict the "preference rating" that user would give to items, aiming at recommending items (movies, music, books, web pages) to users [1]. Recommender systems typically take either of two basic approaches: collaborative filtering or content-based filtering. Other approaches, such as hybrid approaches, also exist [2,3,4].

(1) Content based filtering approach. Content based filtering approach gives recommendation to a certain user by examines properties of the items that have been rated by the user and the description of items to be recommended. In a content based system, we must construct for each item a profile, which is a record or collection of records representing important characteristics of that item that are interested to the users.

(2) Collaborative filtering approach. Collaborative filtering approach makes recommendations to users not on basis of their own preference profile, but based on similarities between the profiles of the active user and other users in the system. Collaborative filtering approach is further categorized into memory based and model based approaches.

(3) Hybrid approaches. Hybrid approaches can be obtained from a combination of collaborative and content based filtering approaches, try to capitalize on the strengths of each approach and increase the efficiency of recommender systems. Seven categories of hybrid recommendation approaches, weighted, switching, mixed, feature combination, feature augmentation, cascade, and meta-level have been introduced by Burke [2].

2.3   Feature Extraction

A feature is a distinctive property (or characteristic, attribute) which is used to distinguish between two objects. Feature extraction is an essential pre-processing step to pattern matching problems. Many feature extraction methods have been proposed for finding representative features of a content profile, such as clustering, classification and regression. Feature values are ratings of features that are provided by the users. Features of an object can be viewed as an n-dimensional vector, and can be represented as a vector space. A pattern is an n-tuple vector of n scalars $x_j, j \in [1, n]$, which are called the feature values. A set of feature values used for classification forms a feature vector. Conventional form of a pattern is given by the following expression: $V = [x_1, x_2 ... x_j ... x_n]^T$, where V is known as the feature space; x_j can be Boolean variables for 0 and 1 respectively; n is the dimension of V. After establishing a set of profiles by the recommender system, it is possible to calculate the similarity values among items, and finally chooses a best suited item to an active user.

2.4   Approximate String Matching

Pattern matching in data mining is to find a relatively small pattern in a relatively large text. Patterns and texts can be combinatorial pattern matching or spatial pattern matching. Examples of combinatorial pattern matching are approximate string matching, tree pattern matching, DNA sequence matching and text editor pattern matching. Examples of spatial pattern matching are geometric pattern matching and graph pattern matching. The problem of approximate string matching is typically divided into two sub-problems: finding approximate substring matches inside a given string and finding dictionary strings that match the pattern approximately.

3   Research Method

3.1   Target Pattern Recommendation Problem

In the content based recommendation system, there is a set of unclassified historical content profiles and a query profile, characterized by a set of feature-rating pairs.

Definition 1. Historical Content Profile
A historical content profile P_i is a set of binary tuples $\{<f_1, r_1>, \ldots, <f_n, r_n>\}$, where $f_j, j = 1 \ldots n$, are the features that describes the content profile of one specific historical design drawing item; r_j denotes the ratings of f_j. Let $P_i, i = 1 \ldots m$, be a set of historical content profiles. The set of content vectors can be represented by an $m \times n$ binary pattern–feature-rating matrix R_{mn}, where r_{ij} is 1 if the ith pattern has jth feature, and 0 otherwise.

Definition 2. Query Profile
A query profile P_q is a set of feature-rating pair $\{<f_{q1}, r_{q1}>, \ldots, <f_{qn}, r_{qn}>\}$, where $f_{qj}, j = 1 \ldots n$, are the features that describes the query profile of consigner's interested design drawing item.

Suppose we are given m known historical content patterns $P_i, i = 1 \ldots m$, which can be represented as a set of n dimensional content pattern vectors. Given an unclassified query pattern P_q, this can be represented as an n dimensional vector. The vector representation of the query profile P_q and the historical content profile P_i can be used for computing similarity value. The pattern for which the system wants to recommend to the consigner is referred to as the target pattern. The target pattern recommendation problem is formally defined as follows:

Definition 3. Target Pattern Recommendation Problem
Given an $m \times n$ binary pattern–feature-rating matrix R_{mn} rated by the designer and a set of n dimensional query vector P_q rated by the consigner, identify a target pattern P_i such that P_i is best suited to that of the query profile P_q. Then, only the target content profile that has a highest degree of similarity to the query profile would be selected and recommended.

3.2   Cosine Similarity Function for Pattern Matching

In computer science, a similarity measure or similarity function is a real-valued function that quantifies the similarity between two objects. Approximate of similarity matching usually relies on similarity functions $f(v_1, v_2) \rightarrow s$ taking two data instances v_1 and v_2 as inputs and calculating a score value s between 0 and 1. If the value for s surpasses a given threshold α, the values v_1 and v_2 are considered to be representations of the same real world object (entity). This process appears in several data management applications such as approximate querying and data integration.

Direction cosine and Euclidean distance are two types of most commonly used similarity measures. A commonly used similarity measure is cosine similarity [5, 6]. Cosine similarity function has been proven to work well and been one of the most popular similarity measures applied to in information retrieval and pattern matching [7,8].

Definition 4. Cosine Similarity Function

Given vector $A = (a_1, a_2, a_3...)$ and $B = (b_1, b_2, b_3,...)$, where a_j and b_j, $j = 1...n$ are the components of the vector (features of the object) and n is the dimension of the vectors. Dot Product (A, B) is defined as: $A \cdot B = \Sigma^n_{j=1} (a_j \times b_j) = a_1 * b_1 + a_2 * b_2 + a_3 * b_3 + ...$; Norm($A$) is the length of vector A and is defined as: $\|A\| = \text{sqrt}(a_1{}^\wedge 2 + a_2{}^\wedge 2 + a_3{}^\wedge 2 + ...)$. Then the cosine similarity between vector A and B is measure by using the Euclidean dot product formula:

Cosine Similarity$(A, B) = CosSim(A, B) = \text{Cosine}(\theta) = \text{Dot } (A, B) / \|A\|\|B\| = A \cdot B / \|A\| \|B\| = \Sigma^n_{j=1} (a_j * b_j)/(\text{sqrt}(\Sigma^n_{j=1} \ a_j \ ^\wedge \ 2 \ ) \times \text{sqrt}(\Sigma^n_{j=1} b_j{}^\wedge 2)) = (a_1 * b_1 + a_2 * b_2 + a_3 * b_3 + ...) / (\text{sqrt}(a_1{}^\wedge 2 + a_2{}^\wedge 2 + a_3{}^\wedge 2 + ...) * \text{sqrt}(b_1{}^\wedge 2 + b_2{}^\wedge 2 + b_3{}^\wedge 2 + ...))$

As used in positive space, θ ranges from 0 to 90 degrees, cos θ ranges from 1 to 0. The cosine similarity ranges from 1 meaning exactly the same, to 0 meaning exactly dissimilarity, and in-between values indicating intermediate similarity or dissimilarity.

To resolve the target pattern recommendation problem, we use the cosine similarity measure to calculate the degree of similarity between the query pattern vector P_q and each of the known historical content patterns P_i, $i=1...m$. Then, only the content profile that has a highest degree of similarity measure to the query profile would be selected.

4   The Architecture of the Proposed Recommender System

During requirements analysis stage of the interior design drawing development life cycle, existing methods used in developing interior design prototype drawings usually lack sense of realism, as well as require the designer and the consigner several iteration of time consuming design reviews. To solve this problem, this study developed a virtual reality based recommender system for interior design prototype drawing retrieval. The proposed recommender system consists of a virtual reality based query system, a database system and a pattern matching engine. The architecture of the proposed recommender system is summarized in the flowchart in Fig.1.

The database system stores a group of historical interior design drawings. FancyDesigner is a wonderful virtual reality based interior design tool for creating these interior design drawings [9]. A set of characteristics or attributes expressing the design features of these drawings is extracted by referring to the related common industrial standards. Then, ratings of the extracted design features are assigned to the historical design drawings to form a set of historical content patterns (database profiles).

When a consigner needs to find an interior design prototype drawing, he/she could be guided to rate a questionnaire. The query system puts in the consigner's requirements using the questionnaire, usually in the form of a "like" click, and transfers the entries of the questionnaire into a query feature pattern (query profile). The pattern matching engine compares a consigner's query pattern with each of the historical database content patterns in order to select the most suitable candidate pattern.

This selection task is accomplished by employing a pattern matching algorithm using cosine similarity measure function to calculate the degree of pattern matching between the two patterns. When the calculated highest pattern matching value exceeds the predefined threshold value, the target pattern in the database is selected. The target design drawing which is associated to the target pattern is then triggered and displayed to the consigner through the query system. By only clicking on questions of a questionnaire query for a consigner, the consigner is taken to a screen that displays the design drawing recommended from historical database design drawings in response to that questionnaire query.

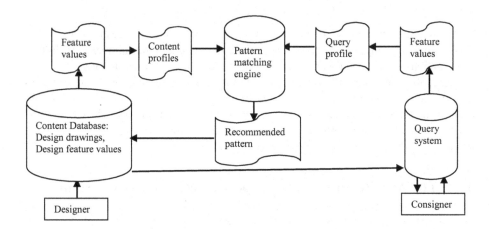

Fig. 1. The architecture of the proposed recommender system

5 A Preliminary Case Study

This case study presents result from an experiment examining the effectiveness and efficiency of the proposed recommender system. Before experiment, twenty-five historical interior design drawings are manually selected and stored to the database.

The experiment of the recommender system consists of a front end data collecting and preprocessing phase and a back end pattern matching phase. In the front end data collecting and preprocessing phase, both designer's feature ratings of the extracted design features and consigner's questionnaire query are transformed into designer's database patterns (content profiles)and consigner's query pattern (query profile) of the recommendation system. In the back end pattern matching phase, the transformed query pattern vector and database pattern matrix is compared in the pattern matching engine of the recommender system. According to the calculation result of the proposed pattern matching algorithm, the database content pattern with the highest degree of similarity matching is selected and an associated drawing is triggered and recommended accordingly. The frontend phase includes extracting design features (Step 1) and establishing designer's content profiles and consigner's query profile

(Step 2). The back end phase matches the consigner's query vector with each vector of the database content pattern matrix through a pattern matching engine (Step 3) and making a recommendation to the consigner (Step 4).

Step 1. Extracting Design Features of Interior Design Drawings

Design feature extraction is defined as the process of uncovering features in historical interior design drawings. In order to classify underlying design feature patterns for historical interior design drawings, it is desirable to extract design features for the drawings.

Before proceeding, several design patterns of historical interior design drawings have been classified and pre-loaded to the database of the recommender system. In this study, we identify a set of design features for classifying the interior design drawings, which are Budget, Style, Design Theme, Color Hue, Chrome, Brightness, Color Scheme, Saturation and Lightness.

Step 2. Establishing Designer's Content Profiles and Consigner's Query Profile

Feature values are ratings of features that are provided by the users (designer and consigner) in the front end phase. Data collecting and preprocessing for the designer's pattern contents and the consigner's questionnaire query is done in this stage. The designer's contents and consigner's query are then transformed into designer's content profiles and consigner's query profile.

(1) Data Collection and Preprocessing for Designer's Content Profiles

The designers not only have decided what kinds of design features that they characterized the historical design drawings, but also have evaluated which drawings possessed which features. Design features decided in our study are: Budget, Style, Design Theme, Color Hue, Chrome, Brightness, Color Scheme, Saturation and Lightness.

Features are classified as "relevant" or "not relevant" to the design drawing by adopting a simple binary rating scale. A designer's preference content profile can be constructed through ratings for indication of "relevant" or "not relevant", inferring features "relevant" or "not relevant" to the design drawing.

Taking pattern P_1 for example, a designer's pattern–feature ratings can be represented as a binary string array:

$$P_1 = [\underset{\text{Budget}}{1000} \quad \underset{\text{Design Style}}{10000000} \quad \underset{\text{Theme}}{00100000} \quad \underset{\text{Color Hue}}{000010000} \quad \underset{\text{Brightness}}{01} \quad \underset{\text{Color Scheme}}{001} \quad \underset{\text{Saturation}}{10} \quad \underset{\text{Lighting}}{10} \quad]$$

A designer's content profile consists of several feature-rating vectors, which can also be represented by a pattern–feature-rating matrix. Each row in the matrix is the vector representation of a design pattern. As shown in Fig. 2, pattern–feature-rating matrix usually is described as an $m \times n$ ratings matrix R_{mn}, where each entry r_{ij} ($1 \leq i \leq m$, $1 \leq j \leq n$) means the rating assigned to the pattern i on the feature j; the row represents m patterns and the column represents n features. Values of the rating matrix come from two values: $r_{ij} = 1$ means that the item i possess the feature A_j; $r_{ij} = 0$ means that the item i does not possess the feature A_j.

$$R_{mn} = \begin{vmatrix} r_{11} & r_{12} & \cdots & r_{1n} \\ r_{21} & r_{22} & \cdots & r_{2n} \\ \vdots & \vdots & r_{ij} & \vdots \\ r_{m1} & r_{m2} & \cdots & r_{mn} \end{vmatrix}$$

Fig. 2. A pattern–feature matrix

(2) Data Collection and Preprocessing for Consigner's Query Profile

When a consigner needs to find a best suited interior design drawing in the historical database, he/she could be guided to rate a questionnaire query. Questions of the questionnaire represents design features of the design query, which are Budget, Style, Design Theme, Color Hue, Chrome, Brightness, Color Scheme, Saturation and Lightness. For collecting consigner's preference ratings, we designed a questionnaire using the design features as query questions. Multiple choice questions regarding the design features of the interior design drawing are designed and asked in the questionnaire. The multiple choice questions allow the respondent to choose a single or multiple click from a list of possible entry options. A consigner's preference query profile can be constructed through questionnaire query with checkboxes for indication of interest as a "like" click, inferring features liked by the consigner.

For explanation, questions asked in the questionnaire are illustrated: "Which of the following entry options of the design feature are requested in your interior design drawing query?" "You can select one or multiple entry options per feature that will help us recommend the best matching design item that interests you."

1.The Decorating Budget is:

☐ ~1M	☐ 1M~3M	☐ 3M~5M	☐ 5M~

2.The Interior Design Theme is:

☐ Greece	☐ Middle east	☐ European	☐ American	☐ Taiwanese	☐ Chinese	☐ Japanese	☐ Korean

3.The Interior Design Color Hue is:

☐ Yellow	☐ Green	☐ White	☐ Blue	☐ Violet,	☐ Pink	☐ Black	☐ Brown	☐ Gray

For example, a consigner's query pattern can be represented as:

$P_q=$ [1000   10000000 01100000 000010000 11   001   11   10   ]
      Budget   Style   Theme   Color Hue Brightness Color Scheme Saturation Lighting

Step 3. Computing Similarity Measures between Patterns

Let $F = \{f_1, f_2, \ldots f_j \ldots, f_n\}$, $j = 1 \ldots n$, be a set of design features; $P = \{P_1, P_2 \ldots P_i \ldots P_m\}$ be a set of content patterns and P_q be a query pattern. Let $P_q = \{r_{q1}, r_{q2}, \ldots r_{qj} \ldots, r_{qn}\}$, $j = 1 \ldots n$, be a set of n dimensional feature values of query pattern P_q, and $P_i = \{r_{i1}, r_{i2}, \ldots r_{ij} \ldots, r_{in}\}$, $i = 1 \ldots m$, $j = 1 \ldots n$, be a set of n dimensional feature values of database patterns. One specific database pattern P_i, whose similarity measure value matches the query pattern P_q, is defined by a cosine similarity function:

$CosSim(P_q, P_i) = \text{Cosine}(\theta) = \text{Dot}\ (P_q, P_i)\ /\ \|P_q\|\|P_i\| = P_q \cdot P_i\ /\ \|P_q\|\|P_i\| = \sum_{j=1}^{n}\ r_{qj}$
$* r_{ij} / (\text{sqrt}(\sum_{j=1}^{n}\ r_{qj}{}^{\wedge}2\ ) \times \text{sqrt}(\sum_{j=1}^{n} r_{ij}{}^{\wedge}2),\ i = 1 \ldots m.$

Each feature in a profile defines a dimension in Euclidean space and the binary entry of each feature corresponds to the value in the dimension. The two feature profiles (feature patterns) P_q and P_l can be transformed into two vectors. When feature patterns are represented as feature vectors, the cosine similarity of two feature vectors is quantified as cosine of the angle between vectors.

P_l = (1, 0, 0, 0, 1, 0, 0, 0, 0, 0, 0, 0, 0, 0, 1, 0, 0, 0, 0, 0, 0, 0, 0, 0, 0, 1, 0, 0, 0, 0, 0, 1, 0, 0, 1, 1, 0, 1, 0) is a set of n dimensional database pattern vector; P_q = (1, 0, 0, 0, 1, 0, 0, 0, 0, 0, 0, 1, 0, 1, 1, 0, 0, 0, 0, 0, 0, 0, 0, 1, 0, 0, 0, 0, 1, 1, 0, 0, 1, 1, 1, 1, 0) is a set of n dimensional query pattern vector. The cosine similarity measure between the query pattern vector P_q and the database pattern vector P_l is calculated as follows:

$CosSim\ (P_q,\ P_l) = (a_{11}\ a_{q1} + a_{12}\ a_{q2} + \ldots\ + a_{1j}\ a_{qj} + \ldots + a_{1n}\ a_{qn})\ / (\text{sqrt}(a_{11}{}^{\wedge}2\ + a_{12}{}^{\wedge}2 + \ldots + a_{1j}{}^{\wedge}2 + \ldots + a_{1n}{}^{\wedge}2)\ \text{sqrt}(a_{q1}{}^{\wedge}2 + a_{q2}{}^{\wedge}2 + \ldots + a_{qj}{}^{\wedge}2) + \ldots + a_{qn}{}^{\wedge}2)) = 8\ /(2.83) * (3.46) = 0.82$

Step 4. Making Recommendation to the Consigner
The pattern matching engine compares a consigner's query pattern with each of the historical database patterns in order to select the most suitable target pattern. The pattern matching engine employs a pattern matching algorithm using cosine similarity measure function to calculate a set of pattern similarity matching values between the two patterns. Once these similarity values are computed, the matching pair of the target pattern that has the highest similarity value and exceeds the predefined threshold value is selected. The target design drawing which is associated with the target pattern is then triggered and displayed to the consigner as a prototype drawing through a virtue reality platform. Thereafter, necessary modification to the recommended prototype drawing can be made. Through communication on the recommended prototype drawing, the gap between the designer's design concept and consigner's design requirements can be bridged

6   Conclusions and Future Work

In this paper, a virtue reality-based recommender system was proposed. The recommender system consists of a virtual reality based query system, a database system and a pattern matching engine. A preliminary case study of the recommender system was made, including a front end data collecting and preprocessing phase and a back end pattern matching and recommendation phase. Results have demonstrated that the virtual reality based recommender system has been shown to greatly help the designer in storing historical drawing items, extracting relevant and useful design features, as well as to direct the consigner from the query system to the historical database to retrieve the most matching design drawing that best suites the consigner's interests or requirements.

The retrieved design drawing can be recommended and displayed to the consigner's query interface as a prototype drawing to proceed with design prototype requirement analysis task. Thereafter, necessary modification to the retrieved prototype drawing can be made. Thus, bridging the communication gap between the designer's design concept and consigner's design requirements can be easier. Applying the proposed recommender system in the interior design industry, it not only avoids haphazard and tedious selection process of the best suited alternative among historical interior design drawings, but also reduces unproductive try-and-error time wastage in design communication. This will increase the effectiveness and efficiency of interior design service.

In the future work, we will develop a user-friendly interface to collect more real-world dataset from participants. Also, we will apply other pattern matching methods to the recommender system and make a comparison among the result that obtained in this study with the results that obtained with other pattern matching methods. Furthermore, we will validate the result of the proposed system so as to achieve a more accurate recommendation and much better user satisfaction.

References

1. Ricci, F., Rokach, L.: Introduction to Recommender Systems Handbook. Recommender Systems Handbook, pp. 1–35. Springer (2011)
2. Burke, R.: Hybrid Recommender Systems: Survey and Experiments. User Modeling and User-Adapted Interaction 12(4), 331–370 (2002)
3. Breese, J.S., Heckerman, D., Kadie, C.: Empirical Analysis of Predictive Algorithms for Collaborative Filtering. In: Proceedings of the Fourteenth Annual Conference on Uncertainty in Artificial Intelligence, Madison, Wisconsin, pp. 43–52 (1998)
4. Pennock, D.M., Horvitz, E.: Analysis of the Axiomatic Foundations of Collaborative Filtering. In: AAAI -99 Workshop on AI for Electronic Commerce at the 16th National Conference on Artificial Intelligence, Orlando, Florida (1999)
5. Larsen, B., Aone, C.: Fast and Effective Text Mining Using Linear-Time Document Clustering. In: Proceedings of the Fifth ACM SIGKDD International Conference on Knowledge Discovery and Data Mining, San Diego, California, pp. 16–22 (1999)
6. Nahm, U.Y., Bilenko, M., Mooney, R.J.: Two Approaches to Handling Noisy Variation in Text Mining. In: Proceedings of the ICML-2002 Workshop on Text Learning, Sydney, Australia, pp. 18–27 (2002)
7. Ye, J.: Cosine Similarity Measures for Intuitionistic Fuzzy Sets and their Applications. Mathematical and Computer Modeling 53(1), 91–97 (2011)
8. Yates, R.B., Neto, B.R.: Modern Information Retrieval. Addison Wesley, New York (1999)
9. FancyDesigner Homepage, http://www.fancydesigner.com.tw/bbs/modules/wordpress/

Adaptation of Social Network Analysis
to Electronic Freight Exchange

Konrad Fuks, Arkadiusz Kawa, and Bartłomiej Pierański

Poznań University of Economics, al. Niepodległości 10, 61-875 Poznań, Poland
{konrad.fuks,arkadiusz.kawa,bartlomiej.pieranski}@ue.poznan.pl

Abstract. The purpose of this article is to explore the opportunities of application of social network analysis method (SNA) into the business environment. After conceptually explaining the idea of business network as well as providing brief overview of the social network analysis method, the authors present an in-depth investigation of network relationships of the members of electronic freight exchange called Trans.eu. This freight exchange is one of the biggest in Europe.

Keywords: SNA, business networks, electronic freight exchange.

1   Introduction

A business network is defined as a set of connected actors that perform different types of business activities in interaction with each other [5]. The actors are also described as "nodes" or positions (occupied by firms, households, trade associations, and other types of organizations) [11]. The interactions (also called relationships) are manifested by "links" that connect nodes with each other. According to this definition, business networks can be understood as: supply chains, supply networks [4, 7], strategic alliances, clusters, franchising systems, etc. Business networks are getting more and more important in the economic life because of their "capacity for regulating complex transactional independence as well as cooperative interdependence among firms" [2]. Business networks can also be approached in terms of activity links, resource ties and actor bonds [3]:

- Activity links regard technical, administrative, commercial and other activities of a company that can be connected in different ways to those of other companies as an inter-firm relationship develops.
- Resource ties connect different types of resource elements (both tangible resources and intangible resources such as knowledge, brand image, etc.) of two or more companies. Resource ties result from how the relationship has developed over time and it represents a resource in itself for a company.
- Actor bonds connect actors (market entities) and influence the way the two actors perceive each other and form their identities in relation to each other. Bonds become established in the inter - firm relationships and reflect the interaction process.

© Springer International Publishing Switzerland 2015   151
D. Barbucha et al. (eds.), *New Trends in Intelligent Information and Database Systems,*
Studies in Computational Intelligence 598, DOI: 10.1007/978-3-319-16211-9_16

One of the major problems when analysing business networks is to determine a network's boundaries. As stated, network analysis has been "relatively mute" on the topic of boundary definition [8]. According to literature, the boundaries of a business network cannot be defined exactly [3]. This is so because the business network is without a centre and without boundaries. Some authors even say that networks have no boundaries [12]. Networks stretch out in all directions and interaction processes may occur in any direction [9].

This theoretic approach makes it almost impossible to analyse business networks. It is extremely difficult to analyse a system that comprises an infinite number of nodes and links. Therefore, an empirical investigation into a given network requires implementation of a method for identifying its boundaries. In this aspect authors present a novel approach to the problem. In most cases network boundaries are understood as self-evident [8]. This way of thinking results in an implicit assumption that each analysed network is reasonably completed [1]. Having rejected this way of justifying boundaries, authors propose two major approaches to binding network actors, namely a nominalist and a realist one [8].

The realist approach could also be described as an 'inside' approach, because it takes into consideration the point of view of actors (nodes) understood as members (parts) of the network in question. It is based on the assumption that network boundaries 'are less or more known to the actors themselves', or, in other words, a network's boundaries are 'consciously experienced' only by the actors composing the network [8]. In contrast to the realist approach, the nominalist approach could be viewed as an 'outside' approach. In this situation a network's boundaries are defined by a researcher (an individual that is outside the network) on the basis of his/her analytical objectives [8]. This approach raises a very important question to what extent an analyst's perception of reality conforms with the reality itself [8]. Regardless of which approach is considered, it must be strongly stated that a network's boundaries are highly context dependent and contingent on the observer/researcher [6].

2   Social Network Analysis

One of the methods which are more and more boldly used in management sciences (in particular in the study of business networks) is network analysis known as the SNA - social network analysis. Initially, SNA mainly referred to an analysis of networks created by humans. It allowed to analyse phenomena occurring in human communities connected to such events as the spread of rumors or information and identifying people who played significant or peripheral roles in a group. Business networks concern companies and the links and interactions between them, in particular. However, companies are created by employees who decide to work with people from other companies. The application of SNA in business networks is to reveal the interactions not only between people but also between enterprises and / or their stakeholders.

SNA provides answers to the following questions: which actors are central, which are located on the periphery, which play a smaller or a greater role? What is the significance of the position of actors in the network for new opportunities, which actors are given a competitive advantage thanks to their position in the structure, and which are limited because of it? Which actors have more power than others? One can learn how information is disseminated among enterprises, how experience is shared, and which entities cooperate with a larger number of partners.

SNA is based on the achievements of the social sciences and on the methods of network analysis and graph theory. Thanks to network analysis, problems expressed by network structures can be formulated and solved. In turn, the graph theory shows the properties and relationships of these network structures in a visual form. SNA allows to look at the network from the perspective of a comprehensive, specific group of companies and an individual participant. One can examine the structure of the network, its strength, characteristics, capacity for certain behaviours, the role and position of the individual participants.

The basis of SNA are numerous measures which have been developed in recent years. The measures that can be applied in the analysis of business networks can be divided into two basic groups - those that are considered from the perspective of the actor, and those that concern the entire network:

a. Actor perspective
   - Centrality degree
   - Betweenness
   - Closeness
   - Eigenvector
b. Network perspective
   - Density
   - Clustering coefficient
   - Small world effect
   - Diameter

2.1 Centrality Degree

The centrality degree is a fundamental measure of centrality. This is the number of connections of a node with other nodes. The greater the centrality, the more important a given node is in the network. It is the number of edges entering and leaving a node. The node with higher in-degree stage is more popular, while the one with higher out-degree stage has a greater impact on other vertices.

2.2 Betweenness

Betweenness is also a measure of centrality. It is the ratio of the number of the shortest paths between any two nodes passing through the given node to the total number of all the shortest paths. The higher the level of betweenness, the more important the node is in the network. This also has its drawbacks. A node with a high degree of betweenness may cause a loss of cohesion of the network. A company which has a high level of betweenness plays an important role in the network because it is the link between directly unrelated cliques in the network (the so-called bridge).

2.3   Closeness

Closeness is another measure of centrality. This is the average length of the shortest paths between the given node and all other nodes, and so the distance needed to reach other nodes from the given node. A node with a high degree of closeness makes it easier to reach other nodes. A company which has a high degree of closeness plays an important role in the network, in terms of communication, diffusion of information, etc.

2.4   Eigenvector

The eigenvector is also a measure of centrality referring to individual network nodes. With the use of this rate, not only the number of links of a given node and its "neighbours" can be measured, but, above all, the quality of those links. In practice, this means that a higher value is represented by a node whose links are characterized by higher quality (e.g. they have a significant position in the network). In business networks, a company with a high value of the eigenvector has relationships with nodes (other entities) which play an important role in the network, for example, they have key resources, they are a source of innovation, they have a large market share.

2.5   Density

Network density is the ratio of the number of the existing relationships to the total number of potential relationships in the network. Density is also often referred to as network completeness or a degree thereof. A directed graph due to the greater number of potential relations has density no greater than the corresponding undirected graph. In business networks, network density decreases along with an increase in the number of nodes - a given entity is not able or not interested in cooperating with all the other entities.

2.6   Clustering Coefficient

In most networks, cliques can be found that, from the network perspective, mean that more than two nodes are interconnected. For three nodes such a formation takes the shape of a triangle (each node is connected to two other nodes), and in the case of a larger number of nodes the structure may be delimited to a fixed number of triangles. The clustering coefficient for a given network determines the participation of cliques in all relationships within the network, i.e. the ratio of all closed triangles (cliques) to the number of unclosed triangles (between the three entities there are two out of the three possible combinations). In business networks, the clustering coefficient can help to identify both positive cooperation (e.g. the occurrence of technological clusters) and negative one (e.g. price fixing agreements restricting competition in the network).

2.7   Small World Effect

A characteristic feature of networks is their large size. It is manifested in the fact that in the context of a specific network a very large number of entities linked directly and indirectly operate. The high number of participants in the network means that, in most cases, they are connected to one another indirectly. However, the network peculiarity means that any two entities can be linked to each other by a relatively small number of connections. It suggests that, in networks, connecting any two entities requires a relatively short path. This phenomenon is referred to as the small world effect. Mathematically speaking, "the small world effect" is the average of the shortest distances between any two entities in the network.

2.8   Diameter

The diameter of a network is determined by establishing the longest of the shortest paths connecting pairs of entities in the network. The diameter is, therefore, determined on the basis of two entities that are most distant from each other (the distance between another randomly chosen pair of nodes is smaller than the diameter).

3   Use of Metrics of SNA in the Electronic Freight Exchange

3.1   Electronic Freight Exchange

The development of information technology has contributed to the emergence of new types of services and business activities. One of such forms is the electronic logistics platform, in particular the electronic freight exchange. The electronic freight exchange is a system in which information is placed about excess or missing cargo transport needs, which enables other parties to find a favourable offer of transport services and a better use of the cargo space. They are primarily intended for forwarders and carriers, but, more and more often, manufacturers and commercial companies are also convinced to use them.

The electronic freight exchange is a complementary solution to other traditional forms of searching for contractors. Its main feature is the fact that mainly single transactions are carried out. Freight prices within the electronic exchange are shaped by the current supply and demand for a given service.

Currently in Europe there are over 100 different electronic freight exchanges. The most complex and popular ones are Teleroute and Timocom and Trans.eu.

Trans.eu was established in April 2004 in Poland. Despite its young age, it quite quickly gained the trust of companies in the TSL. Trans.eu is available in 19 languages and its users are from 24 European countries. The company has branches in 8 European countries. In order to be able to use the system it is necessary to download software from the manufacturer's website, which, once installed on the target computer, allows to report and respond to the posted bids.

3.2   Application of SNA to Electronic Freight Exchange Business Network

For the purpose of our research we've applied two data filters to build up the network. First was date range which was set to year 2012 and second was number of transactions between nodes (strength of business relation). We've assumed that two nodes has strong business relation when there is 10 or more transactions between them in corresponding time frame. Edges are representation of business transactions between companies (nodes). Each business transaction has its initiator, therefore the network is directed.

After filters application we've received directed network of 787 nodes and 1308 edges. Additionally network was divided to 60 independent components: 44 just with two nodes, 9 with three nodes and, 5 with four nodes, 1 with seven nodes and 1 with 643 nodes. From perspective of SNA we've analysed only the "big component" which has 643 nodes and 1218 edges. Network visualization with SNA metrics from the network perspective is presented on Fig. 1. This analysis was made with Cytoscape 3.0.2 software.

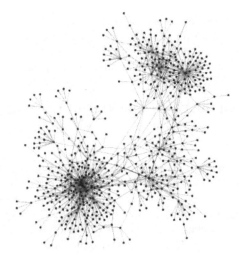

SNA metric	Value
density	0,003
clustering coefficient	0,018
small world effect	3,492
average degree	1,894
diameter	10
number of nodes	643
number of edges	1218

Fig. 1. Network visualization and general SNA metrics

Density metric (0,003) shows that network consists of only 0,3% of possible relations (edges). So when you choose two random nodes it is highly probable that they are connected indirectly. Clustering coefficient has also very low value (0,018), hence only 1,8% of nodes create cliques (clusters). Network diameter illustrates that two most distant nodes are distant from each other with 10 business relations (edges of the network). However, small world effect has value of 3,492. This shows that there are nodes in the network, which significantly shortens the distance between any other two nodes, acting as central points of the network (the so-called hubs).

SNA metrics connected with individual nodes are second part of research presented in this paper. This analysis was made with Gephi 0.8.2 software. As indicated above (see sec. 2) degree is the most important centrality metrics of the SNA. In directed networks there are three metrics connected with degree: aggregated average degree (1,894 – see Fig. 1), in-degree corresponding with incoming transactions (company is transportation service supplier) and out-degree equivalent of outgoing transactions (company is transportation service buyer). Distributions of in- and out-degree is shown on Fig. 2. Both distributions are power-law (see fitted lines on charts), therefore we can assume that analysed network (filtered time and transactions perspective) is resistant to random node removal. This means that if even "fat cat" (node/company important from network perspective) leaves, the network will not fall apart.

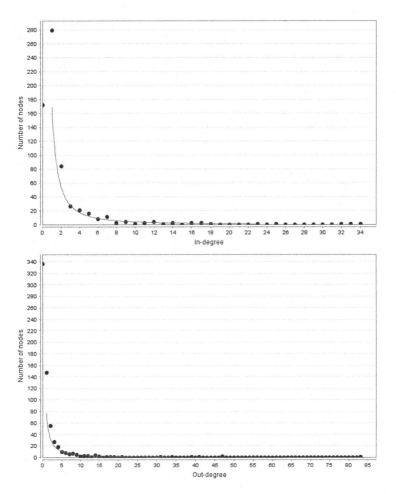

Fig. 2. In-degree and out-degree distribution

Other centrality SNA metrics were additionally filtered for the purpose of this paper, mainly because of clarity of research results presentation. In Table 1 we've presented only nodes with the highest values of combined three centrality metrics: eigenvector centrality, closeness centrality and betweenness centrality. Node is present in Tab. 1 if it has all three metrics in upper 20% values of each metric. Additionally, eigenvector centrality was used as main sorting parameter, because of its importance in analysing business networks [10].

Data shows (see Tab. 1) that the node with highest value of eigenvector centrality is the most important from the whole network perspective, but at the same time its importance is average from the perspective of local clique (closeness centrality). For comparison nodes 8, 13 and 15 are very important for local cliques. Low values of betweenness centrality (node 1 has the highest value in the network) show that there are no central nodes in the network and presented in Table 1 nodes play only role of connecting local cliques. This corresponds with in- and out- degree power low distribution and additionally confirms that analysed network is resilient to random node removal.

Table 1. Individual nodes SNA centrality metrics

ID	Eigenvector Centrality	Closeness Centrality	Betweenness Centrality
1	1	0,398373984	0,018724117
2	0,754654723	0,412631579	0,015256990
3	0,715176124	0,403292181	0,011720866
4	0,122760612	0,431718062	0,007755432
5	0,117681351	0,666666667	0,000310797
6	0,071663436	0,482758621	0,004603237
7	0,047685443	0,487804878	0,000978077
8	0,039024984	1	0,000291600
9	0,022961973	0,4	0,001049762
10	0,022082022	0,424242424	0,000332911
11	0,021652699	0,625	0,000731431
12	0,020671141	0,5	0,002176878
13	0,018441595	1	0,000034020
14	0,015135038	0,526315789	0,000261225
15	0,015113838	1	0,000100093

4   Conclusions

It must be strongly stated that the findings of this research are limited only to the electronic transportation exchange environment. Further research is needed to investigate the potential use of SNA in the analysis of other types of business networks such as: strategic alliances, supply chains etc. Successful adaptation of SNA into the business environment extends the possibilities of the business network analysis/

Acknowledgements. The paper was written with financial support from the National Center of Science [Narodowe Centrum Nauki] – the grant of the no. DEC-2011/03/D/HS4/03367.

References

1. Doreian, P., Woodard, K.L.: Defining and locating cores and boundaries of social networks. Social Networks 16(4), 267–293 (1994)
2. Grandori, A., Soda, G.: Inter-firm networks: antecedents, mechanisms and forms. Organization Studies 16(2), 183–214 (1995)
3. Hakansson, H., Snehota, I.: Developing relationships in business networks. Routledge, Londres (1995)

4. Harland, C.M., Lamming, R.C., Zheng, J., Johnsen, T.E.: A taxonomy of supply networks. Journal of Supply Chain Management 37(4), 21–27 (2001)
5. Holmlund, M., Törnroos, J.Å.: What are relationships in business networks? Management Decision 35(4), 304–309 (1997)
6. Karafillidis, A.: Networks and boundaries. In: International Symposium Relational Sociology: Transatlantic Impulses for the Social Sciences, Berlin (2008)
7. Lamming, R., Johnsen, T., Zheng, J., Harland, C.: An initial classification of supply networks. International Journal of Operations & Production Management 20(6), 675–691 (2000)
8. Laumann, E.O., Marsden, P.V., Prensky, D.: The boundary specification problem in network analysis. Research methods in social network analysis 61, 87 (1989)
9. Prenkert, F., Hallén, L.: Conceptualising, delineating and analysing business networks. European Journal of Marketing 40(3/4), 384–407 (2006)
10. Sanchez, M., Emerson, C.: Essays on economics networks. Dissertation (Ph.D.), California Institute of Technology (2013), http://resolver.caltech.edu/CaltechTHESIS:05312013-135942135
11. Thorelli, H.B.: Networks: between markets and hierarchies. Strategic Management Journal 7(1), 37–51 (1986)
12. White, H.C.: Network switchings and Bayesian forks: reconstructing the social and behavioural sciences. Social Research 62(4), 1035–1063 (1995)

Social Users Interactions Detection Based on Conversational Aspects

Rami Belkaroui[1], Rim Faiz[2], and Aymen Elkhlifi[3]

[1] LARODEC, ISG Tunis, University of Tunis
Bardo, Tunisia
rami.belkaroui@gmail.com
[2] LARODEC, IHEC Carthage, University of Carthage
Carthage Presidency, Tunisia
rim.faiz@ihec.rnu.tn
[3] LALIC, Paris Sorbonne University,
28 rue Serpente Paris 75006, France
aymen.elkhlifi@paris4.sorbonne.fr

Abstract. Last years, people are becoming more communicative through expansion of services and multi-platform applications such as blogs, forums and social networks which establishes social and collaborative backgrounds. These services like Twitter, which is the main domain used in our work can be seen as very large information repository containing millions of text messages usually organized into complex networks involving users interacting with each other at specific times. Several works have proposed tools for tweets search focused only to retrieve the most recent but relevant tweets that address the information need. Therefore, users are unable to explore the results or retrieve more relevant tweets based on the content and may get lost or become frustrated by the information overload. In addition, finding good results concerning the given subjects needs to consider the entire context. However, context can be derived from user interactions.

In this work, we propose a new method to retrieval conversation on microblogging sites. It's based on content analysis and content enrichment. The goal of our method is to present a more informative result compared to conventional search engine. The proposed method has been implemented and evaluated by comparing it to Google and Twitter Search engines and we obtained very promising results.

Keywords: Conversation retrieval, Social user interactions, Conversational context, Social Network, Twitter.

1 Introduction

In the current era, microblogging services gives people the ability to communicate, interact and collaborate with each other, reply to messages from others and create conversations. This behavior leads to an accumulation of an enormous amount of information. Furthermore, microblogs tend to become a solid media for simplified collaborative communication.

© Springer International Publishing Switzerland 2015 161
D. Barbucha et al. (eds.), *New Trends in Intelligent Information and Database Systems*,
Studies in Computational Intelligence 598, DOI: 10.1007/978-3-319-16211-9_17

Twitter, the microblogging service addressed in our work, is a communication mean and a collaboration system that allows users to share short text messages, which doesn't exceed 140 characters with a defined group of users called followers. Users can reply to each other simply by adding @sign in front name user they are replying to. This set of socio-technical features has made possible for Twitter to host a wide range of social interactions from the broadcasting of personal thoughts to more structured conversations among groups of friends [1]. While communicating people share different kind of information like common knowledge, opinions, emotions, information resources and their likes or dislikes. The analysis of those communications can be useful for commercial applications such as trends monitoring, reputation management and news broadcasting. In addition, one of main characteristic of Twitter is that users are not limited to produce contents, they can get involved indirectly in conversations with other users by liking and sharing user's posts. Furthermore, several works [2] have proposed tools for tweets search focused only to retrieve recent tweets. But, these tools are not powerful enough if we want to get tweets about an event or a product. Users are unable to explore results or retrieve more relevant tweets based on content.

This paper proposed a conversation retrieval method which can be used to extract conversation from twitter in order to provide an informative result for users' information needs based on user's content interactions analysis. Comparing with current methods, the new proposed not only extract directly reply tweets, but also relevant tweets which might be retweets or comments and other possible interactions. The method extracts extensive posts beyond conventional conversation.

The rest of the document is organized as follows: we begin by presenting related work in related domains such as forums discussion, Email threads. Then, we focus on more recent works addressing conversation retrieval on microblogging sites. In section 3, we propose our method allows extracting user's content interactions on microblogging sites. The experimentation and evaluation results are detailed in section 4. The final section presents a summary of our work and future directions.

2   Related Work

There are a number of related areas of work, including discussion forums search, email thread detection and conversation search from microblogging sites like Twitter, which is the main domain used in our work. In this section, we will discuss each area in turn.

2.1   Email Threads Reply Detection

Previous research has been focused on using email structure especially emails threads [3,4,5]. Thread detection is an important task which has attracted significant attention [6]. Email is one of the most important tools for treating conversations between people. Generally, a typical user mailbox encloses hundreds of conversations. Few works indirectly address to the problem of thread reply reconstruction. Accorded to [3], the detection of these conversations has been identified as an important task. Clustering the messages into coherent conversations useful to applications, among them, it gives users the opportunity to see a messages greater context they are reading and collating related messages automatically. In[6] the authors suggested a method that allows to assemble messages having the same subject attributes and send them among the same group of people. However, conversations may span several threads with similar (but not exact) subject lines. Furthermore, conversations not include all the participants in all the messages. In the same way,[7] developed an email client extension that makes it possible to clusters messages by topic. However, their clustering approach is focused on topic detection, hence messages belonging to different conversations on the same topic will be clustered together. In addition,[4] recreated reply emails chains, called email threads. The authors suggested two approaches, one based on using header meta-information, the other based on timing, subject and emails content. But, this method is specific for emails and the features cannot be easily extended for microblogs conversation construction.

2.2   Forums Threads Structure Identification

An online forum is a Web application for holding discussions and posting User Generated Content (UGC) in a particular domain, such as sports, recreation, techniques, travel, etc. In forums, conversations are represented as sequences of posts, or threads, where the posts reply to one or more earlier posts. Several studies have looked at identifying the structure of a thread, question-answer pairs or responses that relate to a previous question in the thread. There are many works on searching forum threads that dealt with the reply-chains structure or reply-trees. [8] has concentrated on identifying the thread structure when explicit connections between messages are missing. Despite the fact that replies to posts in microblogging sites, are commonly explicit, it is proved that different autonomous conversations may be developed inside the same replies thread. Furthermore, distinct threads may belong related to macro-conversations. For example, being Twitter hashtags that connect separate threads by common topic. In [9] authors represent the principal differences between traditional IR tasks and searching in newsgroups. They use a measures combination such as author metrics (posts number, number of replies, etc.) and features threads.

2.3   Conversation Retrieval form Microblogging Sites

Conversation retrieval is a new search paradigm for microblogging sites. It results from the intersection of Information Retrieval and Social Network Analysis (SNA). Most of microblogging services provide a way to retrieve relevant information [10], but lack the ability to provide all tweets discussion. There have been few previous researches dealing specifically with conversation detection. In addition, existing conversation retrieval approaches for microblogging sites [11,12,13,14] have so far focused on the particular case of a conversation formed by directly replying tweets. [11] proposed a user-based tree model for retrieving conversations from microblogs. They considered only tweets that directly respond to other tweets by the use of @sign as a marker of addressivity. The advantage of this approach is to have a coherent conversation based on direct links between users. Furthermore, the downside is that this method does not consider tweets that do not contain the @sign. Similarly [12] proposed a method to build conversation graphs, formed by users replying to tweets. In this case, a tweet can only directly reply to other tweet. However, users can get involved indirectly in conversations communities by commenting, liking, sharing user's posts and other possible interactions. In [15] the authors concentrated on different microblogging conversations aspects. They proposed a simple model that produces basic conversation structures taking into account the identities of each conversation member. Other related works [16,17] focusing on different aspects of microblogging conversation, that deal respectively with conversation tagging and topics identification.

3   TCOND: Twitter Conversation Detection Method

We propose an information retrieval method for microblogging sites particularly Twitter based on conversation retrieval concept. Our method combines direct and indirect conversation aspects in order to extract extensive posts beyond conventional conversation. In addition, we defined a conversation as a set of short text messages posted by a user at specific timestamp on the same topic. These messages can be directly replied to other users by using "@username" or indirectly by liking, retweeting and commenting.

3.1   Proposed Method

Our method consists in 2 phases:

- Phase 1: Constructing the direct reply tree using all tweets in reply directly to other tweets.
- Phase 2: Detecting the relevant tweets related indirectly to a same reply tree which might be retweets, comments or other possible interactions in order to extract extensive posts beyond conventional conversation.

Phase1: Twitter Direct Conversations Construction. In this phase, we aim to collect all tweets in reply directly to other tweets. Obviously, a reply to a user will always begin with "@username". Our goal in this step is to create reply tree. The reply tree construction process consists of two algorithms run in parallel recursive root finder algorithm and iterative search algorithm.

Let T0 is the root (first tweet published) of the conversation C and T is a single tweet of the conversation retrieved. Let consider T_i the type of tweet T. A tweet can have three types: root, reply or retweet. The goal of the Recursive Root Finder Algorithm is to identify the conversation root T0 given T. Note that

Algorithm 1. Recursive Root Finder (A:twitter)

Let T be a tweet collected from Twitter (ID tweet)
while (T_i !=root) **do**
    Extract T_i- 1 by matching field "in reply to status id"
end while
A : twitter $=$ A : twitter -1

when the algorithm starts,$|T|$ is not known. Once, the conversation root T0 has been established, the Iterative Search Algorithm is used to seek the remainder of conversation C by searching all tweets addressed to T_i using matching field "in reply to status id". It is run repeatedly until some conditions, indicating that the conversation has ended, are met.

Phase2: Indirect Reply Structure Using Conversational Features. To the best of our knowledge, there has not been previous work on the structure of reply-based on indirectly conversation. Therefore, we define new features that may help to detect tweets related indirectly to a same conversation. The goal of our method is to extract tweets that may be relevant to the conversation without the use of the @symbol. We use the following notations in the sequel:

- t_i is a set of tweets present in direct conversation (tweets in reply to other tweets directly).
- t_j is a tweet that can be linked indirectly to conversation.

The features we used are:

- *Using the Same URL*

Twitter allows users to include URL as supplement information to their tweets. By sharing an URL, an author would enrichment the information published in his tweet. This feature is applied to collect tweets that share the same URL.

$$P1(t_i, t_j) = \begin{cases} 1 \text{ if t contains the same URL.} \\ 0 \text{ otherwise.} \end{cases} \tag{1}$$

- *Hashtags Similarity*

The # symbol, called hashtag, is used to mark a topic in a tweet or to follow conversation. Any user can categorize or follow topics with hashtags. We used this feature to collect tweets that share the same hashtags.

$$P2(t_i, t_j) = \begin{cases} 1 \text{ if t contains the same hashtag.} \\ 0 \text{ otherwise.} \end{cases} \tag{2}$$

- *Tweets Time Difference*

The time difference is highly important feature for detecting tweets linked indirectly to the conversation. We use the time attribute to efficiently remove tweets having a large distance in terms of time compared to conversation root.

- *Tweets Publication Dates*

Date attribute are highly important for detecting conversations. Users tend to post tweets about conversational topic within a short time period. The euclidean distance has been used to calculate how similar two posts publication dates are.

- *Content*

The criterion content refers to the thematic relevance traditionally calculated by IR systems standards. We compute the textual similarity between each element in t_i, t_j taking the maximum value as the similarity measure between two messages. The similarity between two elements is calculated using the well-known tf-idf cosine similarity, $sim(t_i, t_j)$.

- *Similarity Function*

Finally, the similarity between tweets indirectly linked to conversation and tweets which are present in the reply tree is calculated by a linear combination between their attributes.

4 Experiments and Results

The objective of this experimental analysis is to verify the following hypotheses:

- Using a set of conversational features improves the conversation search task in microblogging sites.
- User satisfaction increases when social aspects are included in the search task, i.e., there is a visible and measurable impact on end users.

These two hypotheses are addressed by two separate evaluations. The first consists in an on line user evaluation, where users have been asked to compare the result of queries performed using a traditional search engine and using our method. The second involves several different combination functions to evaluate the criteria applicability for conversation detection.

4.1 Quantitative Analysis

The following experiment has been designed to gather some knowledge on the impact of our results on end-users. For this experiment we have selected three events and queried our dataset using Google[1], Twitter search engine[2] and our

---

[1] www.google.com

[2] Search.twitter.com

method (TCOND). Then, we have asked a set of assessors to rate the top-10 results of every search task, to compare these approaches. In order to measure the results quality, we use the Normalized Discounted Cumulative Gain (NDCG) at 10 for all the judged events. In addition, we used a second metric which is the Precision at top 10. In the following, we first describe the experimental setting, then we present the results and finally we provide an interpretation of the data.

Experimental Settings. The dataset has been obtained by monitoring microblogging system Twitter posts over the period of July-August 2013. In particular, we used a sample of about 113 000 posts containing trending topic keywords using Twitter's streaming API. Trending topics (the most talked topic on twitter) have been determined directly by Twitter, and we have selected the most frequent ones during the monitoring period.

To evaluate our results search tasks we have used a set of 100 assessors with three relevance levels, namely highly relevant (value equal to 2), relevant (value equal to 1) or irrelevant (value equal to 0). The assessors selected among students and colleagues of the authors with backgrounds in computing and social sciences, on a voluntary base, and no user was aware of the details of the underlying systems. Every user was informed of three events happened during the sampling period. For each event we performed three searches:

1. One using Google.
2. One using Twitter Search.
3. One using our method (TCOND).

The evaluators were not aware of which systems had been used. Every user for each search task was presented with three conversations selections, one for each of the previous options with the corresponding top-10 results.

Experimental Outcomes and Interpretation Results. We compare our conversation retrieval method with the results returned by Google and by Twitter search engine using two metrics namely the P@10 and the NDCG@10. From this comparison, we obtained the values summarized in Table 1 where we notice that our method overcomes the results given by both of Google and Twitter. The reason of these promising values is the fact that we combine a set of conversational features and direct replies method to retrieve conversation may have a significant impact on the users' evaluation.

Focusing on the three messages selections, we observe that all conversations obtained with our method receive higher scores with compared to Google and Twitter's selection. According to the free comments of some users and following the qualitative analysis of the posts in the three selections we can see that Google and twitter received lower scores not because they contained posts judged as less interesting, but because some posts were considered not relevant with regard to the searched topic.

Concentration on the three messages selections we observe that all conversations selections obtained with twitter search has higher scores with respect to Google's selection. These results lead us toward a more general interpretation of the collected data. It appears that the social metrics usage have a significant impact on the users' degree interest in the retrieved posts. In addition, the retrieving conversations process from Social Network differs from traditional Web information retrieval; it involves human communication aspects, like the degree interest in the conversation explicitly or implicitly expressed by the interacting people.

Table 1. Table of Values for Computing our Worked Example

	P@10 (Average%)	NDCG (Average%)
Task1		
Google	59.62	56.86
Twitter	65.73	59.71
TCOND	**73.28**	**64.52**
Task2		
Google	57.31	56.02
Twitter	62.78	58.45
TCOND	**67.27**	**62.73**
Task3		
Google	63.21	66.52
Twitter	65.88	68.46
TCOND	**77.27**	**69.33**

4.2   Qualitative Analysis

For this experiment, we worked on our same social dataset mentioned in the previous section. Our goal is to evaluate the usefulness of all proposed criteria on tweets that have not been appeared in direct conversation (created on the basis of our two algorithms detailed in the section 3) to detect tweets which are indirectly related to conversation such as retweets or comments and other possible interactions.

Features Evaluation. In this part, we evaluate the usefulness of single features and combinations of features for indirect conversation detection. To evaluate the applicability of criteria, we experiment with different coefficient weights of the similarity function. The results are given in Table 2 and demonstrate the significance of considering all message features by the similarity function. In addition, the results show that to get higher-quality conversations, the content attribute should be weighted higher than other attributes. However, Table 2 shows also that the dates attribute as almost insignificant for the conversation detection task, and that the content attribute is more important than the time attribute.

The previous experiment supports the hypothesis that a combination of a set conversational feature can improve the results of search tasks in microblogging sites.

Table 2. Conversation Detection Results, Using Different Coefficient Weights

	Precision	Recall
Presence of same URL	0.68	0.61
Hashtags Similarity	0.74	0.86
Time Difference	0.63	0.71
Tweets Publication Dates	0.58	0.64
Content	**0.82**	**0.86**
SameURL+Hashtags Simi	0.69	0.71
SameURL+Content	0.81	0.68
Hashtags+Content	**0.87**	**0.85**
Hashtags Similarity+Time	0.81	0.70
SameURL+Hashtags+Content	**0.88**	**0.86**
Time+Hashtags+Content	0.82	0.76
All	**0.96**	**0.92**

5 Conclusion

This work explored a new method for detecting conversation on microblogging sites: an information retrieval activity exploiting a set of conversational features in addition to the directly exchanged text messages to retrieve conversation. Our experimental results have highlighted many interesting points. First, including social features and the concept of direct conversation in the search function improves the relevance of tweets informativeness and also provides results that are considered more satisfaction with respect to a traditional tweet search task. Future work will further research the conversational aspects by including human communication aspects, like the degree of interest in the conversation and their influence/popularity by gathering data from multiple sources from Social Networks in real time.

References

1. Boyd, D., Golder, S., Lotan, G.: Tweet, Tweet, Retweet: Conversational Aspects of Retweeting on Twitter. In: Proceedings of the 2010 43rd Hawaii International Conference on System Sciences, HICSS 2010, pp. 1–10. IEEE Computer Society Press (2010)
2. Jabeur, L., Tamine, L., Boughanem, M.: Uprising Microblogs: A Bayesian Network Retrieval Model for Tweet Search. In: Proceedings of the 27th Annual ACM Symposium on Applied Computing, pp. 943–948 (2012)

3. Kerr, B.: Thread arcs: an email thread visualization. In: Proceedings of the Ninth Annual IEEE Conference on Information Visualization, vol. 8, pp. 211–218 (2003)
4. Yeh, J.: Email Thread Reassembly Using Similarity Matching. In: The Third Conference on Email and Anti-Spam (CEAS), Mountain View, California, USA (2006)
5. Erera, S., Carmel, D.: Conversation detection in email systems. In: Macdonald, C., Ounis, I., Plachouras, V., Ruthven, I., White, R.W. (eds.) ECIR 2008. LNCS, vol. 4956, pp. 498–505. Springer, Heidelberg (2008)
6. Klimt, B., Yang, Y.: Introducing the Enron Corpus. In: The Third Conference on Email and Anti-Spam (CEAS), Mountain View, California, USA (2006)
7. Cselle, G., Albrecht, K., Wattenhofer, R.: BuzzTrack: topic detection and tracking in email. In: Proceedings of the 12th International Conference on Intelligent User Interfaces, pp. 190–197 (2007)
8. Wang, Y., Joshi, M., Cohen, W., Ros, C.: Recovering Implicit Thread Structure in Newsgroup Style Conversations. In: The Second International Conference on Weblogs and Social Media, ICWSM 2008, The AAAI Press, Seattle (2008)
9. Xi, W., Lind, J., Brill, E.: Learning effective ranking functions for newsgroup search. In: Proceedings of the 27th Annual International ACM SIGIR Conference on Research and Development in Information Retrieval, SIGIR 2004, Sheffield, United Kingdom, pp. 394–401 (2004)
10. Cherichi, S., Faiz, R.: New metric measure for the improvement of search results in microblogs. In: Proceedings of the 3rd International Conference on Web Intelligence, Mining and Semantics, WIMS 2013, pp. 24:1–24:7 (2013)
11. Magnani, M., Montesi, D., Nunziante, G., Rossi, L.: Conversation Retrieval from Twitter. In: Clough, P., Foley, C., Gurrin, C., Jones, G.J.F., Kraaij, W., Lee, H., Mudoch, V. (eds.) ECIR 2011. LNCS, vol. 6611, pp. 780–783. Springer, Heidelberg (2011)
12. Cogan, P., Andrews, M., Bradonjic, M., Kennedy, W.S., Sala, A., Tucci, G.: Reconstruction and analysis of Twitter conversation graphs. In: Proceedings of the First ACM International Workshop on Hot Topics on Interdisciplinary Social Networks Research, vol. 7, pp. 25–31 (2012)
13. Matteo, M., Danilo, M., Luca, R.: Information Propagation Analysis in a Social Network Site ASONAM, pp. 296–300. IEEE Computer Society Press (2010)
14. Magnani, M., Montesi, D., Luca, R.: Conversation retrieval for microblogging sites. Information Retrieval Journal 15, 354–372 (2012)
15. Kumar, R., Mahdian, M., McGlohon, M.: Dynamics of conversations. In: Proceedings of the 16th ACM SIGKDD International Conference on Knowledge Discovery and Data Mining Washington, DC, USA, pp. 553–562 (2010)
16. Huang, J., Thornton, K.M., Efthimiadis, E.N.: Conversational tagging in twitter. In: Proceedings of the 21st ACM Conference on Hypertext and Hypermedia, pp. 173–178 (2010)
17. Song, S., Li, Q., Zheng, N.: A spatio-temporal framework for related topic search in micro-blogging. In: An, A., Lingras, P., Petty, S., Huang, R. (eds.) AMT 2010. LNCS, vol. 6335, pp. 63–73. Springer, Heidelberg (2010)

Measuring Information Quality of Geosocial Networks

Jiří Kysela, Josef Horálek, and Filip Holík

University of Pardubice, Faculty of Electrical Engineering and Informatics,
Studentská 95, 532 10, Pardubice 2, Czech Republic
{jiri.kysela,josef.horalek}@upce.cz,
filip.holik@student.upce.cz

Abstract. Geosocial networks are widespread adopted by tourists for the last years. These networks can replace tourist information center and play important information role in tourism. The main goal of this article is to get amount of coverage of geosocial networks to verify whether there is sufficiently reliable cloud information source for needs of tourism. This paper examines the coverage rate of points of interest used in tourism from Foursquare, Google+ and Facebook in the Czech Republic. Quality of information obtained from these geosocial networks is measured and compared with widely used map applications like Google Maps or local application Mapy.cz.

Keywords: geosocial networks, LBSN, location based services, LBS, Foursquare, Google+, Facebook.

1   Introduction

Geosocial networks, also called location based social networks (LBSN), are a new generation of social networks. Mobility of modern ICT (Information and Communication Technologies) enabled to enrich the data with geographical attributes giving the actual position of the mobile device on the Earth. These data are labeled as geodata and are crucial for usage in LBSN, which is part of location based services (LBS). These LBS provide information of local character. Attributes of mobility and enriched geodata create conditions for LBSN to become a new medium in the area of tourism that will provide users with information that can be subsequently shared. It is possible to share information on LBSN through contributions connected with any given place in the world, a so-called point of interest (POI) [1]. This submission deals with LBSN and with the quality of these cloud information sources, measured in Czech Republic. Section 2 describes specifics of these networks and their current status. Section 3 gives an analysis of completeness of information related to POI provided by LBSN Foursquare, Google+ and Facebook – the most widely used global LBSN. The analysis will examine completeness and sufficiency of information provided by these new media and their potential use in local tourism. The relevance and redundancy of the information acquired will be taken into consideration. The analysis uses POI covering the area of restaurant and bar services according to statistics method EU [6].

© Springer International Publishing Switzerland 2015   171
D. Barbucha et al. (eds.), *New Trends in Intelligent Information and Database Systems,*
Studies in Computational Intelligence 598, DOI: 10.1007/978-3-319-16211-9_18

Information for this submission will be acquired using the API (Application Programming Interface) for individual LBSN and also using field research in defined area – Czech city centers (within a 150 meters radius) of Pardubice and Hradec Králové (both county towns in the east Bohemia with approximately 100 thousand population).

2   LBSN Widespread Adopted in Tourism

LBSN (as well as LBS) provide answers for question like: Where am I? Where are my friends? What is here around me? [2]. There exists a large amount of applications for geosocial networks, but only few of them are used worldwide and has significant quantity of users [8]. Three of them are the following:

2.1   Foursquare

Foursquare is one of the few purely geosocial worldwide networks. An official source [7] actually states that there are approximately 50 million active users and 60 million POI. In Czech Republic, Foursquare is one of the most popular LBSN. POI in Foursquare is called "venue" and made by private users, but also participates global companies acting in the field of tourism (e.g. Lufthansa, Deutsche Bahn and McDonald's). Some businesses through these POI offer various bonuses and discounts for customers, with the goal to ensure loyalty and encourage further visits. The Foursquare campaign of McDonald's [9] resulted in more than 50 articles and blog posts and 600 thousand new fans with a 99 % positive feedback, the campaign increased check-ins of POI McDonald's by 33 % (for the week of the special, check-ins increased by 40 %). Foursquare web application is available on www.foursquare.com or the mobile version on m.foursquare.com. Native Foursquare application is available for all popular platforms:

- Android,
- iOS,
- BlackBerry,
- Symbian,
- Windows.

2.2   Google+

Google+ was originally only a social network, but gradually has changed to geosocial with a new service called as Places, which provides information about POI. Official source [10] actually states approximately 540 million active users. Recently Zagat has been integrated into Google+. Zagat evaluates companies from different perspectives (eg, quality of food, ambience and service) in the range of 0-30 (30 = perfect). A year and a half before Google acquired Zagat. They have been engaged in publishing printed guides in the world of restaurants and their evaluation - so these reviews will appear just in Google+. Web application of Google+ is available on https://plus.google.com or the mobile version on m.google.com. Google+ native application is available for platforms:

- Android,
- iOS.

2.3   Facebook

Originally a social network that due to extension Local Search (until 2012 known as Nearby) acquired properties of LBSN. The fundamental problem Local Search service is its availability only in native mobile applications, potential service thanks to the absence of a web application does not meet this remains only accessible to a limited group of users who own a mobile device running on supported platforms. Native application Facebook with Local Search is available for at least the two currently most widely used platforms, which are:

- Android,
- iOS.

3   Measuring Information Quality of LBSN

3.1   Measured Area of POI

Analysis of completeness will be examined on a base of information obtained inside of two independent defined areas, just in the center of the two different cities. Both of these Czech cities are regional capital in the east Bohemia, with a population of approximately 100 thousand. Defined areas have a circular shape, with a radius of 150 meters incluiding the main square in the old town part of the cities. Inside this circular area information about all restaurants and bars, will be obtained, according to the hospitality statistics method [6]. This area is shown on Fig. 1 and Fig. 2, thanks to own code in JavaScript using Google libraries with Maps visualization.

In the interest of an objective comparison, this analysis will use the information obtained with two independent methods – the first method was a field research by author's personal identification in both defined areas, in the second method the information will be obtained from examined information sources (see section 3.2.).

Pardubice, center (N 50.0385283, E 15.7789706, radius 150 m)

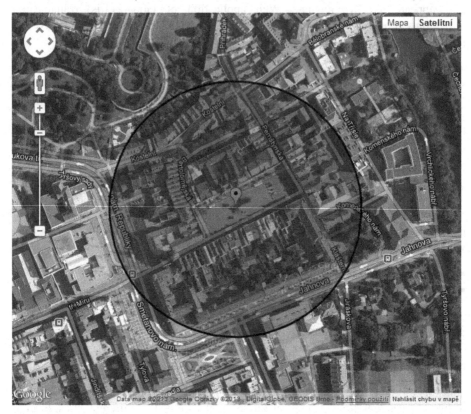

Fig. 1. Map of the analyzed area – Pardubice, historic centre of the city (Source: Google Maps and author, 2014)

Hradec Králové, center (N 50.2092658, E 15.8328122, radius 150 m)

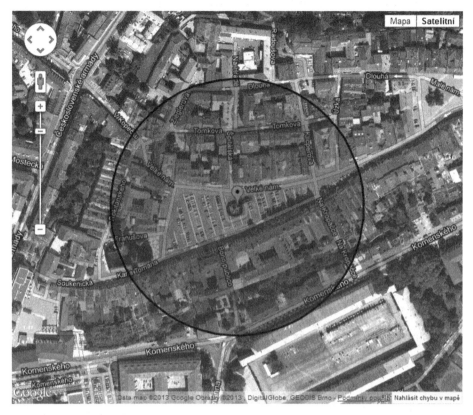

Fig. 2. Map of the analyzed area – Hradec Králové, historic centre of the city (Source: Google Maps and author, 2014)

3.2   Data Mining from LBSN Foursquare, Google+ and Facebook

Due to the unavailability of public information about quantity and quality of POI from geosocial networks, it was necessary to get it any other way, which will provide reliable official results about these networks. For this reason author of the paper obtained needed information using data mining through the API (Application Programmable Interface, described in Table 1) of the selected geosocial network Foursquare, Google+ and Facebook. In case of Google API was found, however, that it provides different results from the web application Google Maps in terms of the number of POI, due to this reason the information was obtained from this application separately. Information was also obtained from a commercial information source Mapy.cz, whose operator is one of the largest Czech company catalog, Seznam.cz. Since the API requires the input position in the long numeric format WGS84 (World Geodetic System 1984) the converter on a long numeric format available at http://www.earthpoint.us/Convert.aspx was used.

Table 1. Parameters for data mining through the API Foursquare, Google+ and Facebook (Source: Foursquare, Google+, Facebook and author, 2014)

Foursquare API [3]	HTTP GET request: https://api.foursquare.com/v2/venues/search?categoryId=P1&ll=P2&radius=P3&limit=P4&oauth_token=P5 API Explorer (https://developer.foursquare.com/docs/explore) request: venues/search?categoryId=P1&ll=P2&radius=P3&limit=P4 <table><tr><td>Pardubice: P1=4d4b7105d754a06376d81259 (coffee, drinks) / 4d4b7105d754a06374d81259 (food) – whole list of the identificators is available on foursquare.com. P2=50.0385283,15.7789706 P3=150 P4=100 P5=*OAuth token*</td><td>Hradec Králové: P1=4d4b7105d754a06376d812 59 (coffee, drinks) / 4d4b7105d754a06374d81259 (food) P2=50.2092658,15.8328122 P3=150 P4=100 P5=*OAuth token*</td></tr></table>
Google+ API [4]	HTTP GET request: https://maps.googleapis.com/maps/api/place/nearbysearch/json?location=P1&radius=P2&types=P3&sensor=false&key=P4 <table><tr><td>Pardubice: P1=50.0385283,15.7789706 P2=150 P3=coffee / bar / food P4=*OAuth token*</td><td>Hradec Králové: P1=50.2092658,15.8328122 P2=150 P3=coffee / bar / food P4=*OAuth token*</td></tr></table>
Facebook API [11]	HTTP GET request: https://graph.facebook.com/search?type=P1&q=P2&center=P3&distance=P4&access_token=P5 Graph API Explorer (https://developers.facebook.com/tools/explorer) request: /search?type=P1&q=P2&center=P3&distance=P4 <table><tr><td>Pardubice: P1=place P2=cafe / bar / restaurant P3=50.0385283,15.7789706 P4=150 P5=*OAuth token*</td><td>Hradec Králové: P1=place P2=cafe / bar / restaurant P3=50.2092658,15.8328122 P4=150 P5=*OAuth token*</td></tr></table>

3.3   Analysis of Information Quality of LBSN Foursquare, Google+ and Facebook and Comparison with Other Information Sources

On the basis of this analysis that compared information acquired using the API of selected LBSN and also using field research in defined area following figures were obtained: Foursquare contains 26 POI covering the area of restaurant and bar services in total in Pardubice, Google+ contains 15 POI in total and Facebook contains 22 POI in total whereas factually there are 28 restaurant and bar services in total in this area (see Fig. 3). In the case of Hradec Králové, Foursquare contains 36 POI covering the area of restaurant and bar services in total in Pardubice, Google+ contains 14 POI in total and Facebook contains 31 POI in total out of 42 existing restaurant and bar services operating in total in this area (see Fig. 4).

The commercial information source Mapy.cz which is run by the biggest Czech company catalogue (Seznam.cz) was also included in the comparison with following results: Mapy.cz contains only 11 POI in Pardubice area and 18 POI in Hradec Králové area. Therefore, it is evident that LBSN offer significantly higher degree of completeness of information – by 54 % for Foursquare, by 14% for Google+, by 39 % for Facebook in Pardubice and by 43 % for Foursquare, by 31 % for Facebook and lower degree of completeness of information by 10 % for Google+ in Hradec Králové. Duplicities and irrelevant POI were eliminated from comparison in Table 2 and 3, because the information was sometimes irrelevant due to wrong GPS location and redundant in relation to given area.

Table 2. Structure of POI in centre of Pardubice – October 2014 (Source: [3], [4], [5], [11] and author, 2014)

Pardubice, centre (N 50.0385283, E 15.7789706, radius 150 m)					
	Foursquare API	Google+ API	Facebook API	Google Maps	Mapy.cz	Total real places
Restaurants POI	7	6	6	5	6	9
Bars POI	19	9	16	6	5	19
Total POI	26	15	22	11	11	28

Table 3. Structure of POI in centre of Hradec Králové – October 2014 (Source: [3], [4], [5], [11] and author, 2014)

Hradec Králové, centre (N 50.2092658, E 15.8328122, radius 150 m)					
	Foursquare API	Google+ API	Facebook API	Google Maps	Mapy .cz	Total real places
Restaurants POI	17	8	12	6	16	20
Bars POI	19	6	19	4	2	22
Total POI	36	14	31	10	18	42

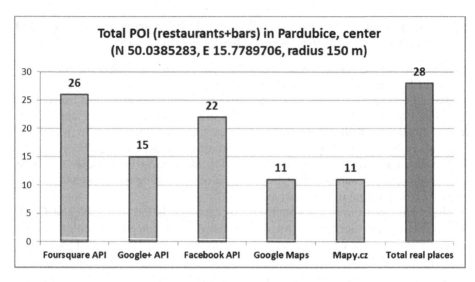

Fig. 3. Comparison of information sources in analyzed area (Source: [3], [4], [5], [11] and author, 2014)

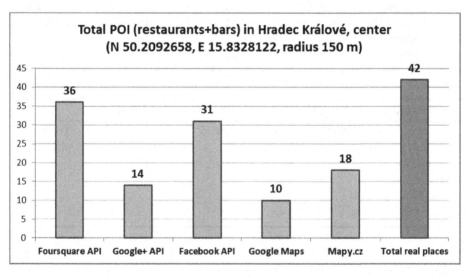

Fig. 4. Comparison of information sources in analyzed area (Source: [3], [4], [5], [11] and author, 2014)

4   Conclusion

This article dealt with LBSN and with the quality of these cloud information sources. Specifics of these networks and their current status was described in section 2. The main point of the submission was given in section 3 – an analysis of completeness of information provided by LBSN Foursquare, Google+ and Facebook. Information from these networks were related to POI in defined area of Czech city centers (within a 150 meters radius) of Pardubice and Hradec Králové (both county towns in the east Bohemia with approximately 100 thousand population). The analysis examined completeness and sufficiency of information provided by these new media and their potential use local tourism. The relevance and redundancy of the information acquired were taken into consideration.

On the basis of this analysis that compared information acquired using the API (Application Programming Interface) of selected LBSN and also using field research in defined area following figures were obtained: Foursquare contains 26 POI covering the area of restaurant and bar services in total in Pardubice, Google+ contains 15 POI in total and Facebook contains 22 POI in total whereas factually there are 28 restaurant and bar services in total in this area (see Fig. 3). In the case of Hradec Králové, Foursquare contains 36 POI covering the area of restaurant and bar services in total in Pardubice, Google+ contains 14 POI in total and Facebook contains 31 POI in total out of 42 existing restaurant and bar services operating in total in this area (see Fig. 4).

The commercial information source Mapy.cz which is run by the biggest Czech company catalogue (Seznam.cz) was also included in the comparison with following results: Mapy.cz contains only 11 POI in Pardubice area and 18 POI in Hradec Králové area. Therefore, it is evident that LBSN offer significantly higher degree of completeness of information – by 54 % for Foursquare, by 14% for Google+, by 39 % for Facebook in Pardubice and by 43 % for Foursquare, by 31 % for Facebook and lower degree of completeness of information by 10 % for Google+ in Hradec Králové. Duplicities and irrelevant POI were eliminated from comparison, because the information was sometimes irrelevant due to wrong GPS location and redundant in relation to given area. As a conclusion for mapped area, LSBN provide significantly higher degree of completeness of information related to POI in the area of restaurant and bar services these days than one of the biggest Czech commercial information source and are therefore applicable in commerce, specifically in local tourism, as a new medium with considerable potential.

References

1. Kysela, J.: Usability of geosocial networks for tourism support (Využitelnost geosociálních sítí pro podporu místního cestovního ruchu). In: 6th International Conference of College of Business and Hotel Management, pp. 204–215. VŠOH, Brno (2013)
2. Kysela, J.: Possibilitites of geolocation web applications for LBS (Možnosti geolokačních webových aplikací pro LBS). In: The Summer School, Interdisciplinary Approaches in Informatics and Cognitive Science, pp. 36–45. University of Hradec Králové, Hradec Králové (2012)
3. Foursquare API, https://api.foursquare.com/v2/
4. Google API,
   https://maps.googleapis.com/maps/api/place/radarsearch/
5. Mapy cz, http://www.mapy.cz
6. Stratilová, Z.: Catering in tourism (Stravovací zařízení v cestovním ruchu). Bachelor thesis, Tomas Bata University in Zlín, Zlín (2008)
7. Foursquare, https://foursquare.com/about/

8. Scellato, S., Mascolo, C.: Distance Matters: Geo-social Metrics for Mobile Location-based Social Networks, `http://perso.uclouvain.be/vincent.blondel/netmob/2011/NetMob2011-abstracts.pdf`

9. Keane, M.: Case Study: McDonald's ups foot traffic 33% on Foursquare Day, `https://econsultancy.com/blog/6582-case-study-mcdonald-s-and-foursquare#i.c7mtm6j2ce5bqy`

10. USA Today, `http://www.usatoday.com/story/tech/2013/10/29/google-plus/3296017/`

11. Facebook API, `https://graph.facebook.com`

Part IV
Cloud Computing and Intelligent Internet Systems

A Secure Non-interactive Deniable Authentication Protocol with Certificates Based on Elliptic Curve Cryptography

Yu-Hao Chuang[1], Chien-Lung Hsu[1], Wesley Shu[2],
Kevin C. Hsu[2], and Min-Wen Liao[1]

[1] Department of Information Management, Chang Gung University,
No. 259 Wen-Hwa 1st Road, Kwei-Shan Tao-Yuan, Taiwan, 333, R.O.C.
{yuhao0512,usnsliao}@gmail.com, clhsu@mail.cgu.edu.tw
[2] Department of Information Management, National Central University, No.300,
Jhongda Rd., Jhongli City, Tao-Yuan County 32001, Taiwan, R.O.C.
{shu,khsu}@mgt.ncu.edu.tw

Abstract. Since the information technology continues to grow and the network applications are getting popular, protecting the privacy of Internet users on the open network becomes a necessary concern. The traditional authentication protocol is not suitable for the requirements of internet security nowadays. That is, it cannot assure that the private information not be revealed during the authentication operation, or be used by malicious terminal service managers for their personal gain in some other business opportunities. Hence, in the dissertation, we propose a deniable authentication protocol based on elliptic curve cryptography (ECC) to satisfy the current public key infrastructure and fulfill the following security requirements of deniable authentication protocols.

1. Each user can prove his/her legitimacy to the designated verifier.

2. The designated verifier cannot prove the identity of the user to the third party even though the verifier provides the testimonials.

Moreover, the proposed deniable authentication protocol is suitable for the mobile devices since it only needs limited computation resources.

Keywords: deniable authentication protocol, privacy, public key infrastructure, elliptic curve cryptography.

1 Introduction

Information technology (IT) is a domain of managing technologies and spans a wide variety of domains including computer hardware, computer software, information system, etc, but are not only limited to such. Hilbert and Lopez introduced the exponential pace of IT change as called a kind of Moore's law [1]. That is, application-specific capacity for computing of machines has roughly doubled every 18 months. At present, the Internet has become an integral part of daily life and breaks the limitations of

© Springer International Publishing Switzerland 2015
D. Barbucha et al. (eds.), *New Trends in Intelligent Information and Database Systems,*
Studies in Computational Intelligence 598, DOI: 10.1007/978-3-319-16211-9_19

time and space. It has penetrated all sectors of our leisure activities: working, shopping, communication, etc. There are several types of online activities offered today such as: online game, online video, online banking, online shopping, e-mail, blog, etc. A staff can use communication tools (such as video conferencing software, wiki systems) to transmit information with his co-workers in the company; in his leisure time, he also can do some recreational activities via Internet.

Recently, since the information technology continues to grow and the network applications are getting popular, the internet has become the most popular communication media. The Internet provides a low-cost space for malicious persons to commit a crime. As the Internet has no borderline and management restrictions, user's privacy has become a major issue for all online activities. Concerns of online privacy lead to unwillingness to provide personal information online (e.g. registered data), rejection of electronic commerce, or even reluctance to surf Internet. Cryptographic techniques are developing to ease the concerns. Thus, how to keep the information transmission secure and to protect confidentiality, integrity, authentication, and non-repudiation of transmitted data in open networks by using cryptography is an important issue.

In traditional cryptographic techniques, the host can identify the validity of a terminal user by using identification algorithms (e.g. Kerberos [2]), and then build a secure channel with the terminal user by using encryption protocols (e.g. secure sockets layer, SSL) to assure transmitted information confidentiality and integrity via Internet. Such kinds of cryptographic techniques should belong to the external security protections. However, if the employee steals users' registered data and then peddles the data to the third party attempting for unlawful gain, the traditional cryptographic techniques (e.g. user identification or encryption) cannot prevent such a crime.

In recent years, different kinds of authentication protocols were proposed by researchers, which are so-called deniable authentication protocols. Compared with the traditional authentication protocols such as [3,4,5,6,7], the deniable authentication protocol has two properties:

(a) The receiver can identify the source of a given message.
(b) The receiver is unable to prove the source of the sender to any third party.

Because deniable authentication protocol possess above properties, it can be applied in various scenarios (e.g. electronic voting systems, secure negotiation systems, and etc.). In 1998, Dwork et al. [8] based on zero-knowledge proof to propose a deniable authentication protocol, in which a time constraint was required and the proof of the knowledge is subject to a time delay during the authentication processes. Meanwhile, based on the computation impossibility of solving the difficult factorization problem, Aumann and Rabin introduced a deniable authentication protocol [9].

In 2001, Deng et al. proposed two kinds of deniable authentication protocols which are based on the computation intractability of solving factoring problem and the discrete logarithm problem respectively [10]. However, both Aumann and Rabin's protocol and Deng et al.'s protocols require a trusted directory to record all public data. To overcome this drawback, Fan et al. based on Diffie-Hellman algorithm [11] to propose a deniable authentication protocol without requiring a trusted directory, in which it can withstand the men-in-the-middle attack using certificates and identify the source of message by employing the signature algorithm [12]. The Diffie-Hellman key exchange protocol is a cryptographic protocol which allows two parties that have no prior knowledge of each other to cooperatively generate a shared key through a public communication channel. It was published by Diffie and Hellman in 1976, which is the earliest practical protocol of key exchange within the field of cryptography [11].

However, Yoon et al. pointed out that a malicious adversary can easily masquerade as a valid verifier without being found to identify the source of message in Fan et al.'s protocol, and hence they further proposed an improvement attempting to overcome this security flaw in Fan et al.'s protocol [13].

Since then, several simple deniable authentication protocols have been proposed [14,15,16,17,18,19], in which all of these protocols are non-interactive. However, in the present mobile communication environment, the computational complexities of above discussed protocols are still high and are not suitable for operating in the mobile devices. At present, several deniable authentication protocols have been proposed continually [13,14,15,16,17], [20]. In Fig. 1, we depict the evolution of deniable authentication protocols.

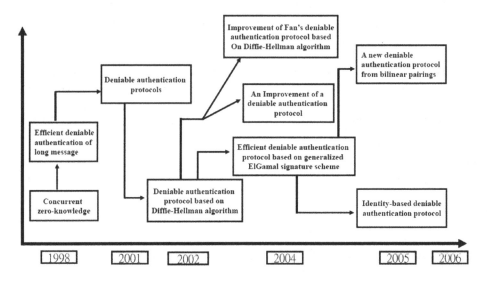

Fig. 1. Deniable authentication protocols

In this paper, we will propose a new deniable authentication protocol based on the Elliptic Curve Cryptography (ECC), in which it not only needs lower computational complexities and communication efforts, but also it can achieve the properties of deniable authentication. The remainder of this paper was organized as follows. Section 2 introduced the proposed deniable authentication protocol with certificates based on ECC and analyzed its security in Section 3. Finally, we gave the conclusions.

2   Deniable Authentication Protocol with Certificates Based on ECC

Elliptic curve cryptography (ECC) was developed by Koblitz and Miller [21] independently, which is a cryptographic approach based on the algebraic structure of elliptic curves over finite field. The primary benefits of ECC are smaller key sizes, less storage, and faster implementations than traditional cryptographic protocols (e.g. RSA [22). That is, an ECC system could provide the same level of security requirements afforded by RSA-based systems by using smaller key sizes. Thus, ECC-based protocols which reduce the storage and transmission requirements are suitable for mobile device applications (e.g. PDA and mobile phone). By these reasons, we design a deniable authentication protocol with certificates, which security is protected under the ECDLP assumption.

In the proposed ECC-based protocol, it is non-interactive and consists of three phases. In this section, we describe in detail the process of each phase of the proposed deniable authentication protocol with certificates based on ECC (DAPCs_ECC for short). In the system initialization phase, a TA is needed to determine some system public parameters as follows: p, q are two large primes, $E \in Z_p$ is a an elliptic curve, a base point $G \in E(Z_p)$ order q, $H(\cdot)$ is a collision-free one-way hash function where $H : \{0,1\}^* \to Z_p$, and the symbol "‖" is concatenate operation of two strings. In the key generation phase, each sender and receiver has to register a private/public key pair as certified by the CA, which are denoted as $x_i \in [1, q-1]$ and $Y_i = x_i \cdot G$, respectively. In the deniable authentication phase, the sender can deniably authenticate a message m to the intended receiver by cooperatively performing the following steps:

1. The sender S chooses a random number $a \in [1, q-1]$ and computes

$$K_{SR} = x_S \cdot Y_R \tag{1}$$

$$r = H(K_{SR}) \oplus a$$
$$C = H(K_{SR} \| a) \oplus m \| H(a \| m)$$

2. After receiving the ciphertext (r, C), R computes

$$K_{RS} = x_R \cdot Y_S \tag{2}$$

$$a' = H(K_{RS}) \oplus r$$

$$m \| H(a \| m) = C \oplus H(K_{RS} \| a') \tag{3}$$

3. R verifies the underlying equation
   $$H(a' \| m) \underset{=}{?} H(a \| m)$$

   If the equation holds, R accepts it. Otherwise, R rejects it.

Note that the sender can attach a timestamp to the message m to withstand the replay attack.

In addition, in order to clearly demonstrate what system functions are performed by which actors, a use case diagram of unified modeling language (UML) has been adopted to illustrate that. The use case diagram of UML can be seen as a type of behavioral diagram which is defined and constructed according to the use-case analysis. The main purpose of a use case diagram is to illustrate a graphical overview of the functionality offered by the system in terms of each actor, their goals, and all dependencies between those use cases (as shown in Fig. 2).

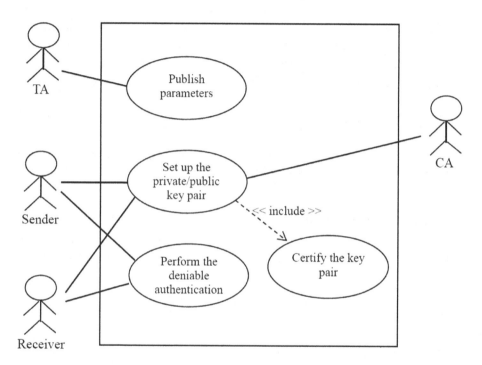

Fig. 2. Use case diagram of UML

3   Security Analysis

In this section, we prove that the DAPCs_ECC can satisfy the security requirements of authentication and deniability.

- *Theorem 1. (Considerations for authentication) In the proposed DAPCs_ECC, the receiver is able to identify the source of the message.*

Proof:

In our protocol, the Eqs. (1) and (2) are compute keys of shared between S and R, where $K_{SR} = K_{RS} = x_S \cdot x_R \cdot G$ based on the Diffie-Hellman key exchange protocol [11]. If someone provides the ciphertext (r, C) and it simultaneously satisfies the Eq. (3), R confirmed the ciphertext (r, C) is constructed by S because the shared key $K_{SR} = K_{RS}$ is only constructed by S and R, granting that intruder E gets all information of communicated between S and R. He is unable to construct a valid ciphertext (r, C) which is protected under the ECDLP assumption. In addition, the message m is encrypted with the shared key K_{RS} and a random number a. An intruder E can get the ciphertext (r, C) by interrupting communication between S and R; however, E is unable to compute the shared key K_{RS} between S and R because the DLP is difficult to solve. That is, E is unable to construct a valid ciphertext (r', C') with m' to satisfy the Eq. (3) without sharing the key K_{RS} to cheat R. Therefore, our proposed DAPCs_ECC can protect against the PIM attack.

- *Theorem 2. (Considerations for deniability) In the proposed DAPCs_ECC, the receiver cannot prove the source of the message to the third party.*

Proof:

Consider the scenario that R can construct a ciphertext (r, C') by himself. In this scenario, R can generate a message m' different from message m, and further constructs the ciphertext (r, C') by a, m' and the shared key K_{RS}, where $K_{RS} = x_R \cdot x_S \cdot G$. According to *Theorem 1,* the ciphertext (r, C') is indistinguishable from the actual message computed by S. Hence, the proposed DAPCs_ECC achieves deniability.

4   Discussions and Conclusions

4.1   Discussions

In this section, we compare some security properties of our proposed deniable authentication protocols with various other deniable authentication protocols. We demonstrate that the Fan et al. and the Wang et al. protocols [12], [14] are not able to withstand the PIM attack nor achieve authentication. Hence, these protocols cannot achieve user authentication as they claimed. Both the Aumann-Rabin [9] and the Deng et al. [10] protocols need a public trust directory to maintain all public data but the others are not. On the other hand, in Dwork et al.'s protocol, a timing constraint was set and proof of knowledge was subject to a time delay during the authentication activities that requires a timing constraint. Consequently, these three protocols are not efficient to apply in practice. In addition, only the Wang et al. [14] and our proposed protocols are non-interactive, which are able to reduce additional communication costs; however, according to above discussion, Wang et al.' protocol cannot achieve the security of user authentication. Moreover, in our proposed protocols, the sender can further insert a timestamp into the authorized message to prevent the replay attack, but others are implicit.

4.2   Conclusions

In recent years, information technologies and networks are developed quickly and consequently the security requirement of a user log into a network is getting more important. Meanwhile, mobile devices replace the traditional device (e.g. desktop) for lots of users to surf the internet. That is, the traditional authentication protocol is not suitable for the requirements of internet security nowadays. In this paper, we proposed a new deniable authentication protocol based on elliptic curve cryptography (ECC) to satisfy the current public key infrastructure and fulfill the following security requirements of deniable authentication protocols. Moreover, the proposed deniable authentication protocol is suitable for the mobile devices since it only needs limited computation and communication resources.

References

1. Moore, G.E.: Cramming More Components onto Integrated Circuits. Electronics 38(8), 114–117 (1965)
2. Kohl, J., Neuman, C.: The Kerberos Authentication Service. Internet RFC 1510 (1993)
3. Huang, H., Cao, Z.: An ID-Based Authenticated Key Exchange Protocol Based on Bilinear Diffie-Hellman Problem. In: ASIACCS 2009:Proceedings of the 4th International Symposium on Information, Computer, and Communications Security (2009)
4. Jun, E.A., Ji, J.D., Lim, J., Jung, S.W.: Improved Remote User Authentication Scheme Using Bilinear Pairings. In: ICHIT 2009: Proceedings of the 2009 International Conference on Hybrid Information Technology (2009)
5. Shieh, W.G., Horng, W.B.: Security Analysis and Improvement of Remote User Authentication Scheme without Using Smart Cards. ICIC Express Letters: An International Journal of Research and Surveys 4(6), 2431–2436 (2010)
6. Shieh, W.G., Horng, W.B.: Cryptanalysis and Improvement of Wen et al.'s Provably Secure Authentication Key Exchange Protocols for Low Power Computing Devices. ICIC Express Letters: An International Journal of Research and Surveys 5(11), 4027–4032 (2011)
7. Yoon, E.J., Lee, W.S., Yoo, K.Y.: Secure Remote User Authentication Scheme Using Bilinear Pairings. In: Sauveron, D., Markantonakis, K., Bilas, A., Quisquater, J.-J. (eds.) WISTP 2007. LNCS, vol. 4462, pp. 102–114. Springer, Heidelberg (2007)
8. Dwork, C., Naor, M., Sahai, A.: Concurrent Zero-Knowledge. In: Conference Proceedings of the Annual ACM Symposium on Theory of Computing, pp. 409–418 (1998)
9. Aumann, Y., Rabin, M.: Efficient Deniable Authentication of Long Message. In: Int. Conf. on Theoretical Computer Science in Honor of Professor Manuel Blum's 60th Birthday (1998), http://www.cs.cityu.edu.hk/dept/video.html (retrieved from)
10. Deng, X., Lee, C.H., Zhu, H.: Deniable Authentication Protocols. IEE Proceeding of Computer and Digital Techniques 148(2), 101–104 (2001)

11. Diffie, W., Hellman, M.E.: New Directions in Cryptography. IEEE Transactions on Information Theory IT-22(6), 644–654 (1976)
12. Fan, L., Xu, C.X., Li, J.H.: Deniable Authentication Protocol Based on Diffie-Hellman Algorithm. Electronics Letters 38(4), 705–706 (2002)
13. Yoon, E.J., Ryu, E.K., Yoo, K.Y.: Improvement of Fan et al.'s Deniable Authentication Protocol Based on Diffie-Hellman Algorithm. Applied Mathematics and Computation 167(1), 274–280 (2005)
14. Wang, Y., Li, J., Tie, L.: A Simple Protocol for Deniable Authentication Based on ElGamal Cryptography. 2005 Wiley Periodicals, Inc. NETWORKS 45(4), 193–194 (2005)
15. Kar, J., Majhi, B.: A Novel Deniable Authentication Protocol Based on Diffie-Hellman Algorithm Using Pairing Technique. In: ICCCS 2011:Proceedings of the 2011 International Conference on Communication, Computing & Security (2011)
16. Lu, R., Cao, Z.: A New Deniable Authentication Protocol from Bilinear Pairings. Applied Mathematics and Computation 168(2), 954–961 (2005)
17. Lu, R., Lin, X., Cao, Z., Qin, L., Liang, X.: A Simple Deniable Authentication Protocol Based on the Diffie-Hellman Algorithm. International Journal of Computer Mathematics 22(3), 1–9 (2007)
18. Shao, Z.: Efficient Deniable Authentication Protocol Based on Generalized ElGamal Signature Scheme. Computer Standards & Interfaces 26(5), 449–454 (2004)
19. Shi, Y., Li, J.: Identity-Based Deniable Authentication Protocol. Electronics Letters 41(5) (2005)
20. Raimondo, M.D., Gennaro, R., Krawczyk, H.: Deniable Authentication and Key Exchange. In: CCS 2006: Proceedings of the 13th ACM Conference on Computer and Communications Security (2006)
21. Koblitz, N.: Elliptic Curve Cryptosystems. Mathematics of Computation 48(177), 203–209 (1987)
22. Rivest, R., Shamir, A., Adleman, L.: A Method for Obtaining Digital Signature and Public Key Cryptosystem. Communications of the ACM 21(2), 120–126 (1978)

An Ad Hoc Mobile Cloud and Its Dynamic Loading of Modules into a Mobile Device Running Google Android

Filip Maly and Pavel Kriz

Department of Informatics and Quantitative Methods,
University of Hradec Kralove, Czech Republic
{filip.maly,pavel.kriz}@uhk.cz

Abstract. This paper introduces Ad Hoc Mobile Cloud (m-cloud) Computing solution, describes its advantages and practical use-cases, e.g. during disaster recovery or mobile network overload during actions involving masses of people. A basic concept of an M-client M-client Server architecture which supports creation and existence of m-clouds is presented further in the text. Principle of a client (an m-client) with variable thickness, a basic building block and thus a necessary component of m-clouds, is discussed. The paper also describes dynamic module loading within an m-client. The Google Android platform has been used for our implementation.

Keywords: Ad Hoc Mobile Cloud, m-cloud, m-client, M-client M-client Server architecture, Android, dynamic module loading, client with variable thickness, LTE Direct.

1 Introduction

Development of mobile communication brings new opportunities for unique cloud forms. In this paper a form of a cloud which is formed by particular mobile devices instead of dedicated data-centers (operated by companies such as Google or Microsoft) like in classic Mobile Cloud solutions [1, 2] is introduced. Mobile device is a part of the Ad Hoc Mobile Cloud and thus it can further provide certain services to neighboring devices. It is necessary to use a suitable architecture to be able to create the Ad Hoc Mobile Cloud. The M-client M-client Server [3] represents such an architecture and an m-client is its basic building block. Ad Hoc Mobile Cloud employing this architecture is called an "m-cloud" in this paper.

The M-client M-client Server architecture enables dynamic forming of m-clouds with the aid of m-clients with dynamic loading of modules. Particular implementation of dynamic module loading is demonstrated at the Google Android platform.

The rest of this paper is structured as follows. In Section 2 the m-cloud and its use-cases are described. In Section 3 we introduce the concept of M-client M-client Server architecture that supports creation and existence of m-clouds. In Section 4 we present main principles of the m-client, the basic building block of any m-cloud. We

© Springer International Publishing Switzerland 2015
D. Barbucha et al. (eds.), *New Trends in Intelligent Information and Database Systems,*
Studies in Computational Intelligence 598, DOI: 10.1007/978-3-319-16211-9_20

demonstrate the implementation of the dynamic module loading in Section 5. Section 6 summarizes our contribution and results.

2   M-Cloud

Mobile Cloud Computing is an emerging trend of last years. Several new approaches have been introduced. Survey [4] summarizes current work on Mobile Cloud Computing, not only on Ad Hoc Mobile Cloud but also on more typical architectures and solutions requiring a server(s) deployed into classic cloud services (IaaS, etc.). Paper [5] describes development and testing of algorithms that allow managing the nodes of the cloud that join and leave the spontaneous ad hoc network. The solution has been implemented within the network simulator and has focused on results of network performance issues related to a number of nodes. Dynamic distribution of particular applications or modules to Android devices is described in [6] tough this work focuses on copy-protection and related copyright issues.

Ad Hoc Mobile Cloud (m-cloud) is composed of autonomous group of mobile devices – network nodes. M-cloud does not usually require access to a central server. We distinguish the following basic states of mobile node in the m-cloud (an "m-client"): first, it has access to the central server (information system) via the Internet and optionally communicates with other m-clients; second, it does not have direct access to the central server and depends on mediated connection with the central server via other m-clients; third, it does not have any access to the central server and uses services provided by other m-clients directly.

2.1   Ways of M-Cloud Forming and Services Offered

Basic services offered by an m-cloud exceed functionalities of classic clouds especially from the mobility point of view.

M-cloud formed of m-clients can offer services such as:

- One powerful distributed system, which is able to offer its computing power for ad hoc analysis or for other purposes.
- Distribution channel for sharing information in a public or private m-cloud independent from mobile network.
- Context-oriented m-cloud. Thanks to the sensors in smartphones context can be detected. Analogically, it is possible to extend this context in the geographical or other dimension with the aid of the m-cloud. Two types of context are distinguished: low-level and high-level [7]. Especially high-level context can be made more accurate thanks to the use of the m-cloud.

M-clouds may be divided according to the way they are formed:

- Geographically defined m-cloud which benefits from geographical location.
- Business-oriented m-cloud is formed by m-clients within one company or organization supporting company's business and improving privacy and security.

Strong dependence of clients on the central server is rapidly decreasing thanks to the solution based on m-cloud. Data can be accessed directly within local networks, even in the field. A form of private m-cloud which enables sharing information without passing them to the third parties (including communication via public network) may also exist. The m-cloud also enables higher availability of data thanks to their distribution among multiple devices.

2.2   M-Cloud Use-Cases

Practical uses of m-clouds can be shown by the following examples:

Offline Rescue Workers. It can happen in the field that some rescue workers are online and the others are offline (disconnected from the mobile network). It is possible to form an m-cloud in which both online and offline mobile devices are connected together using technologies that are independent from the operator's mobile network. Thanks to the m-cloud the map of the search area may be saved into individual mobile devices in a distributed way. Sharing context information about the whole area is another use-case of the context-oriented m-cloud.

Free Roaming Community Network. Costs of mobile connection abroad are too high because of roaming fees. If the traveler wishes to access the Internet from his/her smartphone abroad without extreme financial expense he/she is now forced to look for available hotspots. While Ad hoc networks allow us to share the Internet connection, m-cloud can ensure the sharing of information about roaming community network users and supports the formation of a physical network in particular area [8].

Overloaded or Non-functioning Mobile Networks. The m-cloud supports direct peer-to-peer communication during overload of mobile networks in certain situations (concerts, football matches etc.) or during partial network outages because of natural disasters.

Distribution of Information within Wireless Community Network. Connection of clients into an m-cloud can be used in wireless community networks for distribution of information about the network, its operation, stability, outages, etc. The m-cloud has a substantial role in detection of problems in the wireless network [9].

3   Concept of the M-Client M-Client Server Architecture

The client-server architecture traditionally used in mobile environment fails in the above mentioned cases, especially due to instability of connection with the server. Thus it would be useful to take advantage of communication with the neighboring client. This neighboring client can provide or mediate data or computing power. Such an approach requires a partial shift from the client-server architecture to the peer-to-peer architecture [8].

Implementation of the M-client M-client Server architecture in mobile environment is challenging, particularly due to unpredictable environment and autonomy of mobile devices. Communication technologies such as WiFi, Bluetooth or NFC may cause additional difficulties. Bluetooth in current versions is suitable for communication between two devices within several meters from each other. WiFi technology (or WiFi Direct) can be applied within tens of meters. Both Bluetooth and WiFi suffer from the interference with other technologies in ISM[1] bands. Great potential is in LTE Direct technology[2], which is currently in development. This technology is able to reach the distance up to 500 meters between two mobile devices.

In the following chapters the main aspects of the M-client M-client Server architecture, especially of the m-client itself, are described.

4   Basic Principles of M-Clients

The m-client with dynamic module loading is the basic building block of the mentioned architecture. It is an application running at the mobile device and is implemented as a client with variable thickness. It reconfigures itself on demand so its business logic gets new functions without the need to restart the application. The m-client is able to communicate with the remote server and can also become a local server itself. The architecture is considered decentralized and its advantages are as follows:

- Independence from the server and the rest of the network. Should a task be completed in the field only the neighboring m-clients are required. Neither components of the network nor the server are necessary.
- Making use of information from other (neighboring) m-clients in order to explore their surroundings.

Correctly working dynamic reconfiguration of m-clients depending on certain situations is the basic requirement. These situations are implied by the context information about the environment or by the demand of the user of the mobile device where the m-client is running. The next section deals with the principle of dynamic module loading as a necessary function of the m-client.

5   Analysis of Dynamic Module Loading Implementation

The application consists of the core and the modules. Reference implementation is described at the Android platform. Dynamic loading and activation of application modules without the need of their manual installation is the main issue.

Several circumstances, various issues and limitations have to be considered during the software analysis:

- only Activities defined in the Manifest file may be started
- application (module) is allowed to perform only actions granted by System Permissions defined in the Manifest file
- the Manifest file of the application cannot be changed additionally

---

[1] Industrial, scientific and medical radio bands.
[2] https://www.qualcomm.com/invention/research/projects/lte-direct
    http://www.slideshare.net/yihsuehtsai/lte-direct-in-unlicensed-spectrum

- if the application is divided into multiple Android Application Packages (APKs), it is necessary load individual classes and resources from the corresponding APK

The Manifest file has to be designed in general way in order to enable the application to fulfill all its (future) requirements.

5.1   Modular Solution and the Principle of a Class-Loader

The application should be able to load modules after downloading them from the Internet or from the neighboring mobile device or after reading them from the internal memory card. Considering the above stated limitations of the Android platform every module will have to be implemented as a regular Android Application Packages (APK). All possible System Permissions which particular version of the Android platform offers should be given to the main application (in its Manifest file) to achieve maximum flexibility.

The issue of dynamic loading of modules will be addressed by employing a special class-loader as class-loaders are the components responsible for class loading in Java[3]. In a typical application for the Android operating system running Dalvik run-time class-loaders create the hierarchy described in the Fig. 1. `BootClassLoader` is a top-level class-loader which loads classes of the Android framework. `PathClassLoader[.]` is a system class-loader which can be obtained by calling the `ClassLoader.getSystemClassLoader()` method and is usually being utilized as a parent class-loader for custom class-loaders. The last `PathClassLoader[/data/app/com.example.apk]` class-loader is responsible for loading all classes from `com.example.apk` package thus loading all classes created in the given application.

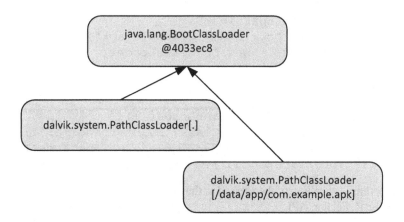

Fig. 1. Class-loader hierarchy

---

3   http://developer.android.com/reference/java/lang/
    ClassLoader.html

Loading classes from the class-loader's point of view is performed in the direction from the parent (top-level class-loader) towards children. Class-loaders also contain references to already loaded classes and every class has the reference to the class-loader by which it has been loaded. Besides the mentioned particular class-loaders the Android platform also offers the DexClassLoader[4] which is intended for dynamic loading of classes from the APK packages. The DexClassLoader's constructor requires the following parameters:

- path to the APK package,
- path to the directory where the class-loader saves the optimized class-files
- optional path to libraries (may be null)
- parent class-loader (mandatory, must not be null)

DexClassLoader will be used in combination with other techniques for dynamic loading of modules in the m-client.

5.2   Dynamic Loading of Modules and Calling the Methods

When implementing dynamic loading of modules it is not possible for particular class-names to be statically referenced in the source code of the main application. It is necessary to use the Reflection API of the Java language for dynamic reference to the individual classes and their methods. The following source code shows dynamic loading of MyInterfaceImpl class from the APK file and calling the callMyInterfaceMethod() method.

```
File apkFolderDir = getDir("apkDir",
                          Context.MODE_PRIVATE);
File dexDir = getDir("dex", Context.MODE_PRIVATE);
File apk = new File(apkFolderDir,
                    "com.example.module.apk");
DexClassLoader dexCl = new DexClassLoader(
    apk.getAbsolutePath(),
    dexDir.getAbsolutePath(),
    null,
    ClassLoader.getSystemClassLoader());
Class<?> clazz = dexCl.loadClass("MyInterfaceImpl");
MyInterface inst = (MyInterface) clazz.newInstance();
inst.callMyInterfaceMethod();
```

Module loading and calling of the method is thus feasible thanks to DexClassLoader class and the Reflection API.

---

[4] http://developer.android.com/reference/dalvik/system/
DexClassLoader.html

5.3   Dynamic Start of Activities

The above described way of dynamic modules loading works well for common classes whose instances are created directly in the application's code. The issue is to start Activities of the Android system. Their instances are not created directly in the application's code but within the Android framework. It is also necessary to define all classes representing Activities in the APK's Manifest file of the main application in advance.

The main part of the code involved in the loading of Activities is the `mClassLoader` private field in `LoadedApk` class which by default contains the reference to the `PathClassLoader`. It is necessary to change the content of this variable with the aid of the Reflection API. It has to contain the reference to the `DexClassLoader` (or its subclass) enabling dynamic loading of modules.

`DynamicDexClassLoader` class, the subclass of the `DexClassLoader` class, has been created in our solution. It loads classes (including Activities) from both dynamically loaded modules and the main application's package. Then `LoadedApkRefChanger` class has been created. It replaces the reference to ClassLoader in the instance of the `LoadedApk` class with the reference to the `DynamicDexClassLoader`. After reference replacement it is possible to create particular Intent that explicitly refers to the `ModuleActivity`. This activity represents the initial Activity which is started after loading of the module. Its name is declared in the Manifest file of the APK package of the main application and thus cannot be changed later. Other possible Activities of the module have to be defined in advance in the Manifest file of the main application. It is recommended to use generic names, e.g. `com.example.intent.action.ACTIVITY_MODULE2`, `ACTIVITY_MODULE3`, etc.

It should be noted that manipulation with references in private variables is possible risk for the application's deployment to another version of the Android platform.

5.4   Results

We have successfully implemented the dynamic module loading according to the described software analysis and design. A simple computationally-oriented application was created as a service (module) to verify our approach. The proof-of-concept implementation has mainly been tested on Samsung I9000 Galaxy with the Android 4.0.4 (API 15) CyanogenMod 9 operating system. Full compatibility of the application has also been proved with versions 2.4.3 (API level 9) and 4.3 (API 18). Compatibility is not guaranteed with the Android KitKat (API 19) version where the application has also partially been tested. Issues related to Resource handling have been found out. The described technique also depends on Dalvik run-time and its adaptation for the new Android Runtime (ART) is the subject of future work.

In the first versions of the application memory leaks have been detected. This happened due to instantiations of new `DynamicDexClassLoader` at every start of the module. To solve this issue a hash-map holding references to the already created `DynamicDexClassLoaders` has been added to the application. This hash-map enables application to reuse instances of `DynamicDexClassLoader`, if necessary, without having to create new instances. This approach effectively avoids possible memory leaks.

6   Conclusion

Ad Hoc Mobile Cloud solution based on m-clients which has successfully been implemented at the Android platform has been suggested. Proof-of-concept application can dynamically download modules from a server and then it is possible to run them. This approach solved one of the main issues regarding the m-client – the basic building block of the M-client M-client Server architecture. This architecture is necessary for the effective building of m-clouds. M-clouds bring new possibilities for the use of mobile technologies not only in emergency situations but also in a company's infrastructure. Openness of the Android Platform's source code is an advantage for developers. Without the possibility to study its code (internal APIs) it would certainly not be possible to create application able to run modules without the necessity to install them in advance. Security policy should be enforced to avoid downloading and running the malicious code (malware). Its definition and implementation will be the subject of further investigation.

Acknowledgements. The authors of this paper would like to thank Pavel Zubr, a student of Applied Informatics at the University of Hradec Kralove, for implementation of the client with variable thickness at the Android platform. They would also like to thank Tereza Krizova for proofreading.

References

1. Reese, G.: Cloud Application Architectures: Building Applications and Infrastructure in the Cloud. O'Reilly (2009)
2. Rittinghouse, J.W., Ransome, J.F.: Cloud Computing - Implementation, Management and Security. CRC Press (2009)
3. Kriz, P., Maly, F., Slaby, A.: Mobile oriented software architecture M client M client server. In: Proceedings of the International Conference on Information Technology Interfaces, Croatia, pp. 109–114 (2011)
4. Fernando, N., Loke, S.W., Rahayu, W.: Mobile cloud computing: A survey. Future Generation Computer Systems 29, 84–106 (2013)
5. Lacuesta, R., Lloret, J., Sendra, S., Peñalver, L.: Spontaneous Ad Hoc Mobile Cloud Computing Network. The Scientific World Journal, Article ID 232419 (2014)
6. Jeong, Y.-S., Moon, J.-C., Kim, D., Cho, S.-J., Park, M.: Preventing Execution of Unauthorized Applications Using Dynamic Loading and Integrity Check on Android Smartphones. Information (Japan) 16(8), 5857–5868 (2013)
7. Chen, G., Kotz, D.: A Survey of Context-Aware Mobile Computing Research. Technical Report: TR2000-381, p. 381. Hannover: Dartmouth College (2000)
8. Kriz, P., Maly, F.: Mobile-oriented scalable cooperative architecture. In: Recent Researches in Computer Science; 15th WSEAS International Conference on Computers, Greece, pp. 375-380 (2011)
9. Kriz, P., Maly, F.: Troubleshooting assistance services in community wireless networks. Journal of Computer Networks and Communications, Article ID 621983, 6 pages (2012)

Methodological Approach to Efficient Cloud Computing Migration

Petra Marešová, Vladimír Soběslav, and Blanka Klímová

Faculty Informatics and Management, University of Hradec Kralove,
Rokitanskeho 62, Hradec Kralove, Czech Republic
{petra.maresova,vladimir.sobeslav,blanka.klimova}@uhk.cz

Abstract. Cloud computing is a technology whose potential is recognized by both private and public sectors in possibilities of savings and high flexibility for the entities which decide to use it. Measurement of cloud computing efficiency and economic assessment of migration is one of the biggest current and future challenges. The aim of this article is to characterize an assessment model of cloud computing with respect to the use in company's practice when evaluating the effectiveness of investments. The model is based on the conditions of the European business environment. Within the framework of this study, attention is drawn to the metric measurements which are good to follow or use in the process of evaluating the effectiveness of cloud computing.

Keywords: cloud computing, economic assessment, measurability, metric measurements.

1   Introduction

Technological developments influence all the spheres of human activities. Within business processes it is particularly an access to information, speed of its processing or ways of communication. In order to increase employees' flexibility and mobility, it is necessary to enable them an access to company's applications, files and services at anytime and anywhere. And cloud computing offers such a solution because it is based on the already existing and tested technologies [1]. At present there exist a number of different views which differ from each other by rigidity and accuracy of setting the boundaries of the term cloud computing. According to Armbrust [2], cloud computing on the one side, includes the applications that are in form of services accessible to users through a communication network, on the other side, it includes all hardware and software tools that use data centres providing these services. The company Forrester [3] expands the above mentioned definition for the standardization of IT tools on the side of suppliers, and self-service principles on the side of users. The agency Gartner proposes a rather compact view on this issue and defines five key pillars of the whole concept [4]: service principle, scalability and elasticity, sharing by more users, measurements according to use, use of the Internet technologies.

© Springer International Publishing Switzerland 2015

D. Barbucha et al. (eds.), *New Trends in Intelligent Information and Database Systems,*
Studies in Computational Intelligence 598, DOI: 10.1007/978-3-319-16211-9_21

Cloud computing is a broad term which for its successful implementation requires a connection of technological and economic-managerial view. The key issue is to describe quantitative and qualitative parameters influencing the process of implementation itself, including the informatics-technical metric measurements. Therefore the aim of this article is to characterize an assessment model of cloud computing with respect to the use in company's practice when evaluating the effectiveness of investments. The model is based on the conditions of the European business environment. Within the framework of this study, attention is drawn to the metric measurements which are good to follow or use in the process of evaluating the effectiveness of cloud computing.

2   Present State of the Research Issues

In 2010 the biggest market of cloud technologies was in the USA. The EU countries came second. This same order is expected again in 2014 [5]. The EU thus ranks below the world's average, which is 34 %. The most common causes of that are that company managers do not sense the need of introducing new technologies, they do not clearly understand the term cloud computing and they are not able to evaluate the effectiveness of this investment in this technology [6]. Although there exist a number of metric measurements which can assess the effectiveness of technologies, they are not used in entrepreneurial practice. The reasons are as follows [7]:

- ignorance of the assessment metric measurements by IT company managers,
- a lack of motivation to apply these metric measurements,
- too technical indicators and a little connected with common entrepreneurial practice of evaluation of the effectiveness of investments,
- they do not provide complete information needed for managers and their decisions,
- insufficient or undervalued information company assets,
- a lack of the assessment of security risks migration of the present technologies into cloud is always solved by a partial area linked with IT.

In connection with the problem of cloud computing efficiency are analysed various issues. For example, Evangelinos and Hill [8] evaluated the performance of Amazon EC2 to host High Performance Computing (HPC). Jureta, and Herssens [9] propose a model called QVDP which has three functions: specifying the quality level, determining the dependency value, and ranking the quality priority. These functions consider the quality of services from the customers' perspective. Kalepu et al. [10] propose a QoS-based attribute model to define the non-functional metrics of distributed services. In the broader context the application performance management (APM) and real user monitoring (RUM) have long been considered "advanced" forms of measurement for information technologies. In the modern economy, the Pareto principle was extended to quality control [11], stating that most defects in production are the result of a small percentage of the causes of all defects. This principle is applied in computer science. Juran introduced the concept of CWQM (Company Wide Quality Management) [12] that was based on three pillars (also known as Juran Trilogy): quality planning, quality control and quality improvement [13].

The above performance models include some shortcomings which are as follows:

- Models focus on how to measure the performance of virtual machines using local experiments. However, the techniques used for measuring the actual resources of cloud.
- Most of the proposed works on performance evaluation do not allow customers to specify the parameters of performance metrics.
- The experiments using the above proposed models do not specify the benchmarks for the performance evaluation.
- In cloud computing architecture, the relationship between the performance monitoring and cost metrics is very important. The proposed models do not link the results of performance monitoring with the service actual cost metrics.

The assessment of the effectiveness of cloud computing is not only the question of the implementation of this technology, but also its operation. Within the framework of the company's strategic management there should be a continuous assessment of the fact if modern technologies meet the company's competitive needs. At this level statistical and dynamic methods of the assessment of the effectiveness of investments are often used. An overview of the most frequent used methods for the evaluation of IT can be divided into two main categories: financial indicators and comprehensive methods of the assessment of the effectiveness of investments. The financial indicators will help to discover if there are a cost and income aspects of cloud computing in the financial expression. These are as follows: return indicators (ROA, ROE, ROI), Internal Rate of Return (IRR), Net Present Value (NPV), Economic Value Added (EVA), Total Cost of Ownership (TCO), Productivity of employees [14, 15, 16].

The second group, within which at first all the impacts of the implementation of this technology are specified, includes general methods which also focus on the qualitative aspects of this issue and they include the financial indicators only in their subsection: Cost benefit Analysis (CBA), Balanced Scorecard (BSC), Porter Value Chain Model, Benchmarking, Total Quality Management (TQM) [16].

With respect to MPS needs, which resulted from the own conducted qualitative and quantitative survey, there is an interest in the method of the evaluation of the effectiveness of cloud computing. This method will use common financial indicators. On the basis of the multicriteria analysis of variations, in which a team of experts from the IT and economic fields participated, the method Cost benefit Analysis was chosen. The following text is focused especially on the metric measurements and indicators which can be appropriately used in this method so that the result of a comparison of all costs and benefits could be formulated numerically.

3   Cost Benefit Analysis for Cloud Computing in European Business Sector

The cost-benefit analysis (CBA) is a systematic approach to estimate the strengths and weaknesses of alternatives that satisfy transactions, activities or functional requirements for business. It is a technique that is used to determine options that provide the best approach for the adoption and practice in terms of benefits of labour, time and cost savings [17]. Its application in the implementation of advanced technologies, information systems or new software (e.g., [18, 19, and 20]) is relatively frequent. The basic terms of this method are as follows: effect resulting from an investment, costs, benefit, and utility.

Effects resulting from an investment – all the impacts on the surveyed entities which the investment activity brings. They can be in form of financial or non-financial (possibly intangible). From the point of view of a certain entity, they can be positive (benefits), negative (costs) or neutral (the entity is not influenced at all).

Costs – all the negative effects on the research entity/ies, or their group. Those are negative effects resulting from the investment [20].

Benefits – all the positive effects on the research entity/ies, or their group. Those are the positive effects resulting from the investment [20].

Utility – establishment of new values (e.g., an increase in the property value) as a result of the project implementation, reducing costs connected with relevant processes as a consequence of the project implementation.

On the basis of the interviews with IT specialists, general steps of the method CBA were modified for cloud computing purposes. They are presented, for example, in [21, 22, 23] They are as follows:

1. criteria specification for the use of cloud computing,
2. characteristics of the current state of IT,
3. characteristics the required services in cloud,
4. decision which entities are affected by the project,
5. determination and possible quantification of all relevant costs and benefits,
6. formulation of all consequences of the cloud computing implementation in financial units,
7. calculation of criterion indicators,
8. analysis of sensitivity and project evaluation.

This method is suitable both for a comparison of the present IT solution with a possibility of cloud computing implementation. It can be also used even within the framework of continuous assessment of functioning of this technology. In the following section particularly the entry parameters are specified. These are necessary to be described for the overall project evaluation. The paper will describe only the steps which are related to quantifiable values.

3.1 Criteria Specification for the Use of Cloud Computing

On the basis of the qualitative survey which was conducted in the selected EU country in 2014, it resulted that companies apart from the possibility of comparing their own current IT costs and cloud computing are also interested in the availability of a certain decision-making tree which would provide them with introductory information on the suitability of the cloud solution. This is the requirement resulting from the present, still ineffective implementation of cloud computing in these countries. Within this step, there is a set of questions which can answer if the cloud computing is suitable for a company and if the cloud will be public, private or hybrid. The questions are divided into two areas. In the first characteristics of a company is provided (e.g., geographic position); in the second there is a specification of company strategic management. With respect to the focus and extent of this article, these questions are not further elaborated.

3.2 Characteristics of the Present IT State

The basis for the implementation of the right form of cloud computing is an identification of possible current insufficiencies of IT centres and the required functionality. In this phase the identification of an implementation model should be made such as public, private, hybrid, community models. The specification of the required functionality should be as exact as possible. The table 1 is just synoptic because the next step is a more precise identification of the functionality of the IT system. Therefore, the provider will have to discover: I/O operations, network speed, network latency, processing.

Table 1. IT entry parameters in a company

Hardware	Description	Unit
Physical servers	Electricity consumption	kWh
UPS, cooling	height	U (4,52 cm)
RAM	Capacity, speed of memory (frequency, latency and speed)	GB, overall benchmark score
HDD	Capacity, performance disk system ()	GB, IOPS
CPU	The number of cores, processor performance	Number of cores, overall benchmark score
UPS, cooling	The power backup time	kVA, minute
		Number of ports,
Structured cabling	Max. transfer rate per unit	Gbit/s.
Network components	Switches, Routers, Access points, converters, Power over LAN, public IP addresses	CZK (currency)
	The total throughput	Mbit/s.

Table 1. (*continued*)

Software	Description	Unit
OS (operating system)	The type	
Database	The type, performance	TB, IOPS

Human Resources	Description	Unit
Number of full-time IT employees	Number of full-time jobs	CZK (currency)
Number of additional staff solving IT strategy	Number of hours / month	CZK (currency)

IT Recovery/year	Description	Unit
Replacement investment HW		CZK (currency)
Investments in SW licenses		CZK (currency)
Failures		CZK (currency),

Security	Description	Unit
Support and updates	The right to update and replacement of Components	CZK (currency)
Authorization	User rights and access management	CZK (currency)
Authentication	Systems for secure authentication, encryption and key exchange	CZK (currency)
Accounting	System event log, audit approaches, security, reporting tools	CZK (currency)
Physical Security	Guarding, fire prevention, prophylaxis, replacement of physical machines	CZK (currency)
Proactive approach	Monitoring, regularly monitored, system updates, event reporting	CZK (currency)

3.3 Characteristics of the Required Services in the Cloud

The key step is connected with the most exact specification of the required services. Therefore not only a person possessing strategic competences, but also an IT specialist must participate in this process. The description of the required final state will not be presented due to the extent of this paper. Further facts which can influence the final state are based on company's characteristics, its orientation, geographical position or company's own requirements on high availability which can influence the costs significantly in Table 2.

Table 2. High availability and downtime, source: [24]

Availability	DPM1	Downtime Per Year (24 x365)		
99.000%	10000	3 Days	15 Hours	36 Minutes
99.500%	5000	1 Day	19 Hours	48 Minutes
99.900%	1000		8 Hours	46 Minutes
99.950%	500		4 Hours	23 Minutes
99.990%	100			53 Minutes
99.999%	10			5 Minutes
99.9999%	1			30 Seconds

Other parameters can be divided into two basic groups: operational and technical criteria. Operational criteria are as follows: client support, compliance with standards, treaties covering the storage, access and transmission of data, packet loss. The operational criteria are related to determining the Service Level Agreement (SLA). SLAs are part of service contracts and are agreements usually between two parties (service provider and customer), which formally define the services. Service contracts use the percentage of service availability as a unit [25]. The technical criteria are as follows: adaptability, availability and use restriction, backup and recovery, response time, elasticity, interoperability, scalable storage, portability. All of these characteristics can be covered by term Quality of Cloud Service (QoCS), which comes from general QoS [26].

3.4 Formulation of All Consequences of the Cloud Computing Implementation in Financial Units

The area of cloud computing has many benefits which are formulated in another than financial form. Provided that there is a market which would set the price of such a product, it is possible to use direct assessment on the basis of market price. However, it cannot be used with many other effects resulting from the use of cloud computing. In that case it is necessary to use other methods. An overview of such methods is demonstrated in Table 3.

Table 3. Overview of valuation methods used in the CBA, source: authors' own source according to [27, 28, 29]

Type of a method according to the source data	Method	
Preferential methods	Contingent valuation method	
	Comparative methods	Method of shadow prices
		Method of asset value
		Method of hedonic prices
		Study of labour markets (a method of wage risk)

Table 3. (*continued*)

Market methods		Travel cost method
		Method of cost prevention
	Cost methods	Consumer behavior study
		Method of cost recovery
		Method of alternative costs
		Method of opportunity costs
	Benefit methods	Method of blanks
		Method of past income

4   Discussion

The most used indicators for the cloud computing evaluation are the indicators of profitability and Total Cost of Ownership (TCO). This method considers not only the initial acquisition price but also the costs connected with the ownership (repairs, consultations, or upgrades. In order the companies could calculate their returns of their investments, they must first know how much their current operations are. This is, however, a problem because the companies do not really know what their costs connected with cloud computing are. Therefore, it seems best to use one of the general methods of the evaluation of investments. The financial indicators there are only their part and general methods include the procedures how to take into account all investment aspects. Apart from the most exploited Analysis of Economic Impacts (CBA) method, they also involve Cost Minimization Analysis (CMA) and Cost Utility Analysis) (CUA) method.

The main purpose of the CBA method is to formulate both the tangible and intangible effects resulting from the use of cloud computing in financial units. A verbal commentary focusing on the individual areas of distortion, which can arise from the use of the proposed methods, should be part of the calculation of indicators within this method. One of the most striking distortions is in step 6 in effort to formulate all the consequences of the cloud computing implementation in financial units. That is why the intangible effects should be commented here also verbally. The last step of the CBA method is also important from the point of distortion, and that is the analysis of sensitivity which enables to discover how much the tested project is sensitive to a change of different factors. This concerns the changes of certain quantities (volume of production, product price, investment costs, project lifespan, or discount rates) in reliance on the changes of the factors which influence these quantities.

5   Conclusion

In the period of the fast development of information technologies and their subsequent use in companies the key issue is effectiveness of such investments. The aim of this article therefore was to characterize the evaluation model of cloud computing with respect to its use in entrepreneurial practice in the assessment of the effectiveness of investments. The model is based on the conditions of the European entrepreneurial environment. In this study attention was also drawn to the metric measurements which are relevant to follow or use in the assessment of effectiveness of cloud computing. Further research will focus on the creation of a specific web application which would be freely accessible and which would enable companies and other entities (after other possible modifications within a testing operation) a primary, independent evaluation of cloud computing potential.

Acknowledgement. The paper is supported by the project of specific science Smart networking & cloud computing solutions and Economical and Managerial aspects in Biomedicine.

References

1. Lacko, L.: Cloud Computing – Current IT Phenomena in Business Economy (2012), http://www.forum-media.cz/res/data/010/001470.pdf
2. Armburst, M., et al.: Above the Clouds: A Berkeley View of Cloud Computing (2009), http://www.eecs.berkeley.edu/Pubs/TechRpts/2009/EECS-2009-28.html
3. Forrester: Cloud Computing for the Enterprise (2009), http://www.forrester.com/imagesV2/
4. Gartner: Gartner Highlights Five Attributes of Cloud Computing (2009), http://www.gartner.com/it/page.jsp?id=1035013
5. Laugesen, N.S.: Cloud Computing Cyber Security and Green IT. The Impact on E-Skills Requirements. In: Final Report, kills_and_cloud_computing_final_report_en (2012), http://ec.europa.eu/enterprise/sectors/ict/files/eskills/e-kills_and_cloud_computing_final_report_en.pdf
6. Marešová, P.: Deployment of Cloud Computing in Small and Medium Sized Enterprises in The Czech Republic. E+M Ekonomie a Management (2014) (in press)
7. Marešová, P., Soběslav, V.: Cost Benefit Analysis Approach for Cloud Computing. E+M Ekonomie a Management (2014) (in review)
8. Evangelinos, C., et al.: Cloud Computing for parallel Scientific HPC. Applications: Feasibility of running Coupled Atmosphere-Ocean Climate Models on. Amazon's EC2. Ratio 2, 2.34 (2008)
9. Jureta, I., et al.: A Comprehensive Quality Model for Service-Oriented Systems. Software Qual. Journal 17, 65–98 (2009)
10. Kalepu, S., et al.: Verity: A QoS Metric for Selecting Web Services and Providers, pp. 131–139 (2003)
11. Juran, J.M.: Juran's Quality Handbook, 5th edn. McGraw-Hill (1999)
12. Juran, J.M.: Quality Control Handbook, 4th edn. McGraw-Hill (1988)
13. Alhamad, M., Dillon, T., Chang, E.: A Survey on SLA and Performance Measurement in Cloud Computing. In: Confederated International Conferences: CoopIS, DOA-SVI, and ODBASE. Hersonissos, Crete, Greece, pp. 469–497 (2011)

14. Uzoka, F.: Fuzzy- Expert System For Cost Benefit Analysis Of Enterprise Information Systems: A Framework. International Journal on Computer Science and Engineering 1(3), 254–262 (2009)

15. Watson, R.T., Kelly, G.G., Galliers, R.D., Brancheau, J.C.: Key IS Issues in Information Systems Management: An International Perspective. Journal of Management Information Systems 13(4), 91-115 (1997)

16. Willcocks, L.: Information Management: The Evaluation of Information Systems Investments. Chapman and Hall (1994)

17. David, R., Ngulube, P., Dube, A.: A Cost-Benefit Analysis of Document Management Strategies Used at a Financial Institution in Zimbabwe: A Case Study. SA Journal of Information Management 15(2) (2012)

18. Harrigan, P.O., Boyd, M., Ramsey, E., Ibbotson, P., Bright, M.: The Development of E-Procurement within the ICT Manufacturing Industry in Ireland. Management Decision 46(3), 481–500 (2008)

19. Gabler, E.: Economics of an Investments: Analysis Helps Agencies Target Limited Transportation Recources to Their Best Uses. Innovations and Technology. Roads & Bridges (2004)

20. Sieber, P.: Analýza nákladů a přínosů metodická příručka. Ministerstvo pro místní rozvoj, http://ww.strukturalni-fondy.cz/uploads/old/1083945131cba_1.4.pdf

21. Boardman, A.E., Greenberg, D.H., Vining, A.R., Weimer, D.L.: Cost-Benefit Analysis: Concepts and Practice, 3rd edn. Prentice Hall, Upper Saddle River (2006)

22. Nas, T.F.: Cost-Benefit Analysis. Theory and Application. SAGE. SAGE. Publications, Thousand Oaks, California (1996)

23. Kunreuther, H., Cyr, C., Grossi, P., Tao, W.: Using Cost-Benefit Analysis to Evaluate Mitigation for Lifeline Systems, National Science Foundation. Earthquake Engineering, Research Centre Program (2001)

24. CISCO: Delivering High Availability in the Wiring Closet with Cisco Catalyst Switches, http://www.cisco.com/c/en/us/solutions/I

25. Bauer, E., Adams, R.: Reliability and Availability of Cloud Computing. Wiley-IEEE Press, Hoboken (2012)

26. Khazaei, H., Mi, J., Mi, V.B.: Performance Analysis of Cloud Computing Centers&Rdquo. In: Proc. Seventh Int', 1 ICST Conf. Heterogeneous Networking for Quality, Reliability, Security and Robustness (QShine), pp. 936–943 (2010)

27. Birol, E., Karousakis, K., Koundouri, P.: Using Economic Valuation Techniques to Inform Water Resources Management: A Survey and Critical Appraisal of Available Techniques and an Application. Springer, Netherlands (2006)

28. Boardman, A.E., Greenberg, D.H., Vining, A.R.: Cost-Benefit Analysis: Concepts and Practice, 3rd edn. Prentice Hall, Upper Saddle River (2006)

29. Brent, J.R.: Applied Cost-Benefit Analysis. Edward Elgar Publishing (2006)

High Level Models for IaaS Cloud Architectures

Ales Komarek, Jakub Pavlik, and Vladimír Soběslav

Faculty of Informatics and Management, University of Hradec Kralove,
Rokitanskeho 62, Hradec Kralove, Czech Republic
{ales.komarek,jakub.pavlik.7,vladimir.sobeslav}@uhk.cz

Abstract. This paper explains how ontology can be used to model various IaaS architectures. OpenStack is the largest open source cloud computing IaaS platform. It has been gaining wide spread popularity among users as well as software and hardware vendors over past few years. It's a very flexible system that can support a wide range of virtualization scenarios at scale. In our work we propose a formalization of OpenStack architectural model that can be automatically validated and provide suitable meta-data to configuration management tools. The OWL-DL based ontology defines service components and their relations and provides foundation for further reasoning. Model defined architectures can support simple all-in-one architecture as as well as large architectures with clustered service components to achieve High Availability.

Keywords: IaaS, OpenStack, Meta-data, Ontology, Service-Oriented Architecture, Configuration Management.

1 Introduction

Nowadays IT infrastructure is the key component for almost every organization across different domains, but it must be maximally effective with the lowest investment and operating costs. For this reason, cloud Infrastructure as a Service (IaaS) is gradually being accepted as the right solution regardless hosting as private, public or hybrid form. Lots of key vendors had tried to develop own solutions for IaaS clouds during several years ago, but infrastructure is being too complex and heterogeneous. Different vendors means different technology, which caused vendor lock-in and limitations in migrations for future growth. In addition, every organization has different requirements for hardware, software and its use.

Based on the idea of openness, scalability and standardization of IaaS cloud platform NASA together with RackSpace founded in 2010 project called Open-Stack, which is a free and open source cloud operating system that controls large pools of compute, storage, and networking resources across the datacenter. It is the largest open-source cloud computing platform today [2]. Community is driven by industry vendors as IBM, Hewlett-Packard, Intel, Cisco, Juniper, Red Hat, VMWare, EMC, Mirantis, Cannonical, etc. In terms of numbers the OpenStack community contains about 2,292 companies, 8,066 individual members and 89,156 code contributors from 130 different countries [3]. These figures confirm that OpenStack belongs to the largest solution for IaaS cloud.

© Springer International Publishing Switzerland 2015
D. Barbucha et al. (eds.), *New Trends in Intelligent Information and Database Systems*,
Studies in Computational Intelligence 598, DOI: 10.1007/978-3-319-16211-9_22

OpenStack is a modular, scalable system, which can run on a single personal computer or on the hundreds of thousands servers as e.g. CERN [4] or PayPal [5].

Lots of vendors and wide community mean lots of ways how OpenStack can be deployed. Each vendor tries to extend core functions and write new service backends to fit their business goals. The actual system consists of many modules and components designed with plugin architecture that allows custom implementations for various service backends. These components can be combined and configured to match available software and hardware resources and real use-case needs.

Each implementation has its own component combination and use some form of configuration management tool to enforce the service states on designated servers and possibly other network components. These tools require data that covers configuration of all components. However, there is not best practise or recommendations how to build suitable OpenStack cloud for different use cases. Detecting component inconsistencies manually is painful and time consuming process. Companies need standardization and validation process for their specific infrastructure requirements, which can help them automate whole implementation, operating and future upgrades of controlling software or physical hardware.

Our project goals are to find a solution to the following issues:

1. Propose high level architecture model definition (Logical model)
2. Implement service that transforms architecture model to solution level model for configuration management tools (physical realization)
3. Provide way how to define and validate architecture based on available hardware resources and target use case.
4. Automate the whole process from high level modelling to actual enforcement on targetted resources.

This paper is focused on designing and creation of high level architecture model, where we propose a formalization of OpenStack service architecture model, based on the approaches developed in classic knowledge representation domain, especially Service-Oriented Architecture by OpenGroup. Component definition is encoded in an ontology using the standard OWL-DL language, which enables sharing of knowledge about configurations across various systems. Reasoning can be used on the specification to automate validation of configuration changes.

When dealing with hundreds of components with thousands of properties and relations, keeping track of changes throughout its life cycle is very challenging. Current approaches are ad hoc, even OpenStack Fuel (Mirantis OpenStack deployment tool [6]) has severe limitations, there exists no standard for specifying common OpenStack architectural model. The question how to convert the proposed OWL-DL schema to metadata format that configuration management tools can process is discussed. We are working on external node classification service that uses graph database to serialize the OWL ontology with REST API that configuration management tools can use as metadata provider. This can streamline the process of adopting new services and service backends in predictable way.

In Section 2 we describe service architecture models, which contain OpenStack moduls, deployment, etc. After we explain ontology models for IaaS cloud in Section 3. Finally we show how to implement high level model into ontology in Section 4.

2   Service Architecture Models

This section describes modularity and complexity of OpenStack IaaS platform including core and supporting services. However, there is not so much place for detail description of all components, since the main idea is contained in Sections 3 and 4. The goal of this section is to show that OpenStack modules are independent services, which can be implemented in many different of ways.

2.1   IaaS Achitecture Core Models

OpenStack is complete Infrastructure as a Service platform. It allows to create virtual servers on virtual networks using virtual block devices.

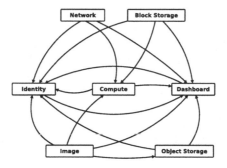

Fig. 1. Logical Model of Icehouse OpenStack service achitecture

Further versions of OpenStack introduce more complex services that use basic services to provide for example Data processsing, Database, Message Queue or Orchestration. All services or modules within OpenStack architecture are independent and have pluggable backends or drivers. This allows vendors to develop plugin for their resources, that can be accessed and managed by the OpenStack API.

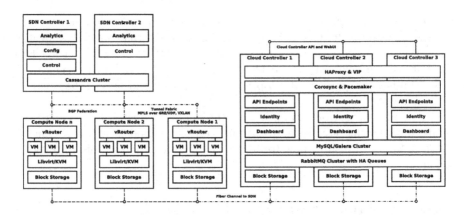

Fig. 2. Locality 2 Architecture

Fig. 1 shows the core modules of OpenStack included in Icehouse release. Each module is briefly described including a serveral backends or plugins.

Identity - Keystone is an OpenStack project that provides Identity, Token, Catalog and Policy services for use specifically by projects in the OpenStack family. *Backends/plugins*: sql, ldap

Image - Glance service provides services for virtual disk images. Compute service uses image service to get the starting image of the virtual server. *Backends/plugins*: dir, Swift, Amazon S3

Compute - Nova service is designed to provision and manage large networks of virtual machines, creating a redundant and scalable cloud-computing platform. *Backends/plugins*: KVM, Hyper-V, VMware vSphere, Docker

Network - Neutron is an OpenStack networking project focused on delivering networking as a service. It makes hard to deploy advanced networking services because of wide range of plugins. *Backends/plugins*: Nova flat networking, OpenVSwitch gre/vxlan, OpenContrail, VMware NSX, etc.

Volume - Cinder provides an infrastructure for managing volumes in OpenStack. It uses storage drivers for volumes direct mapping into virtual instances through FibreChannel or iSCSI. *Backends/plugins*: LVM driver, SAN driver, EMC VNX, IBM Storwize, CEPH, Gluster, etc.

OpenStack is not only about its core services, but there are many services at infrastructural level that are essential as well.

High Availability Cluster software is responsible for clustering OpenStack services and creation High Availability in active/active or active/passive mode. *Backends/plugins*: corosync/pacemaker, keepalived

Communication Service is messaging between components of same OpenStack module. *Backends/plugins*: RabbitMQ, QPid, ZeroMQ

Database Services is responsible for storing persistent data of all modules. *Backends/plugins*: MySQL/galera, PostgreSQL

2.2   Use Cases

We participate in operations of several real OpenStack deployments in Central Eastern Europe. Each of them is different, which means that uses different backends in modules. Because of lack of space we decided to show only one use case. Fig. 2 shows the logical architecture of TCP Virtual Private Cloud.

3   IaaS Service Ontology

What is an ontology? An ontology is a specification of a conceptualization. According to Gruber [8] an ontology defines a set of representational primitives with which to model a domain of knowledge. These primitives are typically classes (or sets), attributes (or properties), and relationships (or relations among class members). The definitions of the representational primitives include information about their meaning and constraints on their logically consistent application.

In general, any enterprise application can benefit from use of ontologies. They are used in field of semantics-based health information systems, interoperable reference ontologies in biology and biomedicine as summarised in Open Biological and Biomedical Ontologies[13].

The formal definition of cloud computing ontology was introduced by Youseff [15]. It maps the complete domain of Cloud computing from software to hardware resources. It is divided into 5 layers of services.

1. Servers (physical and virtual)
2. Core Infrastructure Services (DNS, NTP, config management)
3. Storage (NAS and SAN)
4. Network (Routers, Switches, Firewalls, Load Balancers)
5. Facilities (Power, Cooling, Space)

The scope of IaaS Service Ontology covers the level 1 and 2 with core infrastructure services and servers running OpenStack services. The levels 3 and 4 with network and storage devices will be adopted in further versions of the ontology. The level 5 will be implemented as last as no services are directly configured.

3.1   Ontological Standards

The ontologies define the relations between terms, but does not prescribe exactly how they should be applied. Following ontologies serve as the starting point for creating new ontologies including the IaaS Service Ontology.

Service-Oriented Architecture. The SOA ontology specification was developed in order to aid understanding, and potentially be a basis for model-driven implementation of software systems. It is being developed by Open Group and was updated to version 2 in april 2014. The ontology is represented in the Web Ontology Language (OWL) defined by the World-Wide Web Consortium (W3C). The ontology contains classes and properties corresponding to the core concepts of SOA [10].

OSLC Configuration Management. OSLC Configuration Management Resource Definitions [11] is a common vocabulary for versions and configurations of linked data resources. It provides suitable classes and properties from configuration management domain.

Dublin Core Metadata Initiative. The Dublin Core Metadata Intiative terms provide vocabularies for common resource definition. It is foundational meta-data vocabulary for many other schemas and covers the basic properties.

3.2 Ontology Serialization Formats

There are several ways how to serialize ontology. In it's core ontology representation is a linking structure that forms a directed, labeled graph, where the edges represent the named relation between two resources, represented by the graph nodes. This graph view is the easiest possible mental model for ontologies and is often used in easy-to-understand visual explanations. They differ by reading and writing speed.

RDF/OWL-DL Documents. Ontologies are stored in Web Ontology Language (OWL) defined by the World-Wide Web Consortium (W3C). OWL has three increasingly expressive sub-languages: OWL-Lite, OWL-DL, and OWL-Full [9]. The sub-language OWL-DL provides the greatest expressiveness possible while retaining computational completeness and decidability. RDF is a standard model for data interchange, it as features that facilitate data merging even if the underlying schemas differ, and it specifically supports the evolution of schemas over time without requiring all the data consumers to be changed. The format is used by ontology editors.

Graph Databases. Graph databases can store ontologies very well as they have graph format very similar to RDF format which is standard format of any XML based graph database, just very different implementation. Graph database uses graph structures with nodes, edges, and properties to represent and store data. A graph database is any storage system that provides index-free adjacency. This means that every element contains a direct pointer to its adjacent elements and no index lookups are necessary.

3.3 Plain Meta-data Serialization

The domain of cloud computing services can be mapped not just by ontologies but in a less formal data structures. The most common examples are YAML or JSON files with plain or nested data structures. The schema is enforced by documentation and no semantic validation can be used.

Hierarchical Databases. The more complex meta-data can be stored in hierarchical databases. These systems allow to define service parameters through class inheritance, which can be overridden. Hierarchical classes can be featured as sets, commonalities, or as roles. You can assemble your infrastructure definition from smaller bits, eliminating duplication and exposing all important parameters to a single location. Within hierarchical databases parameters can reference other parameters in the very hierarchy that are actually assembling.

4 Ontology Usage

The journey to mapping high level models in ontologies was long. It was a process of describing everchanging OpenStack cloud architectures over past 2 years. The OpenStack foundation has released 4 major versions with 12 minor versions over that period. The number of managed software components grew from 5 to 17. At the beginnings we started mapping meta-data models in separate files containing complete meta-data set of all services for each server or device. The meta-data was encoded in YAML format. This approach was fine at earlier versions of OpenStack we tested as just the basic set of OpenStack services existed at the time.

Simple meta-data in YAML format

```
service_name:
  service_role:
    data_parameter1: data_value1
    data_parameter2: data_value2
    data_parameter3:
    - data_value3a
    - data_value3b
    object_parameter:
     object1:
       data_parameter1a:
```

The next step was storing the service meta-data in hierarchical databases. The meta-data was split into separate files. The final meta-data for given resources is assembled using the two simple methods. The deep merging of several service definition fragments and parameter interpolation where parameters can be referenced. The reclass [17] was used to implement the desired data manipulation behaviour.

Meta-data in YAML format, parameter interpolation

```
service_name:
  service_role:
    data_parameter1: data_value1
    data_parameter2: {service_name:service_role:data_paramer1}
    object_parameter:
     {otherobject:object1}
```

This was elegant solution that could easily model growing number of services, but still lacked mechanisms to validate given data or encapsulate semantics. It helped to determine the domain and scope of new ontology. The ontology should use existing standards and provide vocabulary to define OpenStack services, core services and later network and storage resources. The ontology should provide mechanisms to validate schema for integrity issues, as missing parameters, disjoint services or values out of proper value domain.

All OpenStack services can be very well described by ontology as they communicate over common message bus, serialize their state and expose services through interfaces in a same way. The resource within ontology have data properties that are derived mostly from Dublin Core metadata terms. Other extensively used standard is OCLS Configuration Management resource definitions and Service-Oriented Architecture Ontology. Resources can have object property types that describe more complex relations. Following example shows excerpt from Glance image service definition on the the controller node:

Meta-data in OWL-DL format

```
<owl:Class rdf:about="#CinderVolumeService">
  <rdfs:subClassOf>
    <owl:Class rdf:about="#VolumeService"/>
  </rdfs:subClassOf>
  <owl:disjointWith>
    <owl:Class rdf:about="#"/>
  </owl:disjointWith>
  <owl:disjointWith>
    <owl:Class rdf:about="#NovaVolumeService"/>
  </owl:disjointWith>
</owl:Class>
```

4.1   Implementation Details

Initial work on creating our Ontology was done in Protege, open-source ontology editor and framework for building intelligent systems. It took some time to evolve ontology classes and create necessary dictionaries [16] to cover first IaaS architectures. Fig. 3 shows the current ontology service architecture.

The ontology is transformed into graph database using our python-bases service named django-ENC that can read and write ontology from OWL-DL XML files created by Protege and communicates with neo4j graph database through REST API. The graph databases are part of family of NoSQL databases and offer much better performance at any volume of data.

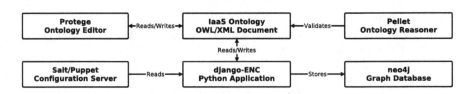

Fig. 3. Ontology Service Architecture

The django-ENC service use web framework Django to provide web services and asynchronous task queue Celery to perform time consuming tasks like ontology assertions and synchronizations between XML and graph database. The

HTTP REST API that can be consumed by configuration management tools like Salt or Puppet through their External Node Classification interface. The meta-data passed to configuration management tools is valid for level 1, 2 and 3 of unified cloud computing ontology [15].

We have successfully tested service status enforcement of several complete OpenStack installations by SaltStack configuration management tool with meta-data acquired from Ontology Service API. The deployment process is not yet fully automated as there is need of setting up network and storage resources manually (only servers are provided), but the progress in both configuration management tools and network and storage will allow better automation of these components by in-place agents or access protocols like SSH in the future.

5   Conclusions

We have managed to do the first steps in formalization of IaaS Architecture high level models. The representation of models, the ontologies, can be used to create and validate meta-data for individual OpenStack cloud installations. The ontology provides schema for the meta-data for each installation so the overall service integrity is ensured.

We created a python-based web service django-enc that use data from the ontology to generate the suitable meta-data for configuration management tools. The service provide simple interface for manipulating the ontology as well as interfaces for ontology editors. The ontology defines the basic services of OpenStack Havana and Icehouse versions. New components and service backends can be easily defined and included.

We plan to expand ontology from virtual and physical servers to network and storage resources by better adoption of configuration management tools. Ontology model is suitable for software agent processing and their rational decisions. It is possible to define agents that will maintain the state of services according to the high-level model. The more parts of the process are modelled and their deployment automated the more manageable the whole system becomes.

Acknowledgements. The paper is supported by the project of specific science Smart networking & cloud computing solutions (FIM, UHK, SPEV 2015).

References

1. NIST: The NIST Definition of Cloud Computing (2011),
   http://csrc.nist.gov/publications/nistpubs/800-145/SP800-145.pdf
2. OpenStack.org: OpenStack Open Source Cloud Computing Software (2014),
   http://openstack.org
3. Stackalytics: OpenStack community contribution in Kilo release (2014),
   http://stackalytics.com
4. Information-technology.web.cern.ch: OpenStack Information (2014),
   http://information-technology.web.cern.ch/book/
   cern-private-cloud-user-guide/openstack-information

5. Openstack.org: PayPal - OpenStack Open Source Cloud Computing Software (2014), http://www.openstack.org/user-stories/paypal/
6. Wiki.openstack.org: Fuel - OpenStack (2014), https://wiki.openstack.org/wiki/Fuel
7. Ivanov, I., van Sinderen, M., Shishkov, B. (eds.): Cloud Computing and Services Science. Springer Science, New York (2012)
8. Liu, L., Tamerzsu, M. (eds.): Encyclopedia of Database Systems, Springer Science, New York (2009)
9. W3C.org: Web Ontology Language (OWL) (2004), http://www.w3.org/2004/OWL
10. The Open Group: Service-Oriented Architecture Ontology, Version 2.0, Open Group, New York (2014)
11. OASIS: Configuration Management Resource Definitions (2013), https://tools.oasis-open.org/version-control/browse/wsvn/oslc-ccm/trunk/specs/config-mgt.html
12. Dublin Core Metadata Initiative: DCMI Metadata Terms (2012), http://dublincore.org/documents/dcmi-terms/
13. Berkeley Bioinformatics Open Source Project: The Open Biological and Biomedical Ontologies (2014), http://www.obofoundry.org/
14. W3C.org: Ontology Driven Architectures and Potential Uses of the Semantic Web in Systems and Software Engineering (2001), http://www.w3.org/2001/sw/BestPractices/SE/ODA/
15. Youseff, L., Butrico, M., Da Silva, D.: Toward a Unified Ontology of Cloud Computing. In: Grid Computing Environments Workshop GCE 2008, IEEE, USA (2008)
16. de Oliveira A.L., de Almeida F.R., Guizzardi, G.: Evolving a Software Configuration Management Ontology. IEEE (2009)
17. Martin Krafft: reclass - Recursive external node classification (2013), http://reclass.pantsfullofunix.net/

Lower Layers of a Cloud Driven Smart Home System

Josef Horálek, Jan Matyska, Jan Stepan, Martin Vancl, Richard Cimler,
and Vladimír Soběslav

University of Hradec Králové, Faculty of Informatics and Management,
Hradec Kralove, Czech Republic
{josef.horalek,jan.matyska,jan.stepan.3,
martin.vancl,richard.cimler,vladimir.sobeslav}@uhk.cz

Abstract. This presented article introduces utilization of a cloud solution as a
part of inovative smart home system (HAUSY - Home Automation System).
This solution is based on highly modular structure, which ensures a unique
solution for individual implementations. In scope of the presented solution, a
universal architecture not requiring a specialized hardware for its utilization is
suggested. In the first part, the original three-part smart home system
architecture and the functionality of individual layers are introduced. The
second part of the article introduces a specific solution for functionalities and
communication between individual end nods and microcontrollers. The
suggested solution is unique for its option of personalizations of user access and
priorities, comfortable interface for defining rules of autonomous operations,
and the option of connecting external systems via a unified API with the option
of using cloud.

Keywords: Smart home, comunication protocols, cloud soloutions, automation,
high modularity.

1 Introduction

We live in the age of smart houses and their dynamic development. This trend was
already described in detail in other study papers, for example in [1, 2, 3, 4] let it be in a
form of a smart home system, home automatization, or ubiquitous homes. Economic or
energetic requirements for smart home system realization are not discussed in this
paper. These impacts will be closely looked at in future study papers based on testing
the final version of systems in model deployment. A part of the future solution can be,
for example, implementation of SmartGrid systems for intelligent consumption
control. These issues are partially discussed in [1], [5].

A Smart house is a house built of modern materials with low-energy consumption
and which contains hardware and software tools enabling automatization of common
tasks and thus effectively affecting operation costs and improving the comfort of living
[2], [6]. The most common hardware tools are modules connected to the control unit
via either a wire or wireless links. These modules are distributed around the house.
Simple programs in these modules and the logic in the controlling unit are then what is

© Springer International Publishing Switzerland 2015
D. Barbucha et al. (eds.), *New Trends in Intelligent Information and Database Systems,*
Studies in Computational Intelligence 598, DOI: 10.1007/978-3-319-16211-9_23

generally understood as software. These tools should also facilitate house control. The house is then possible to be controlled by traditional tools, such as buttons or switches, and at the same time use, for example, touch panels distributed around the house, smartphones, tablets, or computers and laptops. It should also be possible to control a smart house remotely. There are hundreds of manufacturers offering complex solutions (both modules and the control unit) for smart houses. However, almost all solutions on the market nowadays have one of the following faults [7].

Incompatibility - when the costumer decides for a solution by one manufacturer, there is no option of combining that solution with products by other manufacturers. Each system utilizes its own proprietary protocols. A module by one manufacturer will thus not communicate with a control unit by another. There are open protocols and standards, but each manufacturer implements them significantly differently.

Expandability - adding more modules to an already built smart house means to contact the supplier who must perform the installation. Also, it is very common for the user to not be able to change the settings of the control unit in detail, e.g. change the functionality of switches in the house.

Reliability - in case the control unit malfunctions, the house becomes completely uncontrolable. However, such situation can lead to putting house residents in jeopardy.

For these reasons a complex system eliminating such problems was designed for control and automatization of a smart house.

2   A Model of Architecture of the Designed System

Against established standards, when there is only one control unit servicing all modules in the house, automatization system HAUSY is designed as modular and is divided into a three-layered architecture, thanks to which it is resistant to outage of individual devices (Fig. 1).

Instead of the control unit based on a specialized hardware a server is used. The server can be either a physical PC, or a solution such as IaaS (Infrastructure as a Service), or as PaaS (Platform as a Service) available in an independent 24/7 mode, respectively. A local ethernet network leads from the server through a switch into the subsystems. If both recommended IaaS or PaaS technologies are implemented, it is necessary to create a tunnel between individual subsystems and the server in cloud environment. Thanks to this modular solution it is both possible and appropriate to have a server using suitable cloud technology implemented and thus administer the whole system simply and control it with all advantages that the cloud solution provides. The forementioned subsystems are single board minicomputers on ARM architecture. In each logical unit of the house just one subsystem is present. The subsystem then communicates with the modules, referred to as nodes, through a bus. In case of central server outage because of either power outage or a computer component malfunction, the house can still be controlled.

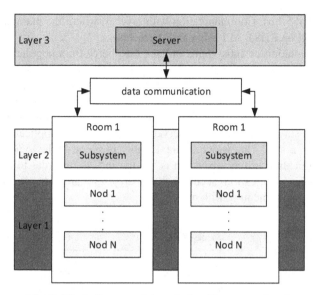

Fig. 1. Block diagram of the solution (Source: authors)

2.1 Layer 1 and 2: Hardware and Subsystem Layers

In the lowest hardware layer there are devices called 'nodes'. These nodes are sensors and actuators. Sensors can be of various types, for example sensors measuring temperature, humidity, air pressure, etc. It is also possible to connect motion detectors, voltage detection, switches, buttons, door control propulsion, etc. The hardware layer is realized by 8 bit PIC single chips by Microchip. Nodes are, by principle, passive elements, which perform no decision on their own, but which only listen to the bus and react to commands sent from the second layer. The first layer can be considered as functional unit.

The second layer, labeled as the subsystem (Fig. 2), is technically realized as a high-performance computer on ARM platform, but out of principle it is possible to use a computer of any other architecture with enough performance capability and compatible hardware. Main function of the subserver is to communicate with nods using an industrial RS-485 bus. The subsystem sends requests for measuring particular values, such as switch position (on/off), temperature value, or motion detection, to the nods and allocates these values to individual nods for parameterization, e.g. turning on an appliance.

For purpose of efficiency the subsystem is divided into two parts. The first part communicates with individual nods and the second part communicates with the central server. Because a fast response is required while communicating with nods, to approximate the response time to real life the software in subsystem communicating with nods was created in C++ language. The part communicating with the central server has been created in Java language. The communication of these two parts of a subsystem is realized via a TCP socket. Individual sockets are sent in JSON format through the RS-485 bus. Communication with layer 3, where the main control system is located, is built on REST API using an Ethernet bus.

Real implementation of subsystems utilizes Raspberry Pi, a popular minicomputer. Two applications are implemented on the subsystem. One application, written in Java, communicates with a server and in default modes it passes information from a lower server application and other way around. There is not complete information about all nodes in the database of this application, but only information about nodes directly connected to the subsystem. In case of server malfunction, this application can assume conrol of its own nodes. That way the functionality in scope of the logical unit is kept. The second application in the subsystem is written in C++ in order to reach maximal performance. This application accesses Raspberry Pi hardware directly and it is responsible for communicating with nodes.

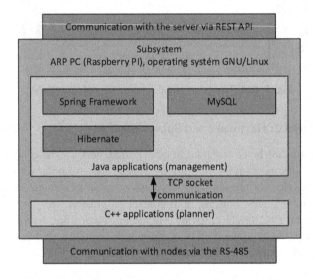

Fig. 2. Block diagram of the Subsystem (Source: authors)

2.2   Layer3: Control and Decision Making Layer

As it was already mentioned, the central server consists of either a high-performance computer of x86 architecture or of a cloud service, on which the main service is implemented. Its main purpose is to receive information about individual nodes and send them commands based on pre-defined rules.

Here, users can define their own rules or control the system manually if needed. Another function of the server is connecting new nodes or subsystems which are automatically assigned IP addresses and makes them available for the user to define new rules. The whole system is thus highly modular. In case of recommended cloud solution, the server enables to change the system independent on user location, by which the cloud solution is used in its full potential.

Each subsystem is responsible for its nodes and in case of failure of communication with the central server, the subsystem remembers last instruction from the central server and follows them until the communication is re-established. Every status is saved into a database and thanks to that it is possible to perform static operations above saved data. It is possible to search for energy leaks or to monitor energy usage (water, gas, electrocity), which the user can observe in a form of well arranged graphs.

System control is then realized by a responsive web application. The system supports two levels of user laws. The first level is the administrator level, which enables to perform all actions, and the second level is the user level, which is in principle restricted and the administrator personalizes it by allowing chosen actions. This restricted role is then appropriate for, for example, children who do not have the possibility to define their own rules and control, for example, the garage door.

A web application written in Java runs on the server (Fig. 3). It utilizes the most common technologies for creating web and enterprise applications. The purpose of this application is to provide the user with an interface in which they can control all parts of the house, scan statistics, and programme the house itself. After launching the system for the first time, the lower layers perform a detection of existing nodes and assign them unique addresses. 'Nodes' is a summary name for all sensors, actuators, and control units. The user can assign the nodes an alias to make them more arranged (e.g. a node with an id=10 can be labelled as light_kitchen1 and a node with id=11 switch_kitchen1). The user then assigns a particular node an action that should be performed dependant on either time or measured values of another node via a graphic tool. All information is stored in a MySQL database. The server communicates with the subsystem via a REST interface with the JSON format for data exchange.

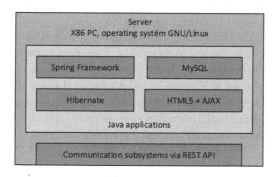

Fig. 3. Block diagram of the Server (Source: authors)

3   Node Implementation

Four principally different node categories can be defined in the designed system.

Category 1 - Immediate, for Reading - nodes from which only their current status can be read. A node can be labelled as immediate, because its current status is always available. It is enough for the subsystem to send a request and the node immediately reports its value. For reading that means that the status of the node can only be read. Such node can be, for example, a button or a switch.

Category 2 - Immediate, for Reading and Writing - this type of nodes can not only write their status, but also read it. A typical example of such node is the light. The subsystem sends a new status (turn on or turn off) and then it can verify whether the light is indeed on.

Category 3 - Immediate, for Writing - nodes that do not send their status upon request, but are used only to write their status.

Category 4 - Lagged, for Reading - all above mentioned nodes have an immediate status available. However, there are sensors (thermometers, moisture meter, etc.) which require certain time to measure the requsted status. If the subsystem was to wait for the node to measure its status and send it, it would not be able to serve other nodes. That might cause uncomfortable delay for the users. For example, if there were nods of a button, lights, and a thermometer in one room and the subsystem would wait for measuring of the temperature, it would not be able to control the other nods. So if the user wanted to turn on the light with a button, they would have to wait until the system finishes measuring the temperature. First, the subsystem sends a request for measuring a new status to the node and after the designated time lapses, the subsystem sends a request for sending the status. In the meantime it can attend to other nodes.

Further information used in the node-subsystem communication is the data width and data amount. Data width states the number of bits possible to write in or read from a node (depending on node category) and the number of data then becomes an information about how much of such data the node requires. To identify possible types unambiguously, each node is assigned a unique number ID. An example can be the following:

Switch: ID = 1, data width = 1 bit, data amount = 1 (only transmits information 0 = off, 1 = on), or RGB light: ID = 2, data width = 8 bit, data amount = 3 (range 0 - 255 for red, green, and blue colors).

Individual physical implementations of nodes contained saved information about the ID type and the number of channels. One node can therefore contain more devices of the same sort. Abovementioned switch node can physically be one switch (i.e. one channel) or up to eight switches and each nod can have a maximum of eight channels.

3.1   Node Communication and Microcontrollers

Nodes must be able to communicate with the subsystem, have their address saved, be able to perform operations with sensors, actuators, and controllers, and they need to be equipped by a microcontroller. Selected model for the introduced solution is PIC 16F690 by Microchip. It is a model of an eightbit series which is equipped by 4 kilobytes of program memory, 256 bytes of operational memory, and 256 bytes of EEPROM memory. It also contains 12 analog digital convertors, and 3 inner timers. It uses UART (Universal Asynchronous Receiver/Transmitter), to which a converter to a suitable bus is connected, to communicate with the subsystem. It is necessary to define which bus will be used to send data to make the nodes able to communicate with the subsystem. Buses like I2C, SPI, or microWire, which have microcontrollers implemented directly in them, exist. However, these buses are designed for communication of components inside the device for short distances in centimeters. It is necessary to deploy some appropriate bus enabling communication for longer distance for communication between devices (nodes and subsystems). Because data from individual nodes will be of maximum size of just few bytes, there is no need to request high transmit speeds and tens of kilobits per second are sufficient.

A suitable variant is the RS-485, which is not a bus, unlike the abovementioned standards, but only a norm defining the voltage level on conductors and specifying the way of communication between devices, i.e. the physical layer. It is used for a Master/Slave type of communication, where one device, the Master, controls the communication and more devices of the Slave type respond after a request. It is possible to use a connection with one pair of twisted conductors (called A and B) and communicate in Half-Duplex mode. Then the Slave devices are in a permanent state of receiving and they switch into a sending mode after a request. Immediately after that they go back to a receiving mode. The Master is, on the other hand, in a transmitting mode and after sending a request to a Slave unit it switches into a receiving mode and awaits a response. The second variant is two pairs of twisted conductors (called TxA, TxB, RxA, and RxB) communicating in Full-Duplex mode. On the first pair the Master is permanently in the transimitting mode, and on the second pair the Master is in the receiving mode. On the first pair all Slave units are awaiting a request. When one Slave unit receives a request, it responds on the second pair. This system is the main advantage of the Full-Duplex mode, when it is no longer necessary to switch between modes. Because switching between modes takes only several nanoseconds, it is more suitable to use the Half-Duplex mode. The RS-485 can reliably transmit data in speed of 20 Mbps for the distance of 10 meters. The distance of 1200 meters permits connecting nodes distributed far from the subsystem. Because the RS-485 does not specify the data link layer and communication protocol, a protocol for communication between nodes and the subsystem was designed. Datagrams sent via the bus always appear in the following format   (Table 1):

Table 1. Datagram format

Number of Byte:	Description:
1	The boot character
2	The number of Treasury data
3	The actual data (1 to n bytes)
n+1	The checksum CRC8
n+2	Termination character

The first byte of the data itself is then a number, which clearly identifies the type and the direction of data transmit (from the subsystem to the node and vice versa). The data itself must always contain at least the first byte, which identifies the type of the datagram.

Microcontrollers are programmed in a simplified version of C language and compiled by MikroC PRO compiler. After switching on (for the first time or after a power outage) the microcontroller checks whether there is an address saved in its EEPROM memory. If yes, the microcontroller goes to a neverending loop and according to its type it waits for specific datagrams from the subsystem, otherwise it goes to an addressing loop. There also is a special function contained which is run after receiving the first data byte and turns on an internal timer in case a communication error caused by interference occurs on the bus. If the datagram is not received in its entirety in 2 miliseconds, the already received data is erased and the microcontroller waits for another first byte.

Main purpose of the designed planner is to convert node data, discover new nodes, measure node status, and report node and communication errors. The main part is an executable method located in a Launcher class. The planner also contains a Connector class, which connects to a Java application (i.e. a higher class) as a client via a TCP socket. It sends and recieves data in the JSON format. Requests of the higher layer are delegated by the Delegator, which delegates data to the Planner class, which contains a planner for nodes on the RS-485 bus. The Delegator then serves as a preparation for system expansion by wireless nodes, which would delegate nodes and data between the Planner and WirelessPlanner classes based on the node communication type (Fig.4). Because the Planner, Delegator, and Connector classes need to run in one instance, the instance for these classes is created after the application is launched. Subsequently, the Launcher class delegates a Delegator reference to the Connector class and vice versa. It also delegates a Planner reference to the Delegator class and vice versa. Then, an init() method is called in the Connector class. A TCP socket is created by the Java application and the following data are received: a list of existing node types, the address of the subsystem, and a list of existing nodes.

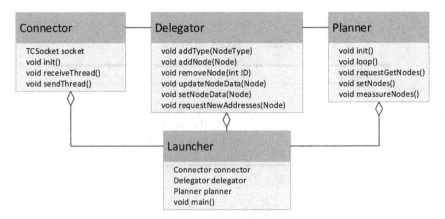

Fig. 4. Subsystem scheme (Source: authors)

With the first launch of the system, no nodes are delegated, only the address and the list of default node types is delegated. Then, receiveThread() and sendThread() methods are launched, both in separate threads. The sendThread() method waits for data from the Delegator class, performs conversion into the JSON format, and forwards data into a higher layer via a socket. The receiveThread() method waits for the JSON data from the higher layer and forwards them to the Delegator class. Data that can be send by the higher layer can be the following: a new status which should be set on the node, a new type of the node, and an address of a node that should be removed. After creating communicational threads, the init() method is called from the Launcher class to the Planer class. It opens UART serial interface, which is built in the Raspberry Pi, and calls a loop() method, which controls the planning process. This is realized via an endless loop, in which time of individual passings throught the loop is measured. If the time is longer than 100 seconds, it is nullified. Also, the following actions are performed based on the current time value: status reading from immediate nodes for reading, status writing onto immediate nodes for reading and writing (applies only to nodes where the status has changed since the last passing), status writing onto immediate nodes for writing (applies only to nodes where the status has changed since the last passing), status reading on nodes for reading and writing, requests for status measuring of delayed nodes, status reading on delayed nodes, search for new nodes.

4   Conclusion

In this article a brand new three-layer system concept for a smart home was introduced. Newly designed communication protocols and Layer 1 along with nodes principles were then introduced in detail. The whole system is then designed as highly modular a provides wide spectrum of extensions. To reach maximal universality, end nodes are fully customable in scope of a particular type. These types can be, for example, a digital node, an analog-input node, a servo-driving node, etc. How many available input/output channels and in what way they will be used is entirely up to the user. API, which will be specified and described in detail after the final version of the system will be finished, being implemented during development. Thanks to these interfaces it will be possible to access the system even from the outer environment in form of, for example, a remote control, a wall terminal, or a monitoring aplication in a smartphone. This will permit the system to be connected via an external control application deploying elements of mechanical learning and neural networks. Other applications, which can fully cooperate thanks to a supervisor (e.g. surveillance camera system, alarm), can run on the server collaterally and fully use its cloud implementation.

Acknowledgements. The paper is supported by the project of specific science Smart networking & cloud computing solutions (FIM, UHK, SPEV 2015).

References

1. Wu, C., Liao, C., Fu, L.: Service-Oriented Smart-Home Architecture Based on OSGi and Mobile-Agent Technology. IEEE Trans. Syst., Man, Cybern. C. 37, 193–205 (2007)
2. Jiang, L., Liu, D.-Y., Yang, B.: Smart home research. In: Proceedings of 2004 International Conference on Machine Learning and Cybernetics (IEEE Cat. No.04EX826) (2004)
3. Ryan, J.: Home automation. Electron. Commun. Eng. J. 1, 185 (1989)
4. Kao, Y., Gu, H., Yuan, S.: Personal Based Authentication by Face Recognition. In: 2008 Fourth International Conference on Networked Computing and Advanced Information Management (2008)
5. Ha, Y., Sohn, J., Cho, Y., Yoon, H.: A robotic service framework supporting automated integration of ubiquitous sensors and devices. Information Sciences 177, 657–679 (2007)
6. Rui, C., Yi-bin, H., Zhang-qin, H., Jian, H.: Modeling the Ambient Intelligence Application System: Concept, Software, Data, and Network. IEEE Transactions on Systems, Man, and Cybernetics, Part C (Applications and Reviews) 39, 299–314 (2009)
7. Kao, Y., Gu, H., Yuan, S.: Integration of Face and Hand Gesture Recognition. In: 2008 Third International Conference on Convergence and Hybrid Information Technology (2008)

Cloud – Based Solutions
for Outdoor Ambient Intelligence Support

Peter Mikulecky

Faculty of Informatics and Management, University of Hradec Kralove, Czech Republic
peter.mikulecky@uhk.cz

Abstract. Intelligent (or smart) environments are usually developed for various indoor applications, capable, e.g., of supporting independent living of seniors or handicapped people. Taking into account recent achievements in ambient intelligence, wearable technologies, and wireless sensor networks, especially focused on the area of environmental monitoring, a need for intelligent support of various human activities also in outdoor environment is evident. Such solutions, focused on intelligent support of human activities in natural outdoor environment, have been still very rarely developed and published. A way to useful outdoor applications may lead through exploitation of "large-scale ambient intelligence" approaches respecting the distributed character of outdoor applications. Here also cloud – based solutions seem to be naturally utilizable.

The purpose of the paper is, after reviewing several existing solutions in related areas, to discuss possibilities for cloud-based solutions in intelligent environments oriented on human activities support in outdoor spaces. The approach is illustrated using scenario-based explanation.

Keywords: Intelligent environments, outdoor environments, cloud-based solutions, scenarios.

1   Introduction

New possibilities of outdoor oriented applications of Ambient Intelligence (AmI) approaches and technologies can be viewed also as a clear result of recent achievements in the area of wide area wireless sensor networks, as well as achievements in sophisticated smart communication devices.

Intensive effort has been devoted recently to the area of wireless sensor networks and their important applications. A wireless sensor network is usually a combination of low-cost, low-power, multifunctional miniature sensor devices consisting of sensing, data processing, and communicating components, networked through wireless link. In a typical application, a large number of sensor nodes are deployed over an area with wireless communication capabilities between the network nodes.

Recent technological advances in sensors facilitate wireless sensor networks that are deeply embedded in their native environments. Wireless sensor networks are highly suitable for various applications, such as health monitoring and guidance [1], home monitoring [2], traffic pattern monitoring and navigation, plant monitoring in

© Springer International Publishing Switzerland 2015   229
D. Barbucha et al. (eds.), *New Trends in Intelligent Information and Database Systems,*
Studies in Computational Intelligence 598, DOI: 10.1007/978-3-319-16211-9_24

agriculture [3], and others. It was mentioned already in [4] that environmental monitoring is recently a strong driver for wireless sensor networks development and for outdoor ambient intelligence applications design.

The current technological and economic trends enable new generations of wireless sensor networks with more compact and lighter sensor nodes, processing power, and storage capacity. In addition, the ongoing proliferation of wireless sensor networks across many application domains will certainly result in a significant cost reduction.

According to [5], cloud computing aims to deliver a network of virtual services so that users can access them from anywhere in the world on subscription at competitive costs depending on their Quality of Service (QoS) requirements. Indeed, this is a challenge for a number of outdoor oriented applications of Ambient Intelligence.

If wireless sensor networks technology is matured enough to enable large-scale planting of sensors in an outdoor area, and these sensors are sophisticated enough to have merged functionalities, communication abilities, and very low energy consumption, why not to use them also for ensuring a number of functionalities that are recently typical for indoor applications of Ambient Intelligence approaches. The idea is to build an intelligent environment based on a multi-agent architecture over the wireless sensor network, including mobile agents representing individual persons appearing in that area equipped with a suitable mobile communicating device, and using a suitable cloud service allowing all the entities in the environment a kind of mutual communication and information sharing about interesting or useful events, about possible endangering situations, or just about possible weather change that is approaching.

In the Section 2 a review of a selection of recent related works in various subareas of Ambient Intelligence employing some cloud service for their solutions is presented. Section 3 is devoted to two scenarios for outdoor applications of Ambient Intelligence, describing the possible solution for some practical outdoor problems. The last parts describe some directions for further research and bring the concluding remarks.

2   Related Works

In this part of the paper, several researches that are oriented somehow on cloud service exploitation in a relatively typical Ambient Intelligence application will be mentioned. These are mainly oriented on Ambient Assisted Living area, usually focused on personal health monitoring. The area of personal health monitoring can be a driver also for outdoor applications; however, all depends on a suitable infrastructure in the environment considered.

Advances in computing and wireless communication have allowed the development of novel distributed computing methods, middleware and applications that support ubiquitous and ambient intelligence systems. A good source for that is [6], a special issue of a journal, which has aimed to explore the recent state-of-the-art developments in theory and application of distributed solutions aiming to support the design, implementation and evaluation of these systems. Some of the works selected for the journal have stressed the importance of cloud computing based solutions to enable ubiquitous systems in very diverse application domains, such as learning, e-health or environment protection.

There are a lot of relevant publications presenting possibilities how to integrate cloud services with ubiquitous and ambient intelligence systems, among others. We can mention here, e.g., [7, 8, 9, 10, 11, 12, 13].

Advancements in mobile technology have allowed mobile devices, such as smartphones and tablets, to be used in a variety of different applications. With the fact that mobile phones (but recently also smart wearable devices, as smart watches) are now a way of life amongst people of all ages due to its ubiquitous nature, it is now becoming more feasible than ever to use mobile technology also for medical applications. A user can simply connect a health monitor to a mobile phone via Bluetooth to develop his or her own personal health monitor and management system [14].

There is currently a strong need to advance the field of health informatics. As the world population ages due to increased life expectancy, this places pressure on the government to fund spending associated with the ageing population, especially in terms of health spending. Consequently, the demand for cutting the cost of healthcare has increased and there is now a growing need for the remote caring of patients at home, in particular, for the elderly and the physically disabled. By leveraging the capability of mobile technology as well as Cloud Computing, one can then develop a health monitoring system where the patient can be assessed by doctors in a remote location from the comfort of their own home. As it is pointed out in [14], there are an abundant number of mobile apps available today for mobile health tele-care. Building on advances in Cloud Computing, it is possible to go beyond the mobile health applications, to enable secure sharing of tele-care data in the Cloud. The Cloud, as an enabler for mobile tele-care, can help provide effective treatment and care of patients due to its benefits such as on-demand access anywhere anytime, low costs and high elasticity, concludes Thilakanathan and others [14].

Recently, there have been works focused on the integration of mobile and cloud technologies for health monitoring. Some of them could be used also for various outdoor applications. Fortino et al. [15] introduced the BodyCloud architecture which enables the management and monitoring of body sensor data via the Cloud. It provides functionality receive and manage sensor data in a seamless way from a body sensor network (BSN). BodyCloud also comprises of a scalable framework that allows support for multiple data streams required for running concurrent applications.

The scheme by Pandey et al. [16] integrates mobile and cloud technologies (MCC, see [17]) with electrocardiogram (ECG) sensors to enable remote monitoring of patients with heart-related problems such as cardiac arrhythmias. The patient connects the sensors to their body and then run an application on a mobile device. The application connects to the sensors via Bluetooth. It will then periodically upload data to the Cloud. The user can then download graphs from the Cloud which represent the user's health states. The scheme also implements middleware in the Cloud. There are web services for users to analyze their ECG, draw graphs, etc. The system is effective as it allows the user to adopt the Pay-As-You-Go methodology every time they require services to analyze their health data.

Bourouis et al [1] in their paper described a new Mobile Health Monitoring system architecture that used a wireless body area sensor networks (WBASN) to collect and send data to the cloud server through GPRS/UMTS. The system uses a cloud service to extract patient data information. These sensory parameters are fed into a neural network engine running as a cloud service that fuses information from multiple disparate sensors to determine whether the patient is in a state that requires some intervention of a doctor or other qualified staff.

Zapater [18] proposes a novel and evolved computing paradigm that: (1) provides the required computing and sensing resources for patient care scenarios, (2) allows the population-wide diffusion, (3) exploits the storage, communication and computing services provided by the cloud, and (4) tackles the energy-optimization issue as a first-class requirement. This computing paradigm and the proposed multi-layer top-down energy optimization methodology were evaluated in a scenario for cardiovascular tracking and analysis, and the obtained results were highly promising. The solution implements Mobile Cloud Computing (MCC) [17] and deals with the typical challenge of distributing efficiently the computation on the cloud [19]. The use of MCC is becoming popular to automate health-care systems [16].

In order to maintain quality healthcare services, it is essential to have an intelligent, highly resourced AAL system that is efficient, responsive, and adequately secures patient health. Forkan et al [20] proposed CoCaMAAL – a cloud platform that offers a high-level abstraction, and its services can be accessed easily via mature web service protocols. Their solution emphasizes a service-oriented architecture that performs context modeling from raw data, context data management and adaptation, context-aware service mapping, service distribution, and service discovery.

A very interesting and relevant paper was published by Vescoukis and others [21], motivated by the fact, that there is a great increase of natural disasters (e.g., forest fires, flooding, landslides) stimulating a big focus in developing smart and intelligent Environmental Information Management (EIM) Systems. These systems should be able to collect, process, visualize and interpret geospatial data and workflows of added-value applications so as to support decision making in case of emergency. It is quite clear that natural disasters pose a great threat to people's lives and their properties while they present a negative impact to the economies. In [21] a novel Service Oriented Information System is proposed that seems to be proper for Environmental Information Management, as well as for planning and decision support in case of emergency. The proposed architecture was designed in close collaboration with real world stakeholders in civil protection and environmental crisis management, and has been implemented as a real system, with applications in forest fire crisis management.

3   The Problem and Its Possible Solutions

In order to support a person's activities outdoor, his/her geographic location must be identified as important contextual information that can be used in a variety of scenarios like disaster relief, directional assistance, context-based advertisements, or early warning of the particular person is some potentially dangerous situations (see, e.g.,

[22] or [23]). GPS can be used here, as it provides accurate localization outdoors, although is not very useful inside buildings. Outdoor to indoor and vice versa activities localization was investigated e.g. in [24] by a coarse indoor localization approach exploiting the ubiquity of smart phones with embedded sensors.

Outdoor acting person's support should provide relevant and reliable information to users who often used to be engaged in other activities and therefore not aware of possible hazardous situations that he or she could encounter in near future. The typical endangering situation and its possible solution using intelligent outdoor environment can be described using the following scenario:

A young man Fred, a keen hiker and mountaineer, started his demanding tour to the top of a high mountain early in the morning. The selected tour is leading through several exacting parts of a national park, in which a modern supporting infrastructure has been installed quite recently. The infrastructure is based on the idea of "large-scale ambient intelligence" with rather dense wireless sensor network monitoring the most important parts of the national park ubiquitously. Fred knows about the features of this service, therefore, in the evening before he started the tour, he subscribed all the necessary service via relevant web pages, learning thus also a couple of important information about weather forecast, his intended tracks condition, opening hours in chalets nearby, etc. The scope of subscribed services covered the instant weather situation information with early warning facilities in the case of a sudden weather change, as well as warning messages about rather unlikely yet potentially extremely dangerous situations, like a forest fire can be. Fred can subscribe also other less important yet interesting or useful messages as, e.g., notifications about interesting places he would possibly passing on his tour, or when the last local bus will leave the stop where Fred intended finishing of his tour.

Being on his tour, Fred wears a smart watch instantly connected with his smart phone, having thus an opportunity to be informed by the underlying intelligent early warning system unceasingly, better to say, if necessary. During the tour, Fred encounters a number of other tourists, partly being subscribers of the service. An elderly lady stopped him suddenly with an urgent request that her husband Jack might have been lost somewhere in the vicinity, but she had had no possibility to find him or to contact him. Fred sent an urgent request for helping support to the underlying monitoring system; the information was stored in the cloud, where all the recently monitored events in the area have been stored. Immediately a mining started through all the recent relevant data in the cloud, and the position of Jack could be located in a couple of minutes. Jack was somehow wounded apparently, therefore a rescue operation started automatically and Jack has been finally transported to a mountain chalet nearby. However, weather started to be a bit unpleasant and Fred got a warning that a storm is very likely to start in a half of hour. He asked for the shortest path to the valley and decided to leave the area using the recommended direction.

A number of similar as well as more elaborated scenarios can be created. These may be oriented, e.g., on a ski resort, where each subscriber of the service will be informed about the best ski track available (best snow, weather, as well as with minimum people skiing), or about potential avalanche threat or other kind of it (if applicable in the environment).

The following, a bit different scenario, can illustrate, how technology enhanced smart ski resort could work:

A senior female Xenya enters the Ski Resort. After purchasing the chip card she wishes to use a cable car in the valley in order to reach the top of the hill. The entrance gate senses her chip card and sends the information about Xenya's starting point into the cloud. After reaching the top, Xenya decides to visit the restaurant there in order to have some refreshments. The entrance gate to the restaurant records her presence there. In the meantime she got a recommendation to her smartphone or smart watch that red slope 11 as well as blue slope 1 is now in best condition and without any crowd. She decided for the blue slope 1 because she wishes to practice downhill on an easy slope first. When skiing down to the slope 1, suddenly information comes that there is a failure on the cable car serving for skiers on slope 1. She deflect her original course towards the red slope 11 aiming thus to prevent herself from increased waiting time for lifts in the slope 1 area. She is practicing downhill for three – four times on the slope 11 using the cable car and subsequent lift there. The sensors incorporated into the entrance gates of those lifts are recording her presence as well as her times for downhill rides. The system calculates from these data that Xenya is not very skillful and as the sensor network incorporated in the slope 11 is starting to signalize downgrading of the snow surface quality, she decides to stop using the slope 11 and enters the chalet nearby for a lunch. The system is aware of her presence in the chalet, and after finishing the lunch some easy relaxing slope is recommended. However, the weather starts to be worse with possible wind and snow, and the system warns Xenya about the fog ascending from the valley. She decided to leave, so system assisted her in finding the optimal journey down using a combination of cable cars and downhill.

All the information is collected in a cloud, where it is available for all the subscribers of this service as well as for subsequent analysis, if necessary. In the urgent case, the system will proactively broadcast an early warning message to the user, offering her/him related navigation services supporting escape from recent dangerous situation.

There are a number of papers that are devoted to various solutions for tourist assistance, mainly oriented on context-aware tourist navigation on their routes. The usual approach (see, e.g. [25]) is based on deployment of intelligent agents, which collectively determine the user context and retrieve and assemble a kind of simple information up to multi-media presentations that are wirelessly transmitted and displayed on a Personal Digital Assistant (PDA). However, these tourism oriented applications are usually deployed for the navigational purposes, without having capabilities of warning the user from potentially dangerous situations that can appear during their routes.

Another couple of related papers are recently devoted to the theme parks visitors and to solutions of recommending systems for visitors in order to get as much from their visit as it is possible. In this area the interesting papers seem to be [26] or [27].

Among most interesting and already existing outdoor applications certainly [28] have to be mentioned, attempting to prevent children from potentially dangerous situations in an urban environment. The research described there was focused on designing a ubiquitous kid's safety care system capable to dynamically detect possible dangerous situations in school routes and promptly give advices to kids and/or their parents in order to avoid or prevent from some possible dangers. To detect the dangerous situations, it is essential to get enough contexts of real environments in kids' surroundings. This is based on two basic assumptions: (1) a big number of sensors, RFIDs, tags and other information acquisition devices are pervasively distributed somewhere in and near school routes, and (2) a kids should carry or wear some devices that can get surrounding context data from the above pervasive devices.

Some other solutions, still without any cloud service support, can be found, e.g., in [29] and [30]. The first is devoted to assist workers in a coal mine in China, contributing to the higher security of their work; the second one describes an interesting case of a wireless sensor network supporting the safety of geophysicists working near a living volcano. Enhancing the both application with a suitable cloud service could be a challenging and possible useful task.

Various other proposals, solutions, as well as related contemplations can be found in [31, 32, 33, 34, 35].

4   Further Research

The environmentally oriented wireless sensor networks [4] are matured enough to become a basis for more complex and intelligent support of various outdoor activities, for instance those described in our scenarios. Nevertheless, the potential of wireless sensor networks alone cannot be adequate for successful development of intelligent outdoor activities support. It seems that the following directions of related research should be pursued:

- designing of adequate agent-based architectures that will be powerful enough to ensure desirable functionalities of a large-scale ambient intelligence system over an underlying sophisticated wireless sensor network;
- such a wireless sensor network must be large enough to be able of monitoring the most important parts of the outdoor area; this might be still very costly;
- the desired monitoring should cover not only possible dangerous situations that are usually relatively transparent (seismoacoustic signals, smoke, water on unusual places, avalanches, etc.), but also sources having potential of causing some unexpected threat, e.g., herds of animals starting to runaway suddenly;
- the agent-base architecture should collaborate very closely with a suitable cloud serving as a storage for various sources of data and information from all the parts of the agent architecture, ensuring the desired functionalities as monitoring the people in the environment, collecting data about weather conditions, watching the possible sources of threats (debris fields, water flows, etc.), and so on;
- the data collected in the cloud should be mined instantly, or in various important cases on request with a specific goal as, e.g., to localize a lost person;
- the warning messages must be delivered to as many endangered people in the area as it is possible, that is, not only to the service subscribers; therefore the means of delivering these messages should be as various as it is possible (also very classical ones, as public radio, etc.), utilizing not only smart phones, but in a short perspective also a palette of wearable devices (smart watches, glasses, etc.).

Any integration of these functionalities with cloud services will enhance the original systems significantly, giving them another dimension, enabling an instant access to potentially endangered people in the particular outdoor environment to important if not vital information, possibly contributing to their rescue. Any simulation of possible AmI and/or cloud-based solutions for various kinds of outdoor environments would be very interesting and promising. One possibility could be using a powerful, yet for still for indoor environment applied simulation architecture AmISim proposed by Garcia-Valverde, Serrano, and Botia [36].

5   Conclusions

Outdoor environments are a great challenge for Ambient Intelligence applications because advances in wireless sensor networks technology, as well as recent advancements in smart and wearable communication devices (smart phones, smart watches, Google glasses, etc.). For instance, proposed intelligent ski resort, but also other outdoor applications, could be useful applications with a great commercial potential. However, there is a permanent development in a number of related research and technology areas, let us mention such areas of research as crowd mining,   crowd sensing, or wearable sensors and devices as the most promising ones for being employed for some new interesting solutions in the development of intelligent outdoor environments. Near future will certainly bring a number of new ideas in this direction.

Acknowledgments. This work was partially supported by the Czech Scientific Foundation research project No. 15-11724S *DEPIES - Decision Processes in Intelligent Environments*.

References

1. Bourouis, A., Feham, M., Bouchachia, A.: A New Architecture of a Ubiquitous Health Monitoring System: A Prototype of Cloud Mobile Health Monitoring System. Int. J. Comp. Sci. 9, 434–438 (2012)
2. Kannan, S., Gavrilovska, A., Schwan, K.: Cloud4Home — Enhancing Data Services with @Home Clouds. In: 31st International Conference on Distributed Computing Systems, pp. 539–548. IEEE Press, New York (2011)
3. Mikulecky, P.: User Adaptivity in Smart Workplaces. In: Pan, J.-S., Chen, S.-M., Nguyen, N.T. (eds.) ACIIDS 2012, Part II. LNCS, vol. 7197, pp. 401–410. Springer, Heidelberg (2012)
4. Corke, P., Wark, T., Jurdak, R., Hu, W., Valencia, P., Moore, D.: Environmental Wireless Sensor Networks. Proc. IEEE 98, 1903–1917 (2010)
5. Buyya, R., Yeo, C., Venugopal, S., Broberg, J., Brandic, I.: Cloud Computing and Emerging IT Platforms: Vision, Hype, and Reality for Delivering Computing as the 5th Utility. Future Gen. Comp. Syst. 25, 599–616 (2009)
6. Ochoa, S.F., López-de-Ipiña, D.: Distributed Solutions for Ubiquitous Computing and Ambient Intelligence (Editorial). Future Gen. Comp. Syst. 34, 94–96 (2014)

7. Cimler, R., Matyska, J., Sobeslav, V.: Cloud-Based Solution for Mobile Healthcare Application. In: 18th International Database Engineering & Applications Symposium, pp. 298–301. ACM Press, New York (2014)

8. Choi, M., Park, J., Jeong, Y.-S.: Mobile Cloud Computing Framework for a Pervasive and Ubiquitous Environment. J Supercomput 64, 331–356 (2013)

9. Garg, S.K., Versteeg, S., Buyya, R.: A Framework for Ranking of Cloud Computing Services. Future Gen. Comp. Syst. 29, 1012–1023 (2013)

10. Grønli, T.-M., Ghinea, G., Younas, M.: Context-aware and Automatic Configuration of Mobile Devices in Cloud-enabled Ubiquitous Computing. Pers. Ubiquit. Comput. 18, 883–894 (2014)

11. Liu, R., Wassell, I.J.: Opportunities and Challenges of Wireless Sensor Networks Using Cloud Services. In: ACM CoNEXT 2011, Art. 4, ACM Press, New York (2011)

12. Lomotey, R.K., Deters, R.: CSB-UCC: Cloud Services Brokerage for Ubiquitous Cloud Computing. In: 5th International Conference on Management of Emergent Digital EcoSystems, pp. 100–107. ACM Press, New York (2013)

13. Mahjoub, M., Mdhaffar, A., Halima, R.B., Jmaiel, M.: A Comparative Study of the Current Cloud Computing Technologies and Offers. In: 1st International Symposium on Network Cloud Computing and Applications, pp. 131–134. IEEE Press, New York (2011)

14. Thilakanathan, D., Chen, S., Nepal, S., Calvo, R., Alem, L.: A Platform for Secure Monitoring and Sharing of Generic Health Data in the Cloud. Future Gen. Comp. Syst. 35, 102–113 (2014)

15. Fortino, G., Parisi, D., Pirrone, V., Di Fatta, G.: BodyCloud: A SaaS Approach for Community Body Sensor Networks. Future Gen. Comp. Syst. 35, 62–79 (2014)

16. Pandey, S., Voorsluys, W., Niu, S., Khandoker, A., Buyya, R.: An Autonomic Cloud Environment for Hosting ECG Data Analysis Services. Future Gen. Comp. Syst. 28, 147–151 (2012)

17. Fernando, N., Loke, S.W., Rahayu, W.: Mobile Cloud Computing: A Survey. Future Gen. Comp. Syst. 29, 84–106 (2013)

18. Zapater, M., Arroba, P., Ayala, J.L., Moya, J.M., Olcoz, K.: A Novel Energy-driven Computing Paradigm for e-Health Scenarios. Future Gen. Comp. Syst. 34, 138–154 (2014)

19. Khan, A.N., Mat Kiah, M.L., Khan, S.U., Madani, S.A.: Towards Secure Mobile Cloud Computing: A Survey. Future Gen. Comp. Syst. 29, 1278–1299 (2013)

20. Forkan, A., Khalil, I., Tari, Z.: CoCaMAAL: A Cloud-oriented Context-aware Middleware in Ambient Assisted Living. Future Gen. Comp. Syst. 35, 114–127 (2014)

21. Vescoukis, V., Doulamis, N., Karagiorgou, S.: A Service Oriented Architecture for Decision Support Systems in Environmental Crisis Management. Future Gen. Comp. Syst. 28, 593–604 (2012)

22. Benikovsky, J., Brida, P., Machaj, J.: Proposal of User Adaptive Modular Localization System for Ubiquitous Positioning. In: Pan, J.-S., Chen, S.-M., Nguyen, N.T. (eds.) ACIIDS 2012, Part II. LNCS, vol. 7197, pp. 391–400. Springer, Heidelberg (2012)

23. Krejcar, O.: User Localization for Intelligent Crisis Management. In: Maglogiannis, I., Karpouzis, K., Bramer, M. (eds.) Artificial Intelligence Applications and Innovations. IFIP, vol. 204, pp. 221–227. Springer, Boston (2006)

24. Parnandi, A., Le, K., Vaghela, P., Kolli, A., Dantu, K., Poduri, S., Sukhatme, G.S.: Coarse In-Building Localization with Smartphones. In: Phan, T., Montanari, R., Zerfos, P. (eds.) MobiCASE 2009. LNCIS, vol. 35, pp. 343–354. Springer, Heidelberg (2010)

25. O'Hare, G.M.P., O'Grady, M.J.: Gulliver's Genie: A Multi-agent System for Ubiquitous and Intelligent Content Delivery. Comp. Communications 26, 1177–1187 (2003)

26. Durrant, A., Kirk, D.S., Benford, S., Rodden, T.: Pursuing Leisure: Reflections on Theme Park Visiting. Comp. Supp. Cooperat. Work 21, 43–79 (2012)
27. Tsai, C.Y., Chung, S.H.: A Personalized Route Recommendation Service for Theme Parks Using RFID Information and Tourist Behavior. Supp. Syst. 52, 514–527 (2012)
28. Takata, K., Shina, Y., Komuro, H., Tanaka, M., Ide, M., Ma, J.: Designing a Context-Aware System to Detect Dangerous Situations in School Routes for Kids Outdoor Safety Care. In: Yang, L.T., Amamiya, M., Liu, Z., Guo, M., Rammig, F.J. (eds.) EUC 2005. LNCS, vol. 3824, pp. 1016–1025. Springer, Heidelberg (2005)
29. Wang, X., Zhao, X., Liang, Z., Tan, M.: Deploying a Wireless Sensor Network on the Coal Mines. In: IEEE International Conference on Networking, Sensing and Control, pp. 324–328. IEEE Press, New York (2007)
30. Werner-Allen, G., Lorincz, K., Welsh, M., Marcillo, O., Johnson, J., Ruiz, M., Lees, J.: Deploying a Wireless Sensor Network on an Active Volcano. IEEE Internet Comp 10, 8–25 (2006)
31. Cech, P., Bures, V.: Recommendation of Web Resources for Academics - Architecture and Components. In: 3rd International Conference on Web Information Systems and Technologies, pp. 437–440. INSICC, Barcelona (2007)
32. Mikulecky, P.: Ambient Intelligence in Decision Support. In: 7th International Conference on Strategic Management and Its Support by Information Systems, pp. 48–57. VSB Press, Ostrava (2007)
33. Mikulecky, P., Cimler, R., Olsevicova, K.: Outdoor Large-scale Ambient Intelligence. In: 19th IBIMA Conference, pp. 1840–1842. IBIMA, Barcelona (2012)
34. Mikulecky, P., Ponce, D., Toman, M.: A Knowledge-based Decision Support System for River Basin Management. In: River Basin Management II, Progress in Water Resources, vol. 7, pp. 177–185. WIT Press, Southampton (2003)
35. Mikulecky, P., Ponce, D., Toman, M.: A Knowledge-based Solution for River Water Resources Management. In: Water Resources Management II, Progress in Water Resources, vol. 8, pp. 451–458. WIT Press, Southampton (2003)
36. Garcia-Valverde, T., Serrano, E., Botia, J.A.: Combining the Real World with Simulations for a Robust Testing of Ambient Intelligence Services. Artif. Intell. Review 42, 723–746 (2014)

Internet of Things Service Systems Architecture

Patryk Schauer and Grzegorz Debita

Institute of Computer Science, Wrocław University of Technology,
Wybrzeże Wyspiańskiego 27, 50-370 Wrocław, Poland
{patryk.schauer,grzegorz.debita}@pwr.wroc.pl

Abstract. Internet of Things (IoT) is growing in popularity, emerging as an innovative paradigm in distributed systems. The key research challenges of the IoT are related with the new approach to designing IoT applications. This article proposes an IoT application designing process based on SOA (Service Oriented Architecture) experiences. Standardization is very important aspect in IoT, it is the reason we compare presented IoT application designing methodology to IoT architecture known in literature. Due to the importance of hardware solutions in IoT systems, the IoT device model, is also proposed, and it is shown how it can be used for describing IoT hub as well as independent node.

Keywords: Internet of Things, Service Oriented Architecture, IoT application designing, IoT device model.

1 Introduction

Internet of Things (IoT) can be called as a new concept of sharing information between objects in computerized world [1]. Nowadays almost all electronic devices can be connected to the Internet. Starting from personal networks used to monitor health parameters, through car area networks that help control traffic and sensor based systems for environment monitoring, ending on huge communication systems for finance operations [2,3]. Differences between classical Internet and Internet of Things came from many aspects in designing and implementing of IoT systems, and also from IoT devices construction. Growing popularity of IoT applications entails increase traffic volume in communication networks. These differences result in fundamental need to redesign applied traffic models and traffic control algorithms. Especially, it is important to relate traffic control process with application scenarios and transmitted content.

Nowadays, the idea of "smart" things, "smart" devices, "smart" networks and "smart" environment becomes more and more popular. This smartness is related to certain aspects of IoT systems. First of all, "smart" devices are able to communicate to each other and share information. Absolutely basic feature is the ability of checking the presence and identifying nearby objects. The other feature of "smart" equipment is autonomy, particularly in terms of power sources. The other distinctive feature of "smart" systems is the ability of self-configuration. By using service composition, known from SOA (Service Oriented Architecture

© Springer International Publishing Switzerland 2015
D. Barbucha et al. (eds.), *New Trends in Intelligent Information and Database Systems*,
Studies in Computational Intelligence 598, DOI: 10.1007/978-3-319-16211-9_25

[4,5]) paradigm, it is possible to build systems that make decisions about taking actions on things, objects and data. These decisions are based on information, which is collected from the environment and also from its description supported by various models of reality. Due to the diversity and multitude of things and processing services, it is necessary to design and build IoT systems based on standardized and common architecture. The main contribution of this paper is to introduce an innovative approach to IoT development, basing on the combination of SOA and IoT systems advantages.

The proposed process of IoT application design assumes definition and description of reality in a formal manner in order to support the designed application in the interactions with real world objects. We propose to describe the interactions between applications, things and the reality as a decision making process in which application components, things, relations and the reality are defined using domain-specific and general ontologies. IoT application is a combination of software and hardware components designed for support of chosen process occurring in reality. IoT applications base on hardware (IoT devices), which can interact with this reality in two ways. They can retrieve information about objects in reality or change their state. In the relevant part of reality there are many processes which can be supported by an IoT application. Usually there is more than one possible way to achieve similar effect, and these ways are called application scenarios.

In IoT application design process we are starting from reality description by means of domain ontology, then we model the decision making processes occurring in reality. Through services composition, taken from SOA approach, we prepare complex services for each scenarios. Note, that the SOA paradigm assumes semantic description of Web services, which means that their input and output data, and their functionalities are also described by the concepts taken from domain ontology. Hence, the descriptions of processes and services (software components) are consistent. On each stage of this processes service requirements appear. They are typically related to IoT devices functionality. Unfortunately, devices usually have previously determined functionalities and methods of realization, such as data formats for generated data. IoT device model presented in this paper partially solves this problem.

Section 2 introduces the idea of IoT application as decision making and support systems. Next, in section 3 we describe IoT application designing process, which is placed in IoT-Architecture [6] in section 4. In section 5 IoT device model is described.

2   IoT Environment

IoT systems can be described as a process operating in real world. In the Fig. 1 the IoT Environment model is proposed. The key idea is that an IoT application is interacting with the reality. Basing on gathered data and model of this reality it is possible to make decisions about future actions.

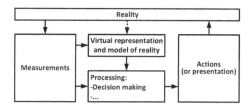

Fig. 1. IoT Environment model

The main feature of IoT application is interaction with reality. Virtual representation and model of the reality is prepared at designing stage. It is based on part of reality in which the application should work or support the work. The most important function of virtual representation is the containment of up-to-date information about objects in the reality, especially their state (values of their parameters). In the reality model the information about objects relations is contained. There are also rules, which define how to process measured data into variables needed in decision making process. These variables can be also extended to estimate the most efficient result concerning process in reality.

In the presented IoT application model the main functionalities of the application fall into three categories: measurements, processing and actions. Processing is the core task of the IoT application. Basing on data provided by measurements and rules stored in reality model, the IoT application can make decisions about actions or support this process by suggesting proper activity to user. Information about the current state of the objects is acquired by measurements. Measurements results can be delivered directly to be processing service or to be supplied with virtual representation and model of reality.

The IoT application interacts with reality by means of IoT devices. There are various types of devices, but they can be grouped into two general categories: measurement devices and actuators. IoT device model is described in section 5.

3 IoT Application Designing

The Fig. 2 is showing the modeling of a process occurring in reality. The detailed answers to the presented questions are used to define the concepts necessary for the description of objects and processes. Objects description contains list of objects with their parameters, and process description contains list of processes, rules of interaction between objects, possible actions on the objects and measurable parameters of the objects. They are all included in the domain ontology.

Modeling process can be started at any step. Depending on the knowledge, if the objects are well known, it is good to start from their description. But in case of different domain, the other parameters from objects description can be important. In that case, it is better to start from describing the application domain. If the domain is not specified, one could start from describing processes in the reality and then decide for the domain of application. The last approach is useful when it is needed to find IoT application domains in the reality.

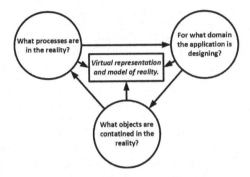

Fig. 2. IoT application modeling process

All steps of IoT application designing process are based on virtual representation and model of reality (Fig. 3). On each step of the process, designing information is taken from the model. To be more precise, on each step the application makes actions basing on that information.

The most important and difficult action is first step, transition from reality description to decision making process description. The expected result of this step is a structure containing tasks of modeled process. In hard cases modeling is making manually, it will be time consuming nevertheless automatisation of this step could be also very difficult. Reality and process descriptions must be complete and consistent. Either way, the goal of this task is to prepare good reality description, according to standardized schema. It is important to build a structure which will be based on commonly known schema. This way of acting simplifies dealing with further steps.

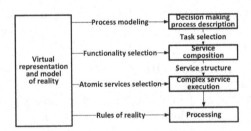

Fig. 3. IoT application designing process

Next step in IoT application designing process is selecting tasks from decision making process graph and transforming it into service graph by means of service composition. Task selection is used only to narrow process description to those ones which are needed in designing application. Service composition uses information about available functionalities of device-controlling services in repository. Mostly, they can be data sources like IoT measure devices functioning in reality. They also deliver functionalities of available data processing services, which can be useful in designing application (processes contain data processing functions). Finally, they can be IoT updating devices (actuators), which can interact with the reality. The final result of service composition is the complex service, represented by its functionality graph, containing nodes which are functionalities and directed, labeled (with data types) edges determining the direction of transferred data.

It's important to notice that in many cases the result of composition would be delivered as many functional graphs corresponding to different application scenarios [7]. Differences might be minimal and could correspond to non-functional, quality, requirements (e.g. road traffic conditions may have highest priority on working days than on weekend). This scenario-optimized complex services can also have extended functionality. The enhanced scenario might include more sophisticated measures and processing (like encryption) than the basic one, especially in eHealth domain, where security and privacy are key issues [8].

The last, but not least, step in application design is complex service execution. During complex service execution there are some things to determine before the execution engine starts complex service. Examplary IoT applications composition and execution platform is ComSS [9] which uses atomic service selection mechanism. This step consists of selecting the atomic service to each of functionalities existing in complex service graph. It is usually thought to be selection of optimal service considering the values of non functional parameters, for example: price or performance. The other parameters which are verified before service start are data formats. For each pair of services which are exchanging data, data formats have to be determined. For example there can be services which need information about temperature. Data source, the IoT device, generates data samples as float number. One of service requirement is to convert float numbers. Of course data types usually are more complex, such as multimedia codecs. Often it happens that format conversion between two services is needed and it is typically provided by the service execution engine.

In "smart" systems application designing does not end on services start. As the services are working, different events may appear. Application has to be prepared to handle this events. In basic case we only need to execute complex service prepared for another scenario. But the common types of events are the ones which came after service failures. In those cases it could be necessary to redesign the application. Reselecting of atomic service usually is not complicated, but when the decision making process has to be rebuild, it could involve re-checking of functionality requirements and/or the repetition of composition process [10].

4   IoT Application Architecture

IoT architecture was widely discussed by specialists in one of the European projects [6]. In our work we are trying to be coherent with presented in IOT-A assumptions. In this paper the greatest attention is devoted to Functional model presented by IOT-A project (Fig. 4).

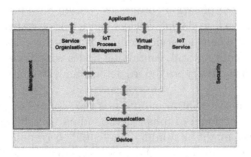

Fig. 4. IoT Functional Model [6]

This model contains nine Functionality Groups (FG) and their interactions. Application FG is a result of the IoT application designing process presented in previous section. As it can be seen in this model, Application FG interacts with four FGs. Functionalities from this FGs are necessary to design and provide IoT applications.

The Service Organisation FG contains functionalities which are required to describe reality and process its description for decision making process description. It's also responsible for formal description of this process. The functionalities of this FG provides as well information about elements of the reality defined as Virtual Entities.

Virtual Entity FG is one of the most interesting. This FG consists of virtual representations of reality elements. In IoT application designing process which was presented in previous section we do not differentiate reality and it's virtual representation - in IoT system instance it is necessary to present reality in the computer understandable form. In IoT application designing process, some parts of Virtual Entities descriptions are passed from reality description to the next steps.

In the SOA context the most interesting is IoT Service FG and IoT Process Management FG. Generally the first one corresponds to services concept, the second one to composition process, especially representing the processes occurring in reality in the form of complex services. In IoT application designing process task selection and functionality selection are inputs to service composition. This steps are related to the IoT Process Management FG. Of course the importance of services descriptions (IoT Service FG) for composition should not be forgotten.

For IoT application designing process another important FG is Management. The last step of our process is services execution. During this step, atomic services are selected with their parameters being set. These decisions have an effect on proper work of IoT systems, inter alia, QoS assurance, resource management, but also on reducing the cost of services [11].

The lowest FG, Communication, correlates to the IoT device which was mentioned above. IoT devices are the basic sources of data and provide a basis for connecting components of the IoT systems. IoT conception rely on possibility of connecting different devices which are integrated with the environment and reality. IoT devices model will be widely discussed in the next section.

5   IoT Device Model

The idea of concentrating device network traffic is well known in the literature of telecommunications [12]. The first telecommunications traffic concentrators were used in systems with circuit switched calls (switched circuit). Using concentrating device traffic provides tremendous advantage. It helps in dimensioning and modeling network traffic [13]. As described by the authors in the introduction to this article, IoT network is classified as a wide area network. At the same time "Things" are able to communicate in a network by means of different network interfaces. This approach implies that the problems known from the literature related to the sensory systems and methods of modelling traffic in computer and telecommunications networks, are still actual and should be further considered in the context of the IoT [14].

In this article the idea of IoT network traffic concentrator is proposed, which, as it could be seen is much more different than pursued switching circuit. IoT concentrator device is designed to connect two network segments IoT. These segments can be named arbitrarily as local and global segments. Local devices segments are herein understood as closely associated with the detectors, sensors, actuators, cameras and other devices designed to transmit signals. On the other hand, global segment is associated with information processing of digital data and sending them to the IoT network system (Fig. 5). Every concentrator is designed to send control information to the local segment of the IoT network in order to control actuators. The idea of IoT concentrator relies on the device which, by definition, is able to have multiple inputs and outputs to both network segments [15]. This device is designed to integrate the devices connected into it as services with specific parameters. To be confident that all of the functionality of the device acting as the concentrator, it is required to have a communications interfaces with equipment which is on the local segment and from the upper layer applications in a global segment IoT network. IoT concentrator should have the functionality associated with the adaptation of their digital circuits for applications performing digital signal processing. The authors of this article propose the application of the idea of Universal Hardware Platform (UHP) which is able to adapt appropriately to the requirements which will be imposed via the communication interfaces and services. The idea of this solution is shown below (Fig. 5). UHP hardware implementation is now possible due to the increasing of technological advances in the area of programmable FPGA circuits. Today, single chips containing huge numbers of transistors and capable of building virtually any digital circuit structures are available to the designer.

The concentrator as a device should have the ability of handling multiple data formats and communication interfaces. A multitude of acquired interfaces provides the concentrator with enormous opportunities allowing the integration of multiple devices in a simultaneous method. At this stage device designer

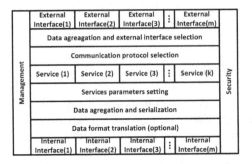

Fig. 5. IoT hardware device architecture – UHP (Universal Hardware Platform)

should be aware that some of the equipment have a fully standardized data formats related to communication in form of simple frames of communication such as RS232, RS485, SPI, I2C etc. While other equipment that have only general recommendations with regard to the format of the frames and signalling must be properly integrated into the concentrator. Specification IEEE 802.15.4. does not define exactly one particular communication protocol. Therefore, the device has two ways of aggregating data. One of them is called directly from the interfaces, assuming that the concentrator knows the data format. The second is a translator function, which is a dedicated function for communication interfaces. The realization of the interfaces and data translation functions may be different - from soft (simple computer program or a shell script), to hard (microprocessor or FPGA programmable digital circuit which performs the functions of translation). Both could come in a form of mixed implementation which could be provided by a person responsible for designing that part of communication. The idea being proposed by the authors assumes that most of standardized methods of communication would be implemented in the form of dedicated microcontrollers. With additional functions related to the implementation of additional digital circuits will be provided in FPGA. The authors have in mind the new innovative methods of communication which are not implemented at the time of design UHP. UHP functionality allows deploying them to the IoT network without replacing the concentrator device. Moreover UHP architecture can be used as a reference to description of any IoT device. It is possible by selecting needed elements from IoT device model presented in the Fig. 5.

The idea that the IoT is different from the standard Internet, so concentrator have to perform an operation of translation and integration data being collected from the local segment equipment and have the ability to control the actuators placed on the local segment by running the corresponding services in the higher layers of the network IoT. The assumption is being made, that the device performing the functions of concentrator should meet the idea of "smart' network. UHP device is able to identify peripherals connected via the internal interfaces. Concentrator has the ability to make devices connected to the local segment of the services at higher network layers IoT. These services are parametrized in a way to reflect the character of environmental data collected (concerning detectors and sensor) and a method for communicating with the device responsible for the control (concerning the actuators). The services should have described parameters such as: (minimum delay, bit rate, etc..). There should be fully established description and characteristics of the internal interface and the appropriate selection of the external interface that will implement suitable demand for bandwidth. The authors define this concept as an appropriate level of QoS and QoE. One of the main tasks of the concentrator is to provide a high level of reliability of the data transmission on the IoT network. Such as a modeling approach allows using accurate IoT traffic models and gives a better way of managing it.

6 Conclusion

This paper was devoted to problem of IoT application designing. The solution was presented, which proposes a process definition based on virtual representation of process functions and model of reality. Proper reality description consistent with the description of services being used and basing on the common domain ontology allows to compose a IoT application which fits the requirements. Following steps in this process are: transformation from reality model to formalized structure of processes description, services composition based on required functionality and service execution with non-functional parameters setting. Referring to the issues of the importance of hardware in IoT systems an IoT device model was discussed. Service composition needs a detailed description of available services and it is another reason of focus given to IoT device model.

IoT application designing approach presented in this paper is only an introduction to IoT service systems. Other important subjects are communication and quality assurance. Abovementioned issues of designing methods, especially consideration devoted to describing reality, processes in the reality, software and hardware based on services are the key elements for the proposed approach, while another important subjects like communication and quality assurance will be addressed in further research.

Acknowledgements. The research presented in this paper was partially supported by the Polish Ministry of Science and Higher Education and the European Union within the European Regional Development Fund, Grant No. POIG.01.03.01-02-079/12 and within European Social Fund.

References

1. Miorandi, D., Sicari, S., Pellegrini, F.D., Chlamtac, I.: Internet of things: Vision, applications and research challenges. Ad Hoc Networks 10(7), 1497–1516 (2012)
2. Lewandowski, J., Arochena, H., Naguib, R., Chao, K.M.,, G.-P.: Logic-centered architecture for ubiquitous health monitoring. IEEE Journal of Biomedical and Health Informatics 18(5), 1525–1532 (2014)

3. Brzostowski, K., Drapała, J., Grzech, A., Świątek, P.: Adaptive decision support system for automatic physical effort plan generation-data-driven approach. Cybernetics and Systems 44(2-3), 204–221 (2013)
4. Świątek, P., Stelmach, P., Prusiewicz, A., Juszczyszyn, K.: Service composition in knowledge-based soa systems. New Generation Computing 30(2-3), 165–188 (2012)
5. Falas, Ł., Stelmach, P.: Web service composition with uncertain non-functional parameters. In: Camarinha-Matos, L.M., Tomic, S., Graça, P. (eds.) DoCEIS 2013. IFIP AICT, vol. 394, pp. 45–52. Springer, Heidelberg (2013)
6. Carrez, F., Bauer, M., Boussard, M., Bui, N., Carrez, F., Jardak, C., Loof, J.D., Magerkurth, C., Meissner, S., Nettstrater, A., Olivereau, A., Thoma, M., Walewski, J.W., Stefa, J., Salinas, A.: Deliverable D1.5 – Final architectural reference model for the IoT v3.0 (July 2013)
7. Groba, C.: Opportunistic service composition in dynamic ad hoc environments. IEEE Transactions on Services Computing 99(PrePrints), 1 (2014)
8. Świątek, P., Klukowski, P., Brzostowski, K., Drapała, J.: Application of wearable smart system to support physical activity. In: Advances in Knowledge-based and Intelligent Information and Engineering Systems, pp. 1418–1427 (2012)
9. Stelmach, P., Schauer, P., Kokot, A., Demkiewicz, M.: Universal platform for composite data stream processing services management. In: Zamojski, W., Mazurkiewicz, J., Sugier, J., Walkowiak, T., Kacprzyk, J. (eds.) New Results in Dependability & Comput. Syst. AISC, vol. 224, pp. 399–407. Springer, Heidelberg (2013)
10. Juszczyszyn, K., Swiatek, P., Stelmach, P., Grzech, A.: A configurable service-based framework for composition, delivery and evaluation of composite web services in distributed qos-aware ICT environment. IJCC 2(2/3), 258–272 (2013)
11. Grzech, A., Świątek, P., Rygielski, P.: Dynamic resources allocation for delivery of personalized services. In: Cellary, W., Estevez, E. (eds.) Software Services for e-World. IFIP AICT, vol. 341, pp. 17–28. Springer, Heidelberg (2010)
12. Habrych, M., Staniec, K., Rutecki, K., Miedzinski, B.: Multi-technological transmission platform for a wide-area sensor network. Elektronika ir Elektrotechnika 19(1), 93–98 (2013)
13. Miedzinski, B., Rutecki, K., Habrych, M.: Autonomous monitoring system of environment conditions. Electronics and Electrical Engineering.–Kaunas: Technologija (5), 101 (2010)
14. Staniec, K., Debita, G.: Interference mitigation in wsn by means of directional antennas and duty cycle control. Wireless Communications and Mobile Computing 12(16), 1481–1492 (2012)
15. Staniec, K., Debita, G.: An optimal sink nodes number estimation for improving the energetic efficiency in wireless sensor networks. Elektronika ir Elektrotechnika 19(8), 115–118 (2013)

An IoT Based Service System
as a Research and Educational Platform

Thomas Kimsey[1], Jason Jeffords[1,3], Yassi Moghaddam[2], and Andrzej Rucinski[1]

[1] University of New Hampshire, Durham, New Hampshire, United States of America
tpt22@wildcats.unh.edu, andrzej.rucinski@unh.edu
[2] International Society of Service Innovation Professionals,
San Jose, California, United States of America
yassi@issip.org
[3] Deep Information Sciences Inc., Portsmouth,
New Hampshire, United States of America
jason@deep.is

Abstract. The Internet of Things (IoT) is a rapidly expanding space for researchers, students, and professionals. Development of IoT applications is hindered by the lack of available design tools, methodologies, and interoperability. The creation of a standards-based, open source hardware and software platform paired with full documentation and educational materials greatly aids in the accelerated development of IoT applications and provides a fertile ground for Science Technology Engineering and Math (STEM) educational opportunities.

Keywords: IoT, Service Science, Internet Architecture, M2M, MQTT.

1 Introduction

Low power Internet enabled embedded systems used in sensor and/or control networks belong to the emerging Internet of Things space. The physical IoT can be defined as a collection of objects connected to an Internet like structure (Fig. 1). An Internet connected weather-monitoring station or an Internet enabled air conditioner are both examples of IoT devices. In conjunction with the growth of Internet-enabled devices, the number of IOT applications is expanding rapidly. However, there is no clear definition of how these systems can be integrated into existing infrastructure and networks.

Since these low power devices cannot run many of the processor intensive protocols standard Internet applications use, new protocols, interworking functions, and architectures must be defined. Often researchers and engineers developing IoT applications must define their own network structure before working on the problem they originally set out to solve. For example, in order to make a prototype solution for data collection and control, how should the architecture of IoT networks be laid out in order to provide required functionality, good network performance, and scalability?

© Springer International Publishing Switzerland 2015
D. Barbucha et al. (eds.), *New Trends in Intelligent Information and Database Systems,*
Studies in Computational Intelligence 598, DOI: 10.1007/978-3-319-16211-9_26

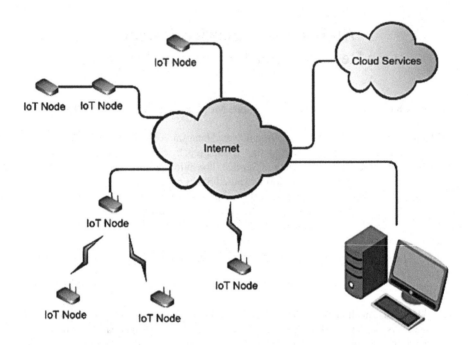

Fig. 1. Example IoT Network

2   Overview

The goal of this paper is to investigate software, hardware, and standards that are applicable to creating a generalized prototype IoT solution. Since the prototype needs to work for many different kinds of applications, a wide range of technologies will be considered. The product of the investigation will be a list of compatible hardware, software, and standards ease and accelerate the development of distributed IoT systems.

Before describing how such a system should be built, further investigation must be done to define what kind of things will be connected to the Internet. The "things" in "Internet of Things" refers to a thing's information and not the thing itself [1]. So any object or "thing" that has some variable data associated with it can be one of the objects associated with the Internet. The effect of this is that a prototype definition must be mindful that an IoT application can take any shape and form.

What does an IoT device look like? Does an IoT node require user input or is it autonomous? The answer is that an IoT node can take many forms and has no explicit declaration of how humans interact with it or how it interacts with the environment [2], [4] (Fig. 2). The only linking factor is their interface to the Internet. In order to accommodate a broad usage description, a re-usable platform solution must be very general, easily allowing for any possible use case [5]. However, if the solution is too general it sacrifices its usefulness. The challenge is to create a system that is both general enough to complement any use, but functional enough to provide a fast and high performance implementation of an IoT node and its associated services.

IOT Nodes

Fig. 2. Example IoT Nodes

One issue facing the expansion of IoT technology is the aging internet and related technologies. Key technologies of the Internet were created in the 1970s and while there have been significant advances and upgrades over the years large portions of the Internet were not built for the current usage or upcomming scale (e.g. IPv4 address exhaustion). As a result, many elements of a large scale IoT system, such as addressing and data storage, will be limited if traditional methods are used [3]. The data collected in IoT systems is collected on nodes and sent to a central database. This data transfer is needed because the data is more valuable when collected in a central location and analyzed as a set rather than individual points [8].

Data rates for generalized IoT systems vary considerably and are constantly increasing. Due to their nature it would not be abnormal to see a system generating Gigabytes of data per second. Traditional database storage engines, such as MySQL's InnoDB, cannot handle the influx of data from a large distributed IoT network [6]. To ensure reliability, performance, and scalability of data storage, a modern storage engine must be selected and used [7]. Using a database that is designed to work better than traditional databases, such as MySQL using InnoDB, with today's network storage, the scaling capabilities of an IoT system can be improved.

3   Network Structure

The physical layout of IoT systems is straightforward. There are a number of network nodes connected to the Internet. The connection from the node to the Internet can be wired or wireless. Also included in IoT networks are cloud services such as data servers, data routing devices, and web servers. The cloud services are responsible for processing and storing data. Lastly, there are user interfaces to the network such as a personal computer.

The largest challenge in creating IoT systems is design and layout of the logical, application layer network topology. The challenges facing IoT networks include the variability of the number of nodes, the data sources and data users are unknown, and node resources are limited. The network has to be able to deal with high volumes of nodes joining and leaving the network, keeping all of the protocols lightweight, and distributing data efficiently without sending data where it is unneeded.

A simple approach to IoT network topology design is a point-to-point connection from each node to the data server (Fig 3). Data is sent between the node and the server via independent TCP/IP connections.

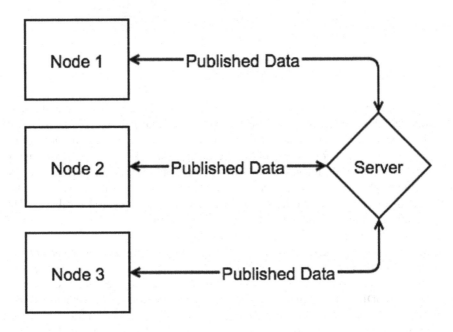

Fig. 3. Simple Point-to-Point Network Topology

The downside of this network topology is keeping track of nodes on the network due to each node's independent connection with the server and the lack of provisions for inter-node communication.

To solve the issue of nodes not being able to communicate with each other, a broadcast network topology could be implemented (Fig. 4). In a broadcast network topology the server forwards data collected from one node to all nodes. In this way nodes are aware of the actions and data from other nodes on the network.

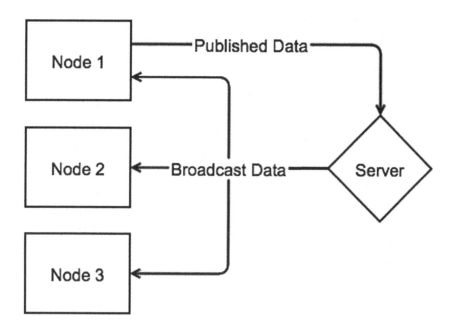

Fig. 4. Broadcast Network Topology

In a broadcast topology each node processes a significant amount of information because every node receives all of the information sent to the server, yet not all of the data is useful to every node. There is also a very high use of network bandwidth since every packet sent to the server gets sent to every node on the network. These two facts make a broadcast network infeasible for medium to large-scale IoT networks.

In order to maintain minimal network traffic and support internode data transfer a publish/subscribe system is implemented (Fig. 5). In a publish/subscribe network nodes publish data under topic identifiers. Nodes interested in receiving certain types of data subscribe to specific topics. When data is published to a server a message broker handles the forwarding of data to only those nodes that are subscribed to the topic the data was published under. To illustrate this concept, consider a network consisting of three nodes and a message broker. Node # 1 publishes data under the topic "Node1/Sensor1" to the message broker. Only Node #3 is subscribed to the topic "Node1/Sensor1" therefore only Node #3 is sent the data by message broker.

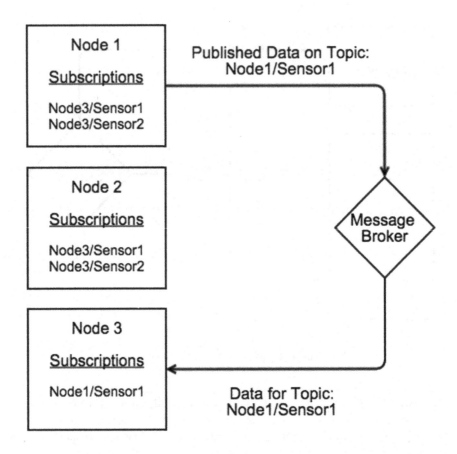

Fig. 5. Publish/Subscribe Network Topology

This system minimizes both network traffic and on-node processing making it an ideal solution for IoT networks.

The key to building a scalable application network topology is to allow nodes to join and leave the network dynamically. This means that all connections are created during run time and as little information as possible about network addresses and resources are stored in firmware. The main building block that enables this type of network is the protocol used. For this system the protocol selected is MQTT, which is a lightweight publish/subscribe messaging protocol.

MQTT is based on a packet format with each data packet having a topic. Data sources can publish data to a topic and data users can subscribe to a topic. The act of subscribing to a topic allows for messages to only be sent to network entities that require that data. Each IOT node on the network communicates with a central MQTT broker. The central broker is responsible for accepting messages from nodes and passing them along to subscribers.

4   Hardware

Open source hardware has been created to aid in the development of IoT systems. The foundation of the hardware is built on a Cypress Programmable System on Chip or PSoC®. The PSoC® was chosen because it is an easy to use development and prototyping platform and it is natively a mixed signal device meaning it handles both analog and digital signals well. PSoC® based systems are especially useful for data collection and control applications. A common problem with the development boards made by Cypress, as well as similar boards, is the lack of a straightforward way to connect to the Internet. To solve this we have developed expansion boards to enable network connectivity.

These expansion boards are open source designs that plug into the expansion header ports on PSoC® development boards. The standard method of connection chosen between a node and the Internet is Ethernet. Ethernet connection provides a stable link and is widely used. The hardware chosen is the WIZnet W5100 Ethernet controller. For point-to-point wireless operation the XBee Pro module produced by Digi International was chosen. These modules are 3.3 volt devices with an output power of 50 mW. Data rates can be configured up to 156 Kilobits per second. The best-case range with these devices is 10 kilometers line of sight with high gain directional antennas. A cellular modem expansion board can be used to provide Internet access in cases where a node is very remote.

This solution is not limited to PSoC® and these expansion boards. Since the designs are lightweight and scalable, the principles and protocols used in this solution can be applied to any IoT node. Implementations could include Arduino, Raspberry Pi, BeagleBone, or any other embedded platform. It can even be easily applied to PC and server based nodes.

These hardware pieces represent building blocks to physical IoT networks. The end result is that this platform is highly scalable, to the point that the only limit to the size of the IoT network is the number of IP addresses available.

5   Educational Uses

The system described in this paper has several educational uses. Firstly the system can be used to demonstrate various engineering topics such as IoT, embedded systems, network communications, and database design. By analyzing and using the system in a classroom or lab setting students gain a hands on connection to a system that spans and integrates several topics. The system can also be used in a project-oriented class where the goal is to develop a full application. Students can use this platform as a building block to create their own application. To demonstrate how a student would use this solution let us consider a student who wants to implement a web based weather station. The student has the sensor package they wish to use and knows that they want the information to end up in a database. The student is able to take the networking hardware discussed earlier, write their application code, and integrate it with the API for the network expansion boards. This allows the student to implement their design quickly and end up with data published to the web easily. Since the network is scalable, adding new sensors and new weather stations does not require redesigning the system. This adds value since now the student's project is highly scalable and nodes can be added and removed on the fly.

At The University of New Hampshire this IoT platform has had significant exposure. It has been presented in Electrical and Computer Engineering (ECE) 401, ECE 562, and ECE 583 and has been used by students to complete their final projects in ECE 711 and ECE 777. This IoT platform is also being used at UNH outside of the classroom. The Programmable Microelectronics Club (A school organization aimed at fostering student driven embedded system development) has been using the platform to educate students on the architecture and uses of an IoT development system. Many of these projects completed by students can be categorized as a smart home/city/state/planet type project, with the goal of creating an IoT application to make human life easier and more productive on a local or global scale. By utilizing this platform both in and out of class, students at The University of New Hampshire have been learning about all aspects of how IoT applications can be built and the issues facing IoT developers.

6   Conclusion

The IoT space is currently expanding rapidly. Many students and researchers are looking to build IoT applications, however, these students and researchers are facing challenges. Laying out the network structure, designing the network hardware, and writing the application code to handle network protocols are tasks that are faced every time an IoT application is developed. Most current solutions offer either a high level solution or a node level solution. For example, the Cisco Internet of Everything covers the high level design including how IoT applications can help tackle challenges facing the world. However, Cisco's solution does not include low-level hardware solutions for node level devices. Another popular solution for IoT development is Arduino. In contrast to Cisco's solution, Arduino provides a node level hardware solution but no high level system solution. The system described in this paper is unique in the fact that it provides both a high level solution including server and network layouts, and hardware and firmware designs for node level development. Having a collection of reusable, open source, predefined structures will facilitate the development of future IoT applications. Lifting the burden of the hardware and networking side of IoT development will foster accelerated growth in the IoT space.

References

1. Huang, Y., Li, G.: Descriptive Models for Internet of Things. In: 2010 International Conference Intelligent Control and Information Processing (ICICIP), pp. 483–486 (2010)
2. Stefanov, R.: Security and Trust in IoT/M2M – Cloud Based Platform. Master's Thesis. Aalborg University Department of Electronic Systems (2013)

3. Tan, L., Wang, N.: Future Internet: The Internet of Things. In: 2010 3rd International Conference on Advanced Computer Theory and Engineering (ICACTE), vol. 5, pp. V5-376–V5-380 (2010)
4. Kranz, M., Holleis, P., Schmidt, A.: Embedded Interaction: Interacting with the Internet of Things. IEEE Internet Computing 14(2), 46–53 (2010)
5. Yashiro, T., Kobayashi, S., Koshizuka, N., Sakamura, K.: An Internet of Things (IoT) Architecture for Embedded Appliances. In: 2013 IEEE Region 10 Humanitarian Technology Conference (R10-HTC), pp. 314–319 (2013), doi:10.1109/R10-HTC.2013.6669062
6. Li, T., Liu, Y., Tian, Y., Shen, S., Mao, W.: A Storage Solution for Massive IoT Data Based on NoSQL. In: 2012 IEEE International Conference Green Computing and Communications (GreenCom), pp. 50–57 (2012), doi:10.1109/GreenCom.2012.18
7. Duan, R., Chen, X., Xing, T.: A QoS Architecture for IOT. Internet of Things (iThings/CPSCom). In: 2011 International Conference on and 4th International Conference on Cyber, Physical and Social Computing, pp. 717–720 (2011)
8. Ding, Z., Gao, X., Xu, J., Wu, H.: IOT-StatisticDB: A General Statistical Database Cluster Mechanism for Big Data Analysis in the Internet of Things. In: 2013 IEEE and Internet of Things (iThings/CPSCom), IEEE International Conference on and IEEE Cyber, Physical and Social Computing Green Computing and Communications (GreenCom), pp. 535–543 (2013)

Part V

Knowledge and Language Processing

Virtual Engineering Objects: Effective Way of Knowledge Representation and Decision Making

Syed Imran Shafiq[1], Cesar Sanin[1], Edward Szczerbicki[2], and Carlos Toro[3]

[1] The University of Newcastle, University Drive, Callaghan, 2308, NSW, Australia
{syedimran.shafiq,Cesar.Maldonadosanin}@uon.edu.au
[2] Gdansk University of Technology, Gdansk, Poland
Edward.Szczerbicki@zie.pg.gda.pl
[3] Vicomtech-IK4, San Sebastian, Spain
ctoro@vicomtech.org

Abstract. This paper presents a knowledge representation case study by constructing Decisional DNA of engineering objects. Decisional DNA, as a knowledge representation structure not only offers great possibilities on gathering explicit knowledge of formal decision events but also it is a powerful tool for decision-making process. The concept of Virtual engineering Object (VEO), which is a knowledge and experience representation of engineering artefacts, is also discussed. In this paper, we present several Sets of Experience of engineering objects used in manufacturing that were collected for the construction of a VEO-chromosome within the VEO-Decisional DNA. VEO is used to enhance manufacturing systems with predicting capabilities, facilitating decision-making in engineering processes knowledge handling.

Keywords: Virtual Engineering Objects (VEO), Decisional DNA, Knowledge Representation, Set of Experience Knowledge Structure (SOEKS).

1 Introduction

In recent years, manufacturing organizations have been facing market changes such as technological advances and the need for short Product Life Cycles [1]. There is also intense pressure from competitors and customer expectation for high-quality products at lower costs. The role of knowledge in industrial design engineering/management has therefore become increasingly important for manufacturing companies to take effective decisions. Knowledge-based industrial design and manufacturing techniques have been used in the past with considerable success. However, they have limitations such as being time-consuming, costly, domain-specific, unreliable in their intelligence, and unable to take previous experience into account [2-4].

Furthermore there is a lot of interest globally in 'Industry 4.0', which is termed as the fourth industrial revolution —following the steam engine, the conveyor belt, and the first phase of IT and automation technology [5]. Industry 4.0 is a powerful concept, which promotes the computerization of traditional manufacturing plants and their ecosystems towards a connected and 24/7 available resources handling scheme through the use of

© Springer International Publishing Switzerland 2015
D. Barbucha et al. (eds.), *New Trends in Intelligent Information and Database Systems*,
Studies in Computational Intelligence 598, DOI: 10.1007/978-3-319-16211-9_27

Cyber Physical Systems (CPS) [6]. The goal is the intelligent factory, which is characterized by adaptability, resource efficiency and ergonomics as well as the integration of customers and business partners in business and value processes [5, 7].

Cyber Physical System (CPS) is emerging as a must have technology needed by industry [8, 9]. CPS is integrations of computation with physical processes [9, 10]. Embedded computers and networks monitor and control the physical processes, usually with feedback loops where physical processes affect computations and vice versa. In the physical world, the passage of time is inexorable and concurrency is intrinsic. Neither of these properties is present in today's computing and networking abstractions [9]. CPS aims to integrate knowledge and engineering principles across the computational and engineering disciplines (networking, control, software, human interaction, learning theory, as well as electrical, mechanical, chemical, biomedical, material science, and other engineering disciplines) to develop new CPS science and supporting technology

In manufacturing, the potential for cyber-physical systems to improve productivity in the production process is vast. Consider processes that govern themselves, where smart products can take corrective action to avoid damages and where individual parts are automatically replenished. Scalable CPS architectures for adaptive and smart manufacturing systems which dynamically enable the continuous design, configuration, monitoring and maintenance of operational capability, quality, and efficiency are, in fact, a requirement for the industry [11]. According to the European commission under the Horizons 2020 programme, the self-learning closing feedback loop between production and design should be included in future factories for optimizing energy expenditure and minimizing waste as a direct relation to the enhancement in control and immediate information processing that a CPS will provide.

The Internet of Things will make a new wave of technological changes that will decentralize production control and trigger a paradigm shift in manufacturing. It is highly likely that the world of production will become more and more networked until everything is interlinked with everything else.

Considering all these factors, this research proposes a novel approach to provide engineering artefacts with an experience-based representation. We introduce the concept of 'Virtual Engineering Object' (VEO), which permits dual computerized/real-world representation of an engineering artefact [12, 13]. VEO is a specialization of Cyber-Physical System (CPS) in terms of that its extension in knowledge gathering and reuse, whereas CPS is only aimed towards data and information management. A VEO is then a generalization of a CPS that by its conceptualization will fall into a homogenous line of thinking aligned to our previous presented work [14, 15].

The concept of Virtual engineering Object uses a standard knowledge representation technique called Set of Experience Knowledge Structure (SOEKS), which comprises Decisional DNA (DDNA)[13] [16]. Decisional DNA is proposed as a unique and single structure for capturing, storing, improving and reusing decisional experience. Its name is a metaphor related to human DNA, and the way it transmits genetic information among individuals through time.

Figure1 summarizes the whole idea of this work, in an industrial domain; manufacturing systems consist of various machines which in turns uses different objects for their working. Virtual/knowledge representation of engineering objects, machines and manufacturing systems will be beneficial in asset, machine and entire system optimization respectively. Intelligent decisions can be made based on intelligent virtual objects and systems.

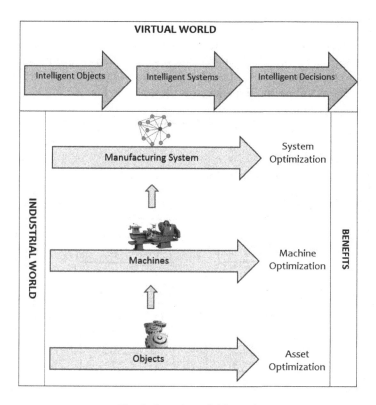

Fig. 1. Overview of this work

The structure of this paper is as follows: section 2 presents the conceptual background and architecture of virtual engineering object. Section 3 describes the case study and proposal to implement the concept of VEO. Section 4 presents the experiments performed and their results. Finally, section 5 presents the conclusions and future work.

2   Virtual Engineering Object (VEO)

A VEO is knowledge representation of an engineering artefact, it has three features: (i) the embedding of the decisional model expressed by the set of experience, (ii) a geometric representation, and (iii) the necessary means to relate such virtualization with the physical object being represented.

A VEO acts as the object's living representation, capable of capturing, adding, storing, improving, sharing and reusing knowledge through experience, in a way similar to an expert in that object [15].

2.1   Architecture of VEO

A VEO can encapsulate knowledge and experience of every important feature related with an engineering object. This can be achieved by gathering information from six different aspects of an object viz. Characteristics, Functionality, Requirements, Connections, Present State and Experience as illustrated in Fig. 2.

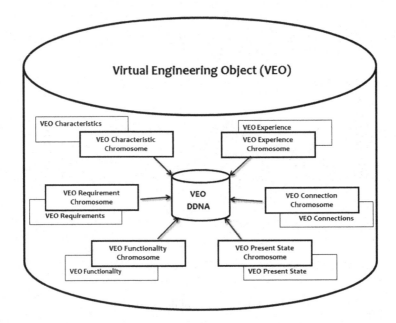

Fig. 2. VEO Structure

The main features of a VEO (shown in Fig. 2) are as follows:

Characteristics describe the set of physical features and expected benefits offered by the artefact represented by the VEO. Not only the information like its geometry dimensions, appearance, weight etc. are captured in this module but also the possible advantages like 'versatility' and the 'ease of operation' can also be achieved from this. Knowledge stored in Characteristics assists in better decision making like which VEO is best suited for a given physical condition and also when more than one VEO of a similar kind are available it helps to decide which is the best in the given situation.

Functionality describes the basic working of the VEO and principle on which it accomplishes its operation. Knowledge related with the functioning and operation of an object like the time consumed, its working boundary limits and the outcome of the process that is performed are stored in Functionality. This module of the VEO assists in storing, selecting and reusing the functional/operational details of the object.

Requirements describe the set of necessities of the VEO for its precise working. Information like type and range of the power source, the space required and the extent of user expertise necessary for operating a VEO can be stored.

Connections describe how the VEO is related with other VEOs. Many engineering objects work in conjunction with other objects. These connecting VEOs may be a "part" or may be a "need" of each other. This module of VEO structure is essential for the scaling up and establishing the interconnections of VEOs in manufacturing scenario.

The Present State of the VEO highlights parameters of the VEO at the current moment. It is like taking a picture and storing information of that particular moment. It also gives an idea about the background of the VEO like its 'reliability' and 'precision' up till now.

The Experience of the VEO stores the information of each decision, operation or event related to the object. This knowledge is dynamic in nature as it changes with each operation. In other words every formal decision related to the VEO is stored in the Experience. This element of the VEO keeps on updating with every activity that is done on the VEO [13].

2.2 Implementation of VEO

For the purpose of implementation of VEO, we integrated it with the Decisional DNA. SOEKS consists of Variables, Functions, Constraints and Rule. Moreover in section 2.1, we also discussed that a VEO structure include elements like Characteristics, Functionality, Requirements, Connections, Present State and Experience. SOEKS for each element of the VEO in the system are created individually. The goal behind this was to provide a more scalable setting, similar to the one that would be found in de-scribing a diverse range of engineering objects. Weights are assigned to the attributes of the variables of an artefact, and then the six sets of SOEKS are generated. These individual SOEKS are combined under an umbrella (VEO), representing experience and knowledge.

3   Case Study

As a case study, we considered a manufacturing set up having three different drilling machines, drilling tools and work holding devices each. Fig. 3 shows the framework for the case study, information and specifications about these above mentioned engineering objects were gathered from standard sources and data is stored according to the SOEKS format. Moreover, every formal decision taken is also stored as a SOE, which leads to the formation of interconnected VEO's.

The objective of this study is not only to develop VEO's for engineering artefacts but also demonstrate that different VEO's connect and forms a network. Furthermore to prove that the experience captured from this VEO network can be reused for better future decision making and efficient utilization of resources (Fig. 3). Effort is made to capture and store all the relevant information of the VEO adhering to the format of the SOEKS.

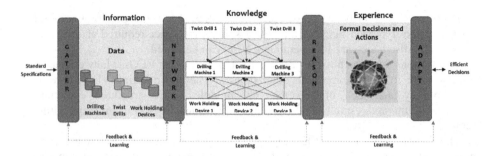

Fig. 3. Framework for case study

First all the necessary information associated to drilling machine was identified and arranged in a standard SOEKS format. CSV files for Characteristics, Functionality, requirements, Present state, Connections and experience were built. Table 1 shows structure of a CSV file for Characteristics. In the similar format other files are developed each representing a category of VEO.

Table 1. CSV file format for VEO Experience

Variables															
OpNo	veo1	veo2	veo3	WPMaterial	WPDepth	Operation	Drill/ToolDia	Feed	CuttingSpeed	DD-F-S	SpindleSpeed	SrfceRough	CoolentUsed	Breakdown	MachiningTime	Date
1	DM1	T1	H1	MS	47	Drilling	5	0.7	760	5-0.7-760	2297	SMOOTH	YES	NO	1.753840413	30/06/2012
2	DM1	T1	H1	MS	48	Drilling	5	0.5	570	5-0.5-570	819	ROUGH	YES	NO	5.023547881	14/08/2012
2186	DM3	T3	H3	CI	49	Boring	30	0.35	90	30-.35-90	2639	ROUGH	YES	NO	1.591511936	23/01/2010
2187	DM3	T3	H3	CI	53	Boring	30	0.3	75	30-.30-75	1081	ROUGH	YES	NO	4.202458041	28/11/2010

Constraints
veo1 = {DM1, DM2, DM3}
veo2 = {T1, T2, T3}
veo3 = {H1, H2, H3}
WPMat = {MS, HSS, CI}
SurfaceFinish = {SMOOTH, ROUGH}
CoolentUsed = {YES, NO}
Breakdown = {YES, No}

Functions
MachiningTime = (60 * SpindleSpeed) / (3.24 * DepthOfHole)

To read these files, a parser was written in Java programming language. Figure 4 shows the simplified JAVA class diagram; parser (veoParserCSV) looks for CSV file, in that file it look for the word 'variables' (table 1), then starts reading (readVariables) the first row under variables. Once all the variables of the first row are read then the parser looks for the word 'functions' (table 1), it reads (readFunctions) all the rows under functions. After that it looks for word 'constraints' (table 1) and read (read-Constraints) all the rows under constraints. This entire information i.e. first row under 'variables', all rows under 'functions' and 'constraints' are stored as one set of experience (SOE). This cycle is repeated for all rows under 'variables', for each row along with functions and constraints, SOEKS are created.

Same procedure is repeated for the all the CSV files. Each file representing a category, collection of SOEKS of same category forms a chromosome of VEO. Collection of different chromosomes forms a Decisional DNA of a VEO. Once the VEO chromosome was constructed, decisional DNA has feature that it can queried [17].

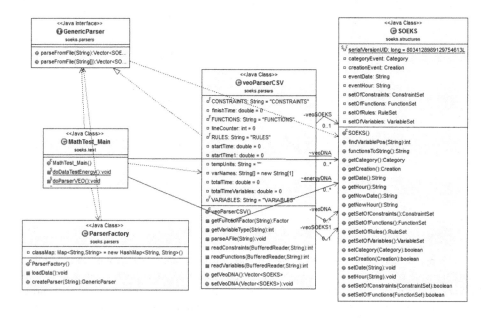

Fig. 4. JAVA Class Diagram for VEO Parser

The VEO Decisional DNA comprised Sets of Experience (SOE) from Characteristics, Functionality, Requirements, Present state, Connections and Experience, having 53 variables, 3 functions and 28 constraints. For testing purposes, we query from the repository of 2256 SOEKS.

A query is used to exemplify this case study. The query is defined in the code as:

```
public static String SIMPLE_QUERY="CLASS variable with the
PROPERTY var_workpieceMaterial EQUALS to HSS &&
var_drillDiameter=15";
```

Notice that this is a value type query. Such query is written in a human-like readable form which means "retrieve all the variables that have the variable name Work piece Material=HSS, Drill Hole Diameter = 15cm". The decisional DNA looks for these two variables in all the SOEKS it has in the memory and returns collection of SOE having both the variables. After all the experiences in the batch are translated using the ideas described before, the inference process can executed to discover new rules according to the categories described above. Then, assuming the existence of more knowledge in the system, under specific conditions and after validation against other experiences, the inference process is able to determine that the values obtained.

The results are shown below.

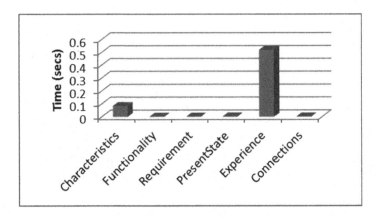

Fig. 5. Time vs VEO Elements

The parsing process of the VEO decisional chromosome was executed, producing a parsing time of 703.0 ms as it can be observed from fig. 5 (with experience CSV file taking 509.0 ms). This is considered a very good time taking into account that those SOE are quite complex due to the amount of variables, functions and constraints, adding up to a total of 84 key features per formal decision event.

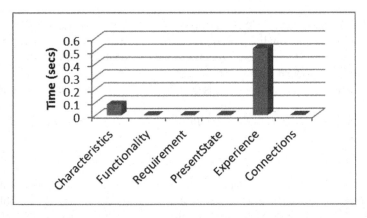

Fig. 6. Time vs SOEKS Elements

The detailed parsing process of the VEO decisional chromosome produced an average parsing time per SOE of 228.0 ms (Variables = 537.0, Functions = 67.0 ms, Constraints = 80.0 ms). It can be noticed from fig. 6 that most of the time is dedicated to the variables due to their large numbers for knowledge construction.

4   Conclusions and Future Work

The main contribution of the work presented throughout this article is the formal definition of mechanisms to implementation of knowledge engineering in the manufacturing field. It explains a practical case of collecting research experience from engineering artefacts, and then, use this information for the construction VEO Decisional DNA. Decisional DNA and the Set of Experience were applied as a knowledge representation structure for gathering the experience. Afterwards, they were used as a tool for decision making processes that can enhance different manufacturing systems with predicting capabilities and facilitates knowledge engineering processes in-side decision making. The relation between CPS and VEO is evident in the sense that a VEO is a specialized kind of CPS system aiming towards the gathering of experiential knowledge and re-use whereas a bare bone CPS is aimed towards data and in-formation gathering and management.

Future work on this proposal includes developing a network of VEO's having a wide variety of engineering objects ranging from simple standalone artefact to a complex multitasking machine. In addition, it is desirable to evaluate the proposed techniques in real-life operational contexts in order to refine, improve, and determine the actual applicability of the proposal presented in this article.

References

1. Verhagen, W.J.C., Garcia, P.B., Reinier, E.C.V., Richard, C.: A critical review of Knowledge-Based Engineering: An identification of research challenges. Advanced Engineering Informatics 26, 5–15 (2012)
2. Danilowicz, C., Nguyen, N.T.: Consensus-Based Partitions in the Space of Ordered Partitions. Pattern Recognition Letters 21(3), 269–273 (1988)
3. Qiu, Y.F., Chui, Y.P., Helander, M.G.: Cognitive understanding of knowledge processing and modeling in design. Journal of Knowledge Management 12(2), 156–168 (2008)
4. Duong, T.H., Nguyen, N.T., Jo, G.S.: Constructing and Mining—A Semantic-Based Academic Social Network. Journal of Intelligent & Fuzzy Systems 21(3), 197–207 (2010)
5. Weber, M.: Industry 4.0., http://www.pt-it.pt-dlr.de/de/3069.php
6. n.a. Zukunftsprojekt Industrie 4.0., http://www.bmbf.de/de/9072.php
7. Böhler, T.M.: http://www.produktion.de/automatisierung/industrie-4-0-smarte-produkte-und-fabriken-revolutionieren-die-industrie/Industrie/
8. Baheti, R., Gill, H.: Cyber Physical Systems. In: Samad, T., Annaswamy, A.M. (eds.) The Impact of Control Technology (2011), http://www.ieeecss.org
9. Lee, E.: Cyber Physical Systems: Design Challenges. University of California, Berkeley, Contract No.: Technical Report No. UCB/EECS-2008-8. Retrieved 2008 06 07 (2008)
10. Lee, E.: Cyber-Physical Systems - Are Computing Foundations Adequate? Position Paper for NSF Workshop On Cyber-Physical Systems: Research Motivation, Techniques and Roadmap; Austin, TX (2006)
11. Garcia-Crespo, A., Ruiz-Mezcua, B., Lopez-Cuadrado, J.L., Gomez-Berbis, J.M.: Conceptual model for semantic representation of industrial manufacturing processes. Computers in Industry 61(7), 595–612 (2010)
12. Shafiq, S., Sanin, C., Szczerbicki, E., Toro, C.: Decisional DNA Based Framework for Representing Virtual Engineering Objects. In: Nguyen, N.T., Attachoo, B., Trawiński, B., Somboonviwat, K. (eds.) ACIIDS 2014, Part I. LNCS, vol. 8397, pp. 422–431. Springer, Heidelberg (2014)

13. Shafiq, S.I., Sanin, C., Szczerbicki, E.: Set of Experience Knowledge Structure (SOEKS) and Decisional DNA (DDNA): Past, Present and Future. Cybernetics and Systems 45(02), 200–215 (2014)
14. Shafiq, S.I., Sanin, C., Szczerbicki, E., Toro, C.: Using Decisional DNA to Enhance Industrial And Manufacturing Design: Conceptual Approach. In: Jerzy, Ś., Leszek, B., Adam, G., Zofia, W. (eds.) Information Systems Architecture and Technology; 22 – Szklarska Poreba, Poland, pp. 23–32. Wrocław: Wrocław University of Technology, Wrocław (2013)
15. Shafiq, S.I., Sanin, C., Szczerbicki, E., Toro, C.: Implementing Virtual Engineering Objects (VEO) with the Set of Experience Knowledge Structure (SOEKS). Procedia Computer Science 35, 644–652 (2014)
16. Sanin, C., Szczerbicki, E.: Decisional DNA and the Smart Knowledge Mangement System: A process of transforming information into knowledge Gunasekaran, A, pp. 149–175. IGI, New York (2008)
17. Sanin, C., Szczerbicki, E., Toro, C.: An OWL Ontology of Set of Experience Knowledge Structure. Journal of Universal Computer Science 13(2), 209–223 (2007)

Finite-State Transducers with Multivalued Mappings for Processing of Rich Inflectional Languages

Ualsher Tukeyev[1], Marek Miłosz[2], and Zhandos Zhumanov[1]

[1] Al-Farabi Kazakh National University, Almaty, Kazakhstan
{ualsher.tukeyev,z.zhake}@gmail.com
[2] Lublin University of Technology, Lublin, Poland
marekm@cs.pollub.pl

Abstract. This paper proposes a processing for rich inflectional languages such as Russian and Kazakh, based on finite state transducers with multivalued mappings. We propose to simplify grammar of inflectional languages and use multivalued mappings. An advantage of finite state transducers with multivalued mappings proposed in this paper is that it automatically generates possible alternatives of words' grammatical characteristics, while in existing rule-based technologies alternatives are written by hand. Ambiguity of grammatical characteristics is solved by comparing alternative grammatical characteristics between adjacent words in the source sentence and matching grammatical characteristics are selected. Here an advantage of proposed method should be noted, it does not require explicit description of matching agreements for grammatical characteristics of adjacent words in a sentence, as it is done in existing rule-based methods.

Keywords: Finite-state transducers, multivalued mappings, computational intelligence, machine translation, natural language processing.

1 Introduction

In the field of languages processing (machine translation in particular) the problem of quality still remains a key topical issue.

In this paper, in line with the development of efficient technologies for languages processing, we propose using of the multivalued mappings theory [1]. The theory of multivalued mappings has been actively developed in the past 30 years, especially in game theory, theory of extreme problems, mathematical economics [2]. In the field of language processing there are examples of using the method of mappings in machine translation [3, 4], but they were not multivalued mappings. Proposed method is closed to finite-state transducers (FST) [5, 6] and especially close to a kind of FST transducers named p-**subsequential** transducers which allow ambiguity of final output strings to be associated with each final state [7]. Also there are many papers using FST for machine translation, in particular, for morphology of agglutinative languages [8, 9, 10, 11]. Approach proposed in this paper is based on simplifying morphology of investigated languages' words for improving of computational processing of machine translation by using of finite state transducers with multivalued mappings.

© Springer International Publishing Switzerland 2015
D. Barbucha et al. (eds.), *New Trends in Intelligent Information and Database Systems,*
Studies in Computational Intelligence 598, DOI: 10.1007/978-3-319-16211-9_28

2   Description of the Method

Using multivalued mappings method for language processing can improve efficiency of language processing algorithms due to the fact that table view of multivalued mappings can significantly increase speed of language processing algorithms, clearly present problems of multiple values at different stages of languages processing, solve them by transformation of multivalued mappings into single-valued mappings. Moreover, the most effective, in authors' opinion, is to use proposed method for rich inflectional languages such as Russian and Kazakh.

When using multivalued mappings apparatus for languages processing, we propose following assumption. Every word of natural language will be represented as a sequence of "stem" and "ending":

$$WORD = STEM + ENDING.$$

With this assumption, for computer representation of language processing we propose to create a table view displaying mapping of word stems from one natural language onto word stems from another natural language in the form of dictionary mapping table. Endings from one natural language (source language) map first onto grammatical characteristics, which then map onto endings from another natural language (target language).

These mappings allow getting corresponding stem ending in the target language for each word in the source language. Joining of stem and ending of target natural language forms required output word.

Further phrases and sentences of target natural language are produced by joining words into a sequence. At this stage a mapping system of phrasal structures from source language onto target language is also required. Further, at the level of target language sentence structure formation a mapping system of sentence structures from source language onto target language is required.

Since multivalued mappings are used at each stage of language processing, then it is natural to have a need to obtain a unique solution, which is possible to achieve by creation and use of conversion system of multivalued mappings into single-valued mappings.

Formally, multivalued mapping is defined as follows.

Let X, Y be discrete spaces, P (Y) is a set of all subsets of Y. Then multivalued mapping F from X onto Y is a correspondence that for each point $x \epsilon X$ assigns an empty subset $F(x) \subset Y$, called an image of point x, i.e. $F: X \rightarrow P(Y)$. We shall call this mapping m-mapping (from X onto Y).

Set Γ_F, a subset of set X×Y, $\Gamma_F = \{(x,y) \mid x \epsilon X, y \epsilon F(x)\}$, is called a graph of m-mapping F. A graph of m-mapping F is a tabular representation of m-mapping F, which is very important and convenient for computer representation of multivalued mappings.

Conversion of multivalued mapping into single-valued mapping.

For this we add an additional set of parameters T to set X:

$$F: X \times T \rightarrow Y, \tag{1}$$

Then multivalued mapping F can be transformed into a series of single-valued mappings

$$\{f_i: X \rightarrow Y\}, \ f_i(x) \in F(x). \tag{2}$$

Based on assumptions made above, for machine translation process of source natural language into target natural language scheme of a multivalued mapping system for stages of morphological analysis and synthesis will be as follows:

$$F_a^m: X_i^k \rightarrow Y_i^k,$$
$$F_s^m: Y_j^k \rightarrow Z_J^k, \tag{3}$$

given that Y_i^k and Y_j^k have to be equal or $Y_i^k \subset Y_j^k$,

X_i^k is a space of source natural language L_i endings for the k-th part of speech; it is an input space for multivalued mapping F_a^m;

Y_i^k is a space of grammatical features for the source language's k-th part of speech; it is an input space for multivalued mapping F_a^m;

Y_j^k is a space of grammatical features for the target language's k-th part of speech; it is an output space for multivalued mapping F_s^m;

Z_j^k is a space of target natural language L_j endings for the k-th part of speech; it is an output space for multivalued mapping F_s^m;

F_a^m is a multivalued mapping of space of endings for the source language's k-th part of speech into space of grammatical features for the source language's k-th part of speech;

F_s^m is a multivalued mapping of space of grammatical features for the target language's k-th part of speech into space of endings for the target language's k-th part of speech.

Let's consider multivalued mapping models for the case when the source language is Russian and the target language is Kazakh. Table segments of multivalued mappings (m-mappings) for these languages' different parts of speech are sown bellow.

Table of m-mapping of Russian noun endings into grammatical characteristics:

$$X_r^n \rightarrow Y^n, \tag{4}$$

where $X_r^n = \{ending\}$, $Y_i^n = \{number, case, declension\}$ is shown below in Table 1. Number: 1-single, 2-plural. Case: 1-nominative, 2-genitive, 3-dative, 4-accusative, 5-instrumental, 6-prepositional. Declension: 1-first, 2-second, 3-third.

Endings' m-mapping tables for other Russian parts of speech are constructed similarly.

Table of m-mapping of grammatical characteristics into Kazakh endings:

$$Y^k \rightarrow X_j^k, \tag{5}$$

where $Y_j = \{POS, number, case, dependence, person, tense\}$, $X_j = \{ending\}$, is shown below in Table 2. Peculiarity of this table is that all kinds of Kazakh parts of speech are presented in a single table. POS: 1-noun, 2-verb. Number: 1-single, 2-plural. Case: 1-nominative, 2-genitive, 3-dative, 4-accusative, 5-locative, 6-ablative, 7-instrumental, Possessive dependence: 1-yes, 0-no. Person: 1-first, 2-second, 3-third, 0-no. Tense: 1-present., 2-past, 3-future.

Table 1. Segment table of Russian language endings' m-mapping

Ending	Number	Case	Declension
а,я	1	1	1
ы, и	1	2	1
е	1	3	1
у, ю	1	4	1
ой, ей, ёй	1	5	1
е	1	6	1
ы, и	2	1	1
-, ей, ь	2	2	1
ам, ям	2	3	1
ы, ей, и	2	4	1
...			

Table 2. Segment table of m-mapping of grammatical features into Kazakh endings

POS	Number	Case	Dependence	Person	Tense	Ending
2	1	0	0	3	3	ады
2	2	0	0	1	3	амыз
2	1	0	0	1	3	амын
2	1	0	0	2	3	асыз
2	1	0	0	2	3	асың
2	1	0	0	0	3	ар
2	1	0	0	0	0	арға

3   Machine Translation Technology

Let's explain machine translation technology multivalued mappings method for rich inflectional languages with an example of a concrete sentence with Russian as a source language and Kazakh as a target language.

We are given a sentence in Russian: *Лук хорошо растет на большом огороде. (Onions grows well in a large garden)*

The machine translation technology based on multivalued mappings method includes several well-known stages. However, some stages are particularly related to the use of the multivalued mappings method.

On the first stage technology divide stem and endings for each word of the sentence using stemming algorithm based on proposed multivalued mapping tables for source language (Russian). Stem + ending: *лук* + - ; *хорошо* + -; *раст+ет; на* + - ; *больш* + ом; *огород* + е.

Then, on next stage the algorithm produces options of grammatical characteristics for source sentence words.

So, word '*Лук (Onion)*' is defined with next grammatical characteristics: **112** (1-nominative case; 1-single number; 2- second declension);

word '*хорошо (well)*' is defined without grammatical characteristics;

word '*растет (grows)*' is defined with two kind of grammatical characteristics: **1311** (1-present tense; 3-third person; 1-single number; 1-masculine) and **3311**(3-future tense; 3-third person; 1-single number; 1-masculine);

word '*на (in)*' is defined with two kind grammatical characteristics: **4-**accusative case; **6-**prepositional case;

word '*большом (a large)*' is defined with two kind of grammatical characteristics: **611** (6- prepositional case; 1-masculine gender; 1-single number); **631** (6- prepositional case; 3-neuter gender; 1-single number);

word '*огороде (garden)*' is defined with three kind of grammatical characteristics: **112** (1-nominative case; 1-single number; 2- second declension; 1-masculine gender); **412** (4- accusative case; 1-single number; 2- second declension; 1-masculine gender); **611** (6-prepositional case; 1-single number; 2- second declension; 1-masculine gender);

Further, parts of speech (POS) of each word are defined: [N *Лук*]; [Adv *хорошо*]; [V *растет*]; [Prep *на*]; [Adj *большом*]; [N *огороде*]. The POS algorithm is rule-based.

On next stage source sentence is chunked: [SP *Лук*] [VP *хорошо растет*] [OP *на большом огороде*], where SP – subject phrases, VP – verb phrases, OP – object phrases. Special grammatical rules defining subject, verb and object phrases are used for chunking.

Translation of words' stems from source language into target language is done. For above sentence words translation defined as: Russian word '*лук*' have two variants of translation on Kazakh 1) '*пияз*' *(onion)*, 2)'*садақ*' *(bow)*;

word '*хорошо (well)*' have on Kazakh '*жақсы*'; word '*растет (grows)* have on Kazakh stem '*өс*';

word '*большом (a large)*' have on Kazakh stem '*үлкен*'; word '*огороде (garden)*' have on Kazakh stem '*бақша*'. For lexicon selection in translation of word on Kazakh a context-vector of ambiguous words is formed, based on which algorithm disambiguates them.

Ambiguity of words' grammatical characteristics is solved by comparing grammatical characteristics of adjacent words (Fig. 1) in the source sentence and matching grammatical characteristics are selected. For example, compare grammatical characteristics of words '*большом (a large)*' and '*огороде (garden)*', the algorithm chooses a word '*большом (a large)*' with grammatical characteristics **611** because they are present in characteristics of considered words. Then, algorithm compare grammatical characteristics of words '*на(in)*' and '*большом (a large)*' and chooses the word '*на (in)*' with grammatical characteristic **6** (prepositional case) because it is present in characteristics ofconsidered words.

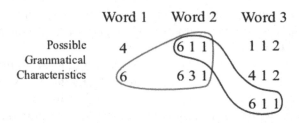

Fig. 1. Choosing of grammatical characteristics for adjacent words

Here we should note an advantage of proposed method, which does not require explicit descriptions of matching agreements for grammatical characteristics of adjacent words in a sentence, as it is done in existing methods.

Another advantage of multivalued mappings method proposed in this paper is automatic generation of possible ambiguity options for grammatical characteristics, while in existing technologies possible options are written manually.

On the next stage morfogeneration of target language words is done based on unambiguously defined stems and grammatical characteristics.

Next stage is chunking transformation of sentence's phrasal structures in the target language. The next step converts syntactic structure of target language sentence. There are two kinds of conversions: on phrase structure level and on sentence level. On each of these levels conversion algorithms from source (Russian) language into target (Kazakh) language are presented.

Advantage of proposed machine translation technology for rich inflectional languages is the ability to automatically generate possible multivalued solutions during translation process for their subsequent conversion into single-valued solutions for the target language.

4   Machine Translation Data Structure

Machine translation data structure is based on the tables of multivalued mappings of endings to grammatical feature and dictionary tables, which created for each part of speech of source and target languages. Below some tables of machine translation data structure are represented: Russian noun endings table, Russian-Kazakh verb dictionary table (Fig. 2-5).

RecNo	id	rus	ch_r	padezh	chislo	rod
			Click here to define a filter			
1	1	ой	3	2	1	2
2	2	ей	3	2	1	2
3	3	ого	3	2	1	13
4	4	его	3	2	1	13
5	5	ых	3	2	2	123
6	6	их	3	2	2	123
7	7	ой	3	3	1	2
8	8	ей	3	3	1	2
9	9	ому	3	3	1	1
10	10	ему	3	3	1	1
11	11	ым	3	3	2	123
12	12	им	3	3	2	123
13	13	ую	3	4	1	2
14	14	ого	3	4	1	1
15	15	его	3	4	1	1
16	16	ое	3	4	1	3
17	17	ее	3	4	1	3
18	18	им	3	4	2	123
19	19	ой	3	5	1	2
20	20	ей	3	5	1	2
21	21	ым	3	5	1	13
22	22	им	3	5	1	13
23	23	ими	3	5	2	123

Fig. 2. Structure of Russian noun endings table (rus-ending, ch_r- POS, padezh-case, chislo-number, rod-gender)

RecNo	id	rus	osnova_rus	kaz	vozvratnyi	vid
<null>	<null>	сыг	<null>	<null>	<null>	<null>
1	<null>	сыграть	сыгра	ойна	0	1

Fig. 3. Structure of Russian-Kazakh verb dictionary table (rus-verb, osnova_rus-stem, kaz-stem, vozvratnyi-reflexive, vid-form)

RecNo	id	rus	kaz	jak	chislo
			Click here to define a filter		
1	1	я	мен	1	1
2	2	ты	сен	2	1
3	3	он	ол	3	1
4	4	она	ол	3	1
5	5	оно	ол	3	1
6	6		біз	1	2
7	7	вы	сендер	2	2
8	8	они	олар	3	2

Fig. 4. Structure of Russian-Kazakh pronoun dictionary table (rus-pronoun, kaz- pronoun, jak-person, chislo-number)

5   Results

Experimental results of proposed approach machine translation models and algorithms for Russian-Kazakh language pair are presented below (Fig. 5, 6).

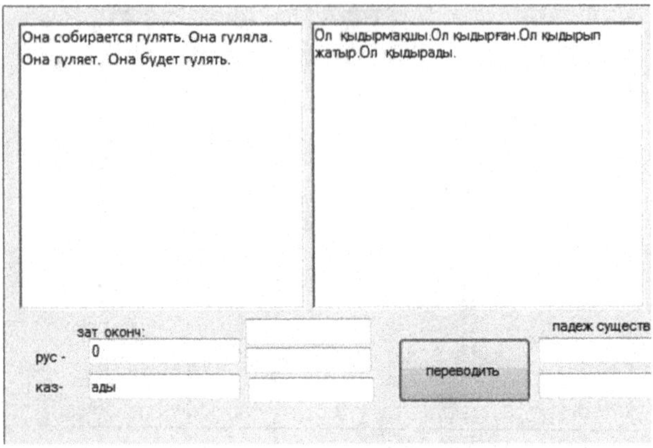

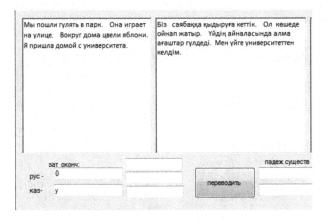

Fig. 5. Example of translation (She is going for a walk. She walked. She is walking. She will walk)

Fig. 6. Example of translation (We went for a walk in the park. She played on the street. Around the house apple trees are blooming. I came home from university)

The results of the comparative analysis of machine translation from Russian into Kazakh language using online translators (Sanasoft, Pragma, Audaru (Soylem)) and our program are shown below (Table 3). For the experiment we used simple sentences.

Table 3. Comparative analysis of machine translation from Russian into Kazakh

Examples on Russian	Sanasoft	Pragma	Audaru (Soylem)	Our program
Добрый день, завтра в субботу 9 июля в 9-00 будем совместно работать (Good afternoon, tomorrow, Saturday July 9, 9-00 will work together)	9 шілдеге мейірбан күнге, сенбіге ертең сонымен бірге жұмыс істейді 9-00	Ізгі күн, ертең в сенбіні 9 шілденің ара 9-00 бір жұмыс істе-боламыз	Қайырлы күн, сенбіге 9 шілде ертең 9-00 жұмыс істейміз	Қайырлы күн, ертең сенбіде 9 шілде 9-00 біріге жұмыс істейміз.
Я ожидаю от школы много знаний и хороших учителей (I expect a lot of knowledge of the school and good teachers)	Мен көп білімдерден және жақсы мұғалімдерден мектептен күтіп жатырмын	Мен көп білімді және жақсы мұғалімдерді мектептен деген күтемін	Мен өнер-білім мектептен көп күтілемін және жақсы мұғалімдер	Мен мектептен көп білім және жақсы мұғалім күтіп жатырмын.

Test results show that for presented simple sentences our program translate well, while online tested programs make morphological and syntactic mistakes in the translation.

6   Conclusion

A distinctive feature of proposed approach is to simplify grammar of source and target languages, which allows using of a tabular method of multivalued mappings. An advantage of finite state transducers with multivalued mappings method for machine translation of complex inflectional languages proposed in this paper is automatic generation of ambiguity alternatives for possible words' grammatical characteristics, while in existing rule-based technologies alternatives are written by hand. Grammatical characteristics ambiguity is solved by comparing grammatical characteristics alternatives of adjacent words in the source sentence and matching grammatical characteristics are selected, which is also an advantage of proposed method, that does not require an explicit description of agreements in matching grammatical characteristics of adjacent words in a sentence, as it is done in existing rule-based methods. Drawback of the proposed approach is increasing number of stems in dictionaries of languages due to fact that word is supposed to consist of a stem and ending, and does not consider prefixes.

Proposed machine translation technology for rich inflectional languages is being implemented for Russian-Kazakh, Kazakh-Russian language pairs. Experimental implementations of machine translation for language pairs listed above show adequately high efficiency of proposed technology.

References

1. Gelman, B.D.: Introduction to the Theory of Multivalued Mappings. Voronezh (2003) (in Russian)
2. Kaczynski, T.: Multivalued Maps As a Tool in Modeling and Rigorous Numerics. Journal of Fixed Point Theory and Applications 4(2), 151-176 (2008)
3. Mitamura, T., Nyberg, E.H.: Hierarchical Lexical Structure and Interpretive Mapping in Machine Translation. In: 6th International Conference on Computational Linguistics – COLING, Nantes, pp. 1254–1258 (1992)
4. Hakkani, D.Z., Tür, G., Oflazer, K., Mitamura, T., Nyberg, E.H.: An English-to-Turkish Interlingual MT System. In: Farwell, D., Gerber, L., Hovy, E. (eds.) AMTA 1998. LNCS (LNAI), vol. 1529, pp. 83–94. Springer, Heidelberg (1998)
5. Koskenniemi, K.: Two-level Morphology: a General Computational Model of Word-form Recognition and Production. University of Helsinki (1983)
6. Beesley, K.R., Karttunen, L.: Finite-State Morphology. CSLI Publications. Stanford University (2003)
7. Mohri, M.: Finite-state Transducers in Languages and Speech Processing. Computational Linguistics 23(2), 269–312 (1997)
8. Oflazer, K.: Two-level description of Turkish morphology. Literary and Linguistic Computing 9(2), 137–148 (1994)
9. Washington, J.N., Salimzyanov, I., Tyers, F.M.: Finite-state Morphological Transducers for Three Kypchak Languages. In: 9th Conference on Language Resources and Evaluation, Reykjavik, pp. 3380–3385 (2014)
10. Salimzyanov, I., Washington, J.N., Tyers, F.M.: A Free/Open-source Kazakh-Tatar Machine Translation System. In: XIV Machine Translation Summit, Nice, pp. 175–182 (2013)
11. Tyers, F.M., Sánchez-Martínez, F., Forcada, M.L.: Flexible Finite-state Lexical Selection for Rule-based Machine Translation. In: 17th Annual Conference of the European Association of Machine Translation, Trento, Italy, pp. 213–220 (2012)

The Implementation of the Perceptual Memory of Cognitive Agents in Integrated Management Information System

Andrzej Bytniewski, Anna Chojnacka-Komorowska,
Marcin Hernes, and Kamal Matouk

Wrocław University of Economics ul. Komandorska 118/120, 53-345 Wrocław, Poland
{andrzej.bytniewski,anna.chojnacka-komorowska,
marcin.hernes,kamal.matouk}@ue.wroc.pl

Abstract. This paper presents the issues related to cognitive agents' perceptual memory implementation - very important module of agent's architecture allows for interpretation of the phenomena occurring in the enterprise's environment. The agents operate in a prototype of an integrated management information system. The main challenge related to perceptual memory is knowledge processing, parallel with symbolic and emergent way. Additionally, a very important issue is using standardized approach due to the need to achieve interoperability. The authors present the use of a topic map standard and semantic network with nodes and links activation levels (slipnet) in order to implement perceptual memory in the Learning Intelligent Distribution Agent architecture.

Keywords: Integrated management information systems, cognitive agents, perceptual memory.

1   Introduction

Integrated management information systems (IMIS) are commonly used by companies and they are characterized by being fully integrated, both in terms of the entire system or applications and in terms of business processes. Note, however, that the properties of contemporary IMIS are becoming more and more inadequate. Apart from collecting and analyzing data and generating knowledge through drawing automatic conclusions based on the results of the analysis, the system should also be able to understand the significance of phenomena occurring around the organization. It is becoming more and more necessary to make decisions based not only on knowledge but also on experience, thus far regarded as purely human domain [3].

In order to accomplish tasks set by IMIS, several cognitive agent programs (cognitive agents) can be used as a multi-agent system. Not only do they enable quick access to information and quick search for the required information, its analysis and conclusions, but also, besides being responsive to environment stimuli, they have cognitive abilities that allow them to learn from empirical experience gained through

© Springer International Publishing Switzerland 2015
D. Barbucha et al. (eds.), *New Trends in Intelligent Information and Database Systems,*
Studies in Computational Intelligence 598, DOI: 10.1007/978-3-319-16211-9_29

immediate interaction with their environments [8]. The cognitive agents allow for generating and executing the decisions.

Taking such action requires putting a number of modules, both for memory and processing, in the agent's architecture. Cooperation among all the modules enables a cognitive cycle to be conducted. The initial phase of the cycle is aimed at receiving and interpreting stimuli from the environment. Receiving stimuli, for which the sensory memory module is responsible for, is a simple process, because it is executed with traditional data reading function (for example, it could be data reading from an internet forum, a customer's opinion on a product). However, the interpretation of stimuli is a more complex process, because it does not allow the use of traditional methods (for instance, data exploration), since they apply only to processing structured knowledge, while a cognitive agent must be able to process unstructured knowledge as well. Besides, traditional methods of data interpretation give little consideration for its meaning (semantics). Therefore an important problem is to implement a perceptual memory module (perceptual memory), responsible for the interpretation of stimuli received from the environments, with the use of tools that consider the aforementioned aspects. The method of implementation of perceptual memory has significant influence on subsequent phases of the cognitive cycle (described in item 3 of the article) and, consequently, on the effectiveness of business process support in a company. Most existing methods of perceptual memory implementation allow for processing knowledge related to business processes only in symbolic way or only in emergent way. Unstructured knowledge, however, must be processed in a hybrid way (parallel in symbolic and emergent way). Moreover, existing hybrid methods (for example use of semantic network with weights of the arts) use mainly their own language (not standardized) of knowledge representation. It is a major barrier for the multi-agent systems designers, because they must design a system based on particular technology (cognitive agent architecture). The interoperability of such system is very difficult to achieve in this case.

The aim of this paper is to present the method of the cognitive agents' perceptual memory implementation in a prototype of an integrated management information system. The main novelty of this approach is the ability to process knowledge both in symbolic and emergent way, and using the existing standards of knowledge representation (a Topic Map) at the same time.

The study has been divided into four parts. First, there is a review of the literature in the field of cognitive agents' perceptual memory implementation; then, a business processes integration in a prototype of Cognitive Integrated Management Information System (CIMIS), is characterized. The third part of the paper presents the Learning Intelligent Distribution Agent (LIDA) architecture running in CIMIS. In the fourth part, it presents the method of perceptual memory implementation with a case study.

2   Related Works

The perceptual memory is implemented in the cognitive multi-agent systems using different approaches and techniques. They mainly depend on the architecture of a cognitive agent. Often the neutral networks are used for implementing perceptual memory of cognitive agents [7]. The work [10] presents a rule-based perceptual memory implementation. The perceptual memory of the Integrated Cognitive-Neuroscience Architectures for Understanding Sensemaking (ICARUS) architecture [11] is implemented as the "percepts" (represented as triple: object, attributes, values) and "beliefs" (describe relation between "percepts", such as the relative position of two buildings). However these approaches allow for knowledge processing only in emergent way or only in symbolic way, while unstructured knowledge, related to business processes, must be processed in both emergent and symbolic way (parallel). The Executive Process Interactive Control (EPIC) [12], instead, has separate perceptual processors with distinct temporal properties for several major sensory (e.g. visual, auditory, tactile) modalities. This approach, allows for knowledge processing parallel in both emergent and symbolic way, however, it requires separate sensory inputs for these processors, therefore processing multimedia environments (for example web sites that contain both text, picture and sound) is very difficult. The work [18] presents the implementation of perceptual memory by using a semantic network. The arcs between nodes can have weights, therefore this approach allows for processing knowledge both in emergent and symbolic ways. Thus this ontological, hybrid (symbolic and emergent) approach is suitable for processing in an IMIS. The main disadvantage of this approach, however, is using not-standardized language (SnePSUL) for semantic network representation.

Considering the existing methods of perceptual memory implementation, the advantages of the method presented in this paper are as follows:

- the possibility of knowledge processing both in emergent and symbolic way, using the semantic net with node and links activation level (the "slipnet"),
- the use of a standardized approach can increase the possibility to achieve interoperability between systems.

3   Business Process Integration Using CIMIS

There are different forms of the IMIS's architectures in the subject literature and in practice, supporting the realization of enterprise's processes [16]. In order to develop a prototype of IMIS (based on cognitive agents), named CIMIS (Cognitive Integrated Management Information System), the following subsystems have been created: fixed assets, logistics, manufacturing management, human resources management, financial and accounting, controlling, business intelligence [2].

Such an architecture enables to optimize internal processes as well as developments in the business environment. Subject bibliography describes a business process in different ways. For instance, [13] defines it as a set of structured and measurable actions designed to make a product for a particular customer or market [5]. According to [17] there are two types of processes: basic (e.g. logistics, manufacturing) and auxiliary (e.g. human resources management, financial and accounting).

Fig 1 presents an example of selected processes carried out in the enterprise along with specification of sub-systems of the CIMIS to support their implementation.

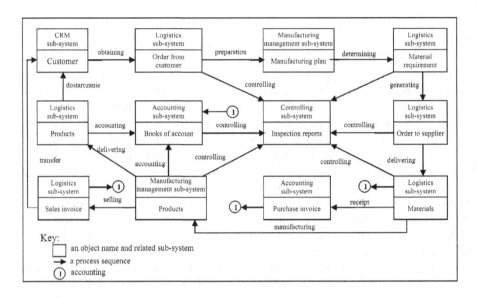

Fig. 1. The example of selected processes carried out in the enterprise along with specifying of sub-systems of the CIMIS to support their implementation

On this example, the execution of processes is initiated through placing orders for ready-made products by customers, since the orders are a basis for preparing production plans and calculating the demand for production materials. The data is necessary to purchase the right amount of materials used in the production process. The last phase is the making of products and delivering them to the customer. Then the cycle is repeated. It is important to run accounting books properly and issue periodical control reports so that all processes are executed efficiently and effectively. Each of the business processes is supported by a particular sub-system of the CIMIS.

The execution of processes is only possible through precise cooperation of sub-systems. Only a running flow of information among the sub-systems allows for efficient and economic execution of business processes in a company.

The CIMIS is developed on the basis of cognitive agents architecture named The Learning Intelligent Distribution Agent (LIDA), which has been characterized in the next part of the article.

4   The LIDA Architecture

The creation of the CIMIS prototype is based on the LIDA cognitive agent architecture [6]. The advantage of this architecture is its emergent-symbolic nature. This makes it possible to process knowledge of both structured (numerical and symbolic) and unstructured (natural language). The LIDA cognitive agent's architecture consist of the following modules [4], [6]:

- sensory memory,
- perceptual memory,
- workspace,
- spatial memory,
- episodic memory,
- declarative memory,
- attentional codelets,
- global workspace,
- action selection,
- sensory-motor memory.

Functioning of the cognitive agent – a cognitive cycle - is divided into three phases: the understanding phase, the consciousness phase and selection of activities and learning phase [6]. These phases have been described in detail in [4], [6]. In this paper we shortly describe the first one due to using perceptual memory in the phase. The understanding phase begins when the stimuli received from the environment activate the codelets of low level features in sensory memory. The codelets are specialized, mobile information processing programs. The outputs of these codelets activate perceptual memory. The perceptual memory represents the knowledge about the environment in the form of semantic network, for example the topic map. In this memory the low level codelets create more abstract occurrences, such as: objects (people, things), categories (for example client, employer, material, product), actions (for example accepting an order, execution of a production task) or events (the emergence of competition, decrease in sales). The results of the perception are transferred to the workspace (it stores the results of perception – topics and associations between currently occurring events in the environment). Next, the consciousness phase begins [6]. The perceptual memory is also the main component of agent perceptual learning (one of the agent's learning mechanisms, the other of them being: episodic learning, procedural learning, attentional learning [6]), which rely on the recognition of new objects, associations, categories. Learning of agents can be implemented as lifelong learning with a teacher or without a teacher.

The next part of the article presents the method of implementation of LIDA agents' perceptual memory used in CIMIS, with the example of CRM sub-system.

5   Perceptual Memory Implementation

The perceptual memory is a very important module of the cognitive agent architecture. It works in the first phase of cognitive cycle. The perceptual memory is defined as the ability to interpret the incoming stimuli through the recognition of their individual characteristics, to make their categorization and to define the relationship between features and categories, at the same time having the ability to learn those skills [1]. Perceptual memory allows to identify well-known objects in the environment or to perceive the deviations from the expected state of the environment. It works as a filter, receiving, on a regular basis, the most important information from the sensory memory, and the results of the perception are transferred to workspace.

In order to implement perceptual memory of cognitive agent in the CIMIS, the topic map standard, introduced by International Organization for Standardization (ISO/IEC 13250:2000), was used. The topic maps are a kind of a semantic network, and they allow to write information of the data ontology and data taxonomy in a semantically ordered manner [15]. The topic map, most often, consists of „parent-child" relations.

5.1   Topic Map Project

The CRM sub-system is engaged in matters connected with ensuring the best company-customer relations and collecting information in customers' preferences in terms of product purchase in order to increase sales [2]. One of the particular goals of the sub-system is to collect and register orders and supervise their execution. Fig 2 presents an example of simplified (due to the volume of the article) topic map (grey topics denote the objects), related to CRM sub-system.

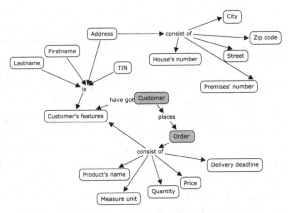

Fig. 2. The example of simplified topic map related to the CRM sub-system

The topic map presented in Fig. 2 presents characteristics related to a customer's order, such as customer's features and attributes of the order. It allows for perception of the orders sent by the customers in a different form, after they have been received by the agent's sensory memory.

5.2   The Example of Implementation of the Topic Map in LIDA

Perceptual memory implementation is based on the developed topic map. The LIDA represents a topic as a node and associations as links. This semantic net with nodes and links, on activation level, is called "slipnet". The configuration of the nodes and links, in design-time, is stored in XML format configuration file. The sensory memory consist of the occurrences relevant to a particular topic.

Fig 3 presents the example of the configuration in accordance with the topic map presented in Fig. 2 (':' denotes a link).

```
<param name="nodes">
    Customer, Customers_features, Firstname, Lastname, TIN,
    Address, City, Zip_code, Street, House_number, Premises_number,
    Order, Product_name, Measure_unit, Quantity, Price, Delivery_deadline
</param>
<param name="links">
    Customer:Customers:features, Firstname:Customers_features,
    Lastname:Customer_features, TIN:Customers_features,
    Address:Customers_features, Address:City, Addres:Zip_code,
    Address:Street, Addres:House_number, Address:Premises_number,
    Order:Product_name, Order:Measure_unit, Order:Quantity,
    Order:Proce, Order:Delivery_deadline, Customer:Order
</param>
```

Fig. 3. Example of configuration of the nodes and links

In order to visualize the content of perceptual memory better, the LIDA agent automatically presents a graphical representation of the configured nodes and links in the form of a semantic network, having characteristics similar to the topic map (Fig 4 – the number of nodes and links was limited for the sake of better figure readability). The features of the nodes and the levels of the link activation (marked in Fig 4 by dots) are determined by the implementation of program code of the class related to a particular node or link. As the result of cognitive agent's learning process, new topic and associations can be generated. They will be stored, in runtime, in agent's perceptual memory and they can be used, despite it has not been previously configured.

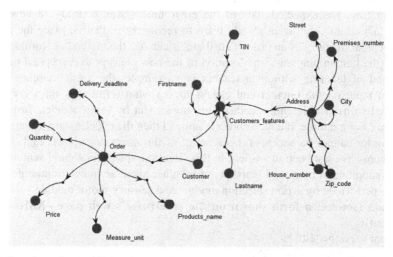

Fig. 4. Sample nodes and links visualization automatically generated by agent of CRM subsystem

5.3   The Example of the Perceptual Memory Functionality – Case Study

The following example presents the functioning of perceptual memory in practice. Assume that sensory memory of the CRM sub-system's agent receive data from environment related to orders from different sources (the agent operates in polish language). Their interpretation in perceptual memory can be as follows:

1.   **Data (source: paper scan):** „My name is Katarzyna Nowak, I order 15 pcs. of Product1 at a price of 20 euro per piece. I ask for delivery to Wrocław, zip code 50-150 Rynek 200 street, until May 20, 2014. I would also like to ask for an invoice – my TIN 111-222-33-44".

Interpretation: The Perceptual memory recognized the Customer object (node) with the following features (attributes): Lastname="Nowak", Firstname="Katarzyna", TIN="111-222-33-44", and address: City="Wrocław", Zip code="50-150", Street="Rynek", House number:="200". The Order subject has also been recognized, with the following attributes: Product's name: „Product1", Measure unit: „pcs", Quantity: „15", Price:"20 euro", Delivery deadline: „ May 20, 2014".

The associations between nodes have also been recognized, e.g. the customer placed an order. Because all of the required objects, attributes and relationships between them have been identified, it is possible to accept the order (carried out by other agent architecture modules).

2. **Data (source: e-mail):** „I order 20 pcs. of product2 at a price of 50 euro. I will pick it up in two days. Requesting discount. My e-mail:mzielinski@com.pl. Best regards, M. Zielinski".

Interpretation: The Perceptual memory recognized the Customer object (node) with the following attributes: Lastname: Zielinski", Firstname:"M.". Whereas the address of the customer has not been recognized. The Order subject has also been recognized, with the following attributes: Product name: „Product2", Measure unit: „pcs", Quantity: „20", Price:"50 euro", Delivery deadline: „in two days". Because a deviation from the expected state of the environment was noticed – a new object (node): „Discount", the agent should learn to recognize it, that is, place the node in the perceptual memory as an order's attribute. It can use the method of learning with a teacher (the human pinpoints the location of the new concept to perceptual memory) or method of learning without a teacher (for example, the agent searches for the Discount topic on the internet and checks to see what terms it is most commonly associated with). The learning methods are implement by using codelets (using Java language). For example, if a new object is noticed then the codelet starts running.

The order cannot be accepted, because all of the necessary objects and links between them have not been identified. In this situation, the agent should sent a request for data supplement by e-mail (carried out by other agent architecture modules). This action is performed by action selection module and sensory-motor module.

3. **Data (source: a form shown on the enterprise's web page –XML format document):**

„<?xml version="1.0"?>
<FEATURES_OF_CUSTOMER>
<LASTNAME>Kowalski</LASTNAME>

```
<FIRSTNAME>Henryk</FIRSTNAME>
<ADDRESS>
     <CITY>Poznań</CITY>
     <ZIP_CODE>50-100</ZIP_CODE>
     <STREET>Krótka</STREET>
     <HOUSE_NUMBER>10</HOUSE_NUMBER>
     <LOCAL_NUMBER>2</LOCAL_NUMBER>
<NIP>333-444-11-22</NIP>
</FEATURES_OF_CUSTOMER>
<ORDER>
<PRODUCT_NAME>Product3</PRODUCT_NAME>
<MEASURE_UNIT>pcs</MEASURE_UNIT>
<QUANTITY>10<QUANTITY>
<PRICE>30</PRICE>
<DELIVERY_DEADLINE>2014-05-31</DELIVERY DEADLINE>
</ORDER>
```

Interpretation: Because sensory memory contains structured data, all of the required objects, attributes and relationships between them have been identified, so the order will be carried out.

It is therefore clear, that the presented method for implementing the cognitive agent's perceptual memory, as a consequence, allows for both structured and unstructured data processing as well as interpretation of this data. The main disadvantage of the presented method is its poor ability to interpret orders which are not well written. For example, if an order contains a typographical or spelling error, the agent will send a request for data supplement to the user, while a human can improve the text of an order. A method of eliminating this disadvantage must be implemented.

6   Conclusions

Proper recognition of items, categories, actions and events occurring both in external environment and inside the cognitive agent depends on the method of implementation of its perceptual memory. Using a topic map for this purpose allows for a semantic representation of structured and unstructured knowledge on economy and management, and learning new topics and relations, which consequently enables the execution of symbol grounding process. The advantage of such approach is the ability to immediately implement a topic map in the perceptual memory of a LIDA cognitive agent program, used in the creation of the CIMIS prototype.

The LIDA cognitive agent is also under implementation as a fundamental analysis tool in a-Trader financial decision support system [9] and as a consensus determining agent in decision support systems[19]. Issues of knowledge integration [14] in these systems arise.

Since a topic map has been recognized as standard, as it was noted before, its employment will lead to interoperability of systems, which is currently widely promoted in many projects financed by the European Commission, including integrated management information systems.

References

1. Bitterman, M.E.: Phyletic difference in learning. American Psychologist 20, 396–410 (1965)
2. Bytniewski, A. (ed.): Architektura zintegrowanego systemu informatycznego zarządzania. Wydawnictwo AE we Wrocławiu. Wrocław (2005) (in Polish)
3. Hernes, M.: A Cognitive Integrated Management Support System for Enterprises. In: Hwang, D., Jung, J.J., Nguyen, N.-T. (eds.) ICCCI 2014. LNCS, vol. 8733, pp. 252–261. Springer, Heidelberg (2014)
4. Cognitive Computing Research Group (April 20, 2014), http://ccrg.cs.memphis.edu/
5. Davenport, T.H.: Process innovation. Reengineering. Work Through Information Technology. Harvard Business Scholl Press (1993)
6. Franklin, S., Patterson, F.G.: The LIDA architecture: Adding new modes of learning to an intelligent, autonomous, software agent. In: Proceedings of the International Conference on Integrated Design and Process Technology, San Diego (2006)
7. Hecht-Nielsen, R.: Confabulation Theory: The Mechanism of Thought. Springer (2007)
8. Katarzyniak, R.: Grounding modalities and logic connectives in communicative cognitive agents. In: Nguyen N. T. (ed.), Intelligent technologies for inconsistent knowledge processing, Advanced Knowledge International, pp. 21-37. Australia, Adelaide (2004)
9. Korczak, J., Hernes, M., Bac, M.: Risk avoiding strategy in multi-agent trading system. In: Proceedings of Federated Conference Computer Science and Information Systems, pp. 1119 – 1126. Kraków (2013)
10. Laird, J.E.: Extending the SOAR Cognitive Architecture. In: Wang, P., Goertzel, B., Franklin, S. (eds.) Frontiers in Artificial Intelligence and Applications, Vol. 171 (2008)
11. Langley, P.: An adaptive architecture for physical agents. In: Proceedings of the 2005 IEEE/WIC/ACM International Conference on Intelligent Agent Technology, IEEE Computer Society Press, Compiegne (2005)
12. Meyer, D.E., Kieras, D.E.: Precis to a Practical Unified Theory of Cognition and Action: Some Lessons from EPIC Computational Models of Human Multiple-Task Performance. In: Gopher, D., Koriat, A. (eds.) Attention and Performance XVII. Cognitive Regulation of Performance: Interaction of Theory and Application, pp. 17–88. Cambridge, MIT Press (1999)
13. Müller, R., Rupper, P.: Process Reengineering. Astrum, Wrocław (2000)
14. Nguyen, N.T.: Advanced Methods for Inconsistent Knowledge Management. Springer Verlag London (2008)
15. Passin, T.B.: Explorer's Guide to the Semantic Web. Manning Publications (2004)
16. Plikynas, D.: Multiagent Based Global Enterprise Resource Planning: Conceptual View. Wseas Transactions on Business And Economics 5(6) (2008)
17. Porter, M.: Competitive Advantage. Free Press, New York (1985)
18. Shapiro, S.C., Rapaport, W.J., Kandefer, M., Johnson, F.L., Goldfain, A.: Metacognition in SNePS. AI Magazine 28, 17 – 31 (2007)
19. Sobieska-Karpińska, J., Hernes, M.: The postulates of consensus determining in financial decision support systems. In: Proceedings of Federated Conference Computer Science and Information Systems, pp. 1165–1168. Kraków (2013)

Consensus with Expanding Conflict Profile

Marcin Maleszka

Wroclaw University of Technology, Wyb. Wyspianskiego 27, 50-370 Wroclaw
Marcin.Maleszka@pwr.edu.pl

Abstract. Consensus methodology is nowadays widely used, but it does not consider the time component of gathering knowledge. We observe how introducing expanding conflict profile influences the behavior of consensus methods for integrating knowledge. This allows observing temporary consensus for partial data, as opposed to only final consensus for all data. We describe some basic notions of how expanding conflict profile changes various parts of consensus methodology. We also observe this approach experimentally, showing that for different data structures temporary consensus is convergent to final consensus. We briefly describe how this approach may be used to work with data-streams.

Keywords: knowledge integration, consensus function, expanding conflict profile.

1 Introduction

In modern information society different tasks related to knowledge management become more and more important. Decision making, information retrieval and knowledge integration are now common occurrence in various applications, often without user input. Multiple methods have been developed to solve different problems occurring in each of those tasks, including consensus methodology used to handle inconsistency of knowledge. Inconsistency is a feature of knowledge which is characterized by the lack of possibility for inference processes. Therefore, solving inconsistency of knowledge is a basic and very essential sub-task in many tasks of knowledge management [8]. Other researchers have previously presented methods for resolving and processing inconsistency on syntactic and semantic level on the basis of mathematical models for representing and measuring it. What those authors did not consider, are the temporal aspects of knowledge inconsistency and consensus. This paper propose some basic notions meant to start addressing this gap.

As a first step towards enhancing consensus methodology with time issues, we propose to use expanding conflict profile. A profile of the conflict is the set of distinct opinions (data, knowledge) from all the agents. Such set is often inconsistent, that is the opinions (knowledge) of some agents are contrary to opinions of other agents. The simple example following the formal definition of the model in [9] gives three states of agent knowledge: positive, negative and neutral. If some agents opinion is positive and others – negative, their opinions

© Springer International Publishing Switzerland 2015   291
D. Barbucha et al. (eds.), *New Trends in Intelligent Information and Database Systems*,
Studies in Computational Intelligence 598, DOI: 10.1007/978-3-319-16211-9_30

are inconsistent and there is a conflict. We propose the expanding conflict profile, where in time knowledge of new agents is added to the conflict profile.

Our proposition allows observing the temporary consensus of agent knowledge (consensus for the current profile), as opposed to only the final consensus (available once all the data has been gathered). We describe briefly how expanding conflict profile influences consecutive parts of consensus methodology, starting from the measurement of data consistency, up to postulates for consensus functions. We then show on simulated data how temporary consensus relates to the final one for different basic knowledge structures. We also briefly describe how this research may be expanded upon and applied to use consensus for data-streams.

This paper is organized as follows: in Section 2 we discuss other research done on time-dependent problems in integration and consensus; in Section 3 we briefly discuss some basic notions that allow changing previously used consensus methodology into time-based one; in Section 4 we show how the theory works in some experimental situations for different data types; we conclude this paper with some final remarks, possible applications and future work aspects in the last Section 5.

2 Related Works

Consensus changing with time is most often considered when using in multi-agent systems, possibly for autonomous robots or network systems [10]. It has broad applications in formation control, attitude alignment, flocking and more. Research focuses, among others, on determining if agents reach consensus in finite time. All research in this group may in fact be considered as related to stability of the multi-agent system [2].

The authors of [4] consider the finite-time consensus problem for leaderless and leader-follower multi-agent systems with and without external disturbances. They show that without disturbances all agents in leaderless system can reach consensus in finite time, and with disturbances a pair of agents can reach similar state. They also propose algorithms for leader-follower systems and verify it both analytically and experimentally.

Similar research was done in [11] and in [3]. In the first, authors construct local non-smooth time-invariant consensus protocols using Lyapunov function, graph theory and homogeneity with dilation to obtain consensus of agents in finite time. They provide both theoretical and numerical analysis of their solution. In the latter, the binary consensus protocol and pinning control scheme are considered, which helps minimize the error relative to each neighbor. Authors, use Filippov solution, set-valued Lie derivative and Lyapunov functions to determine the conditions that guarantee achieving consensus in finite-time. Again, numerical examples are shown to illustrate theoretical results.

In this paper we consider another aspect of consensus over time - how inconsistent data can be integrated. It is an approach first developed to determine median data [1],[7] in formal theory of consensus. Consensus methodology has been proved to be useful in solving conflicts and is effective for knowledge inconsistency resolution and knowledge integration. The general notion of consensus and postulates for consensus choice functions (also called knowledge functions) were defined like: reliability, unanimity, simplification, quasi-unanimity, consistency, Condorcet consistency, general consistency, proportion, 1-optimality and 2-optimality [8]. In other works classes of consensus functions were defined and analyzed referring to postulates. This allowed developing consensus-based integration algorithms for various types of data. What those often efficient solutions do not consider is time. This papers aims to expand this methodology by adding the time component.

3 Basic Notions

We use the notion of consensus for inconsistent knowledge, as given by [8]: in some domain U, for some conflict profile $X \subseteq U$ (a set of inconsistent opinions, data or knowledge), there exists some element $C(X) \subseteq U$, called the consensus, that satisfies a series of select postulates (criteria). This element is the best representative of the whole set X.

In this paper we expand the notion of conflict profile by observing how it changes in time, ie. if new opinions are added or removed from the conflict profile. We use discrete time, where a new time unit starts after the profile was changed. We denote this dynamic profile as $X(n), n \in \{1, \ldots, N\}$. We are therefore able to determine some temporary consensus at this time moment: $C(X(n)), n \in \{1, \ldots, N\}$.

Based on observations of groups of students trying to solve a problem, we make an assumption that the conflict profile $X(t)$ is expanding by addition of new elements (a new agent states his opinion or provides new data) - the previous elements do not change and are not removed. Using the established notions, this is denoted as $X(n+1) = X(n) \cup x$ for some $x \in U$. The real world example that was the basis of this assumption is that a group of student, having difficulties with solving some problem, ask a student outside this group for a his opinion. If this additional opinion is not enough, another student is asked for help. A fallacy may occur here - even the whole group of students may not know the correct solution to the problem. Therefore this approach is best used when the answer depends on the participants (like jury decisions) and not if it is independent of them (like weather forecasts).

We observe how such change in the conflict profile influences the temporary consensus, by comparing the temporary consensus to the final one. In this Section, we do this in analytical form, while in Section 4 we simulate the situation in a virtual environment.

We will consider all the various elements of consensus methodology, as defined in [8], in a concise format and same ordering, presenting only differences from non-dynamic situation.

We may observe that in case of expanding conflict profile the postulates for consistency need to be modified.

First, let us consider the postulate for maximum consistency. Consistency $c(X)$ is a measure used to determine if a conflict profile is homogenous (opinions are identical) or heterogeneous (opinions are different). If the profile is homogenous, then $c(X) = 1$. For expanding conflict profile this means that:

$$\forall_{n \in \{1,...,N\}} : c(X(n)) = 1 = c(X) \tag{1}$$

Consequently, either $X(n) = X$ or $\forall_{n > n_0} : X(n)$ and X are multiples of some $X(n_0)$ where n_0 is a starting time moment. In both cases the profile is not dynamic and cannot be treated as an expanding conflict profile (ie. there is no conflict).

Another postulate that is important to consider is one of the versions of postulate for minimal consistency (P2c). It says that if $X = U$ then $c(X) = 0$. For expanding conflict profile this leads to an observation that if we add elements to X then $c(X(n))$ is non-monotonously decreasing until for $X(N) = X = U$ we have $c(X(N)) = 0$.

We may also analyze how the postulates for consensus behave for expanding conflict profile.

The reliability postulate for consensus function states that some solution always exists for any inputs. For a given time moment, we have some inputs and a single consensus function – if this function satisfies the reliability postulate, then it is satisfied for any such moment. Thus reliability postulate does not change for expanding conflict profile.

The unanimity postulate cannot be considered in the case of expanding conflict profile, as it is based on a homogenous profile ($X = \{k \cdot x\}$).

Similarly, the simplification postulate may be satisfied only in very few situations. It occurs for homogenous profiles (which we consider undesirable due to the previous deliberations) and for very specific non-homogenous profiles (if $X(n_2)$ is a multiple of $X(n_1)$ then $C(X(n_1)) = C(X(n_2))$).

The postulate of quasi-anonymity may be satisfied in an expanding conflict profile and in some cases it may be a desirable situation. If the profile is expanded in each time moment by the same element $\{x\}$ then in infinite time the consensus will also be $\{x\}$. On the other hand, as $N < \infty$, this postulate cannot be fully applied.

The postulate for consistency works as intended for expanding conflict profile. It has the following form:

$$x \in C(X(n_1)) \wedge X(n_2) = X(n_1) \cup \{x\} \Rightarrow x \in C(X(n_2)) \tag{2}$$

The postulate of Condorcet consistency may be used to describe the so-called *trajectory* of the consensus, that is the changes of consensus over time. We have:

$$C(X(n_1)) \cap C(X(n_2)) \Rightarrow C(X(n_1)) \cup C(X(n_2)) = C(X(n_1) \dot\cup X(n_2)) \tag{3}$$

The postulate of general consistency may be used for expanding conflict profile as in the case of constant profile, but the proportion postulate has an interesting form, as:

$$X(n_1) \subseteq X(n_2) \wedge x \in C\big(X(n_1)\big) \wedge x \in C\big(X(n_2)\big) \Rightarrow d\big(x, X(n_1)\big) \leq d\big(x, X(n_2)\big) \tag{4}$$

Therefore we may say that d is non-decreasing over time.

The postulates of 1-Optimality and 2-Optimality are used as in case of constant profile to determine the best value of consensus. They are important for implementation, but are no different than in the standard case.

We may also consider the quality of consensus as an important measure. We have:

$$\hat{d}\big(x, X(n)\big) = 1 - \frac{d\big(x, X(n)\big)}{card\big(X(n)\big)} \tag{5}$$

We know from 4 than $d(x, X(n))$ is non-decreasing. We also have increasing $card\big(X(n)\big)$. As $\forall_{y \in X} : d(x, y)$ is in range $[0, 1]$ with each increase of cardinality by 1, the overall distance increases by $[0, 1]$. Therefore we may say that $\hat{d}$ is non-decreasing.

An important question to this problem is if the following occurs:

$$\forall_{(n_1, n_2): n_1 < n_2} : d\Big(C\big(X(n_2)\big), C\big(X\big)\Big) < d\Big(C\big(X(n_1)\big), C\big(X\big)\Big) \tag{6}$$

We may use a counter-example to show that this does not occur:

First, assume elements of the conflict $A_1 = 0$, $A_2 = 1$, $A_3 = 2$ and $A_4 = 3$ with O_2 to solve the conflict. The final consensus $c(X(N)) = 1.5$.

We will first use A_2 and A_3 in t_1. Then $c(X(n_1)) = 1.5$. The distance to final consensus is 0.

Now in time n_2 we will add A_1 to the temporary conflict profile. Then $c(X(n_2)) = 1$. The distance to final consensus is now $0.5 > 0$.

This counter-example leads to another interesting observation:

$$\forall_{n_1 \in \{1, \ldots, N\}} \exists_{n_2 \in \{1, \ldots, N\}: n_1 < n_2} : d\Big(C\big(X(n_2)\big), C\big(X\big)\Big) \leq d\Big(C\big(X(n_1)\big), C\big(X\big)\Big) \tag{7}$$

In borderline worst case $n_2 = N$ or:

$$d\Big(C\big(X(n_2)\big), C\big(X\big)\Big) = d\Big(C\big(X(n_1)\big), C\big(X\big)\Big). \tag{8}$$

There remain several interesting possibilities that must be considered in future research. We consider the most important of those open research questions as follows:

- The temporary consensus is convergent on final consensus. It should be considered how fast is the convergence and at which point the temporary consensus is close enough to final consensus.
- When observing a stream of data, it is important to know at which point we have enough observations to make some decisions.

In both cases what needs to be verified is how to determine the time n in the following notation:

$$\forall_\epsilon \exists_{n \in \{1,...,N\}} : d\Big(C(X(n)), C(X)\Big) < \epsilon \qquad (9)$$

We intend to present an algorithm for determining this n in future publications.

4   Simulation of Consensus with Expanding Conflict Profile

The mathematical formula in Section 3 are very general and are best described in very specific cases. Due to this we have performed experiments on the behavior of the consensus with expanding conflict profile on simulated data. The experiment was conducted as follows:

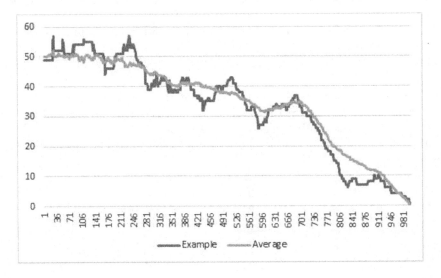

Fig. 1. Distance from final consensus. Simulation for multiple Boolean attribute data.

1. Create a set of $N = 1000$ agents with randomly generated data.
2. Determine final conflict profile $X(N)$ consisting of all agents data.
3. Determine final consensus $C(N)$.
4. Set current number of agents $n = 2$.
5. Create temporary conflict profile $X(n)$.
6. Add data from random unused agents to $X(n)$ until number of used agents is equal n.
7. Determine temporary consensus $C(X(n))$.
8. Calculate distance $d\Big(C(N), C(X(n))\Big)$.
9. If $n < N$, increment n and goto step 5.

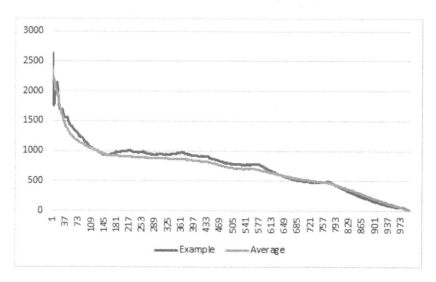

Fig. 2. Distance from final consensus. Simulation for Multiple number attribute data and O-2 consensus choice function.

We have considered the following types of data and consensus choice functions:

– Multiple Boolean attributes. There is one hundred conflict subjects. All domains are binary number $\{0, 1\}$. For each agent agent and for each conflict subject we randomly generate a single binary value. The final consensus for each conflict subject is a binary value, as determined by 1-Optimality. The final consensus overall is a set of hundred binary values. The distance is Hammings.

– Multiple number attributes and O-2 consensus choice function. There is one hundred conflict subjects. All domains are real numbers in range $[0, 100)$. For each agent and for each conflict subject we randomly generate a single natural number in this range. The final consensus for each conflict subject is a real number in this range, as determined by 2-Optimality. The final consensus overall is a set of one hundred real numbers. The distance is Euclidean.

– Multiple number attributes and O-1 consensus choice function. There is one hundred conflict subjects. All domains are real numbers in range $[0, 100)$. For each agent and for each conflict subject we randomly generate a single natural number in this range. The final consensus for each conflict subject is a real number in this range, as determined by 1-Optimality. The final consensus overall is a set of one hundred real numbers. The distance is Euclidean.

Fig. 1, 2 and 3 show the results of the experiment, averaged over a number of runs, and a single experimental run for comparison.

As number of agents used increases, the consensus becomes more similar to the final consensus. In some cases the distance to final consensus was zero with data from just 950 agents, but in some cases only the last run lowered the distance

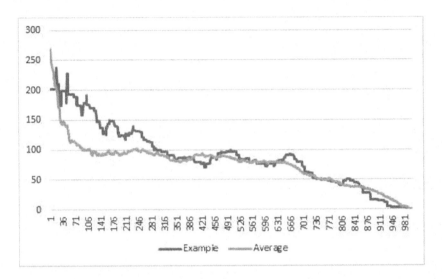

Fig. 3. Distance from final consensus. Simulation for Multiple number attribute data and O-1 consensus choice function.

to zero. This is due to randomly selecting agents in the experimental procedure. It is possible that with good selection of agents, the number of them required to achieve consensus equal to final consensus may be reduced, thus reducing also the running time for many complex algorithms (lower n for algorithms with high complexity $O(f(n))$).

One may observe that the distance does not always decrease, even with averaged data. This clearly shows that while

$$\forall_{n_1 < N} \exists_{n_2 : n_1 < n_2 \leq N} : d\Big(C\big(X(n_2)\big), C\big(X(N)\big)\Big) \leq d\Big(C\big(X(n_1)\big), C\big(X(N)\big)\Big),$$
(10)

it is not always so that $n_2 = n_1 + 1$.

5   Conclusions

This paper briefly describes some basic notions on expanding conflict profile and its influence on determining consensus of inconsistent data. It provides some basic observations and hypotheses on the influence of such changing profile on the process of integration. Section 4 also shows how the temporary consensus behaves in relation to the final one for some basic knowledge (data) structures.

While this research is in very early stages, some interesting applications may be already described. Consider a system that observes the behavior of a car driver in different traffic situations. Such system may for example be mounted in a car of known traffic-offender and monitor him continuously. If one gathered all the observations from a longer period of time, the consensus of them would show if this driver is a good or a bad one. In real world situation we would receive the data as a stream, with periodically added new information. Using the methodology briefly discussed in this paper, we could observe the changing consensus of those data-points. Once a certain threshold is crossed, we know that long-term, the driver will only behave more and more similar to this temporary consensus (but with possible temporary breaks of the trend). Thus further observation is unnecessary.

Similarly we could apply consensus methodology and methods developed for this approach in multiple situations where data arrives as a stream and not as a single packet. We aim to prepare a real-world experiment to verify the veracity of this approach, compared to other methods of analyzing the data-stream.

As stated multiple times previously, this research is only on a very basic level so far. In our future research we intend to develop it more fully, analyzing and testing the behavior of consensus functions for different knowledge structures. We will consider also other types of changes occurring in the system conducting data integration, including other changes occurring in the profile, changes in the consensus function and more.

Acknowledgment. This research was co-financed by Ministry of Science and Higher Education grant.

References

1. Barthelemy, J.P., Janowitz, M.F.: A formal theory of consensus. Siam Journal of Discrete Mathematics 4, 305–322 (1991)
2. Bhat, S.P., Bernstein, D.S.: Finite-time Stability of Continuous Autonomous Systems. Siam J. Control Optim. 38(3), 751–766 (2000)
3. Chen, G., Lewis, F.L., Xie, L.: Finite-time distributed consensus via binary control protocols. Automatica 47, 1962–1968 (2011)
4. Li, S., Dua, H., Lin, X.: Finite-time consensus algorithm for multi-agent systems with double-integrator dynamics. Automatica 47, 1706–1712 (2011)
5. Maleszka, M., Nguyen, N.T.: A Method for Complex Hierarchical Data Integration. Cybernetics and Systems 42(5), 358–378 (2011)
6. Maleszka, M., Mianowska, B., Nguyen, N.T.: A method for collaborative recommendation using knowledge integration tools and hierarchical structure of user profiles. Knowledge-Based Systems 47, 1–13 (2013)
7. McMorris, F.R., Powers, R.C.: The median procedure in a formal theory of consensus. Siam Journal of Discrete Mathematics 14, 507–516 (1995)
8. Nguyen, N.T.: Advanced methods for inconsistent knowledge management. Springer (2007)
9. Pawlak, Z.: An inquiry into anatomy of conflicts. Journal of Information Sciences 108, 65–78 (1998)
10. Ren, W., Beard, R.W., Atkins, E.M.: A Survey of Consensus Problems in Multi-agent Coordination. American Control Conference, Proceedings of the 2005. IEEE, 1859–1864 (2005)
11. Wang, X., Hong, Y.: Finite-Time Consensus for Multi-Agent Networks with Second-Order Agent Dynamics. In: Proceedings of the IFAC World Congress, Seoul, pp. 15185–15190 (2008)

Part VI

Intelligent Information and Database Systems: Applications

Offline Beacon Selection-Based RSSI Fingerprinting for Location-Aware Shopping Assistance: A Preliminary Result

Wan Mohd Yaakob Wan Bejuri[1,2], Mohd Murtadha Mohamad[1], and Raja Zahilah Raja Mohd Radzi[1]

[1] Faculty of Computing, Universiti Teknologi Malaysia, 81310, Malaysia
[2] Faculty of Information and Communication Technology, Universiti Teknikal Malaysia Melaka, 76100, Malaysia
`mr.wanmohdyaakob.my@ieee.org,`
`{murtadha,zahilah}@utm.my`

Abstract. The location determination in an obstructed area can be extremely challenging particularly when the Global Positioning System (GPS) is blocked. When this happens, users will encounter difficulty in navigating directly on-site, especially within an indoor environment. Occasionally, there is a need to integrate with other sensors in order to establish the location with greater intelligence, reliability, and ubiquity. The use of positioning integration may be useful since it involves as many beacons as necessary to determine positioning. However, the implementation of the integration in the mobile devices platform may lead high computation which in turn could increase power consumption. In this paper, an offline beacon selection-based RSSI fingerprinting is proposed in order to lessen the computation task during the location determination process, as it may cause huge power consumption in mobile devices. By reducing the number of beacons that will be processed, the number of RSSI fingerprinting searches of the location in the spatial database also reduced. Lastly, the preliminary results are presented to illustrate the performance of an indoor environment set-up.

Keywords: Global Navigation System, Wireless LAN and Beacon Selection.

1 Introduction

Nowadays, current technologies gain ground (such as: [1, 2]). This includes also location determination technologies. The knowledge of location position in a shopping mall is a common requirement for many people during shopping activities [3, 4, 5]. Considerable research and development has taken place over the recent years with regards to Location-Based Services (LBS), which can now be supplemented and expanded with the help of ubiquitous methods, and perhaps even replaced in the future. The positioning and tracking of pedestrians in smart environments are done differently compared to when conventional navigation systems are used. This is because it is no longer only the passive systems which execute the positioning on demand that

© Springer International Publishing Switzerland 2015
D. Barbucha et al. (eds.), *New Trends in Intelligent Information and Database Systems,*
Studies in Computational Intelligence 598, DOI: 10.1007/978-3-319-16211-9_31

need to be considered. The advancement of location positioning technologies for location-aware shopping assistance can be used to navigate users in shopping mall environments. There are many types of research prototypes which rely on indoor positioning that can be used to locate a user during shopping activities. The system based on GSM signal strengths have been used to determine the floor of a building the user is on. However, accurate tracking within a floor of a building was not achieved. Likewise, short range positioning technologies (Bluetooth and Near Field Communication) were used to narrow down the estimate of the user's location [6]. However, the drawback of this technology is that it requires the installation of markers in each of the environment [7]. In contrast, there is a localisation technique that does not require the installation of additional infrastructure but requires visual data. However, this technique requires pictures of a huge amount of the surrounding environment to determine the location and this is a major drawback to its application in cell phones. In [8], the integration of dead reckoning with WLAN which uses a particle filter to merge observations was shown. A custom rigidly-mounted belt module was used with a WLAN radio and integrated mechanical sensor (such as two (2) axis accelerometer, gyroscope and pressure sensor). However, there was a lack of in-depth clarification of the step counting algorithm or zero-reference determination for gyroscope used. In fact, the sensor mix was in adequate for establishing a global heading in the dead reckoning component if the user was not facing the same direction every time the system is started up. [9] was a research investigating the suitability of FM Radio/WLAN integration in an indoor positioning system platform. The benefit of this method of integration is that the system is able to increase the positioning accuracy in an indoor environment as it involves multiple beacons during the RSSI fingerprinting. The results showed that it achieve a positioning accuracy of 1m. However, as multiple beacons may lead high computation tasks, it would be better if the number of beacons could be reduced to an optimum number. The work in this paper is aimed at reducing the computation task during the RSSI fingerprinting process as it may cause large power consumptions in mobile devices. The structure of the paper is as follows. Section 2 presents the basic concepts related to location-aware shopping assistance, and puts forward the proposed method. Section 3 presents the experiment setup and the details of the preliminary results. In conclusion, a discussion and the future direction of the project are provided in Section 4.

2   Offline Beacon Selection-Based RSSI Fingerprinting

The concept of location-aware shopping assistance relates to positioning determination across all environments [10, 11]. It usually takes a multi-sensor approach, supplementing standalone positioning with other signals, motion sensors and environmental features [12, 13, 14]. In general, the system is based on WLAN/FM radio positioning integration. The reason of this system integration is to ensure the ubiquity of the positioning system, thus enabling the sensors to help each other in determining positioning [15, 16, 17, 18, 19]. In essence, this paper proposes a method of reducing the computation rate during the RSSI fingerprinting by reducing the number of beacons that need to be processed. This may also reduce the number of location searches in the location spatial database. Fig. 1 shows the block diagram of the offline beacon selection-based RSSI fingerprinting. At first, the system will receive the WLAN and FM radio signals simultaneously. However, in an obstructed environment, the signals will deteriorate due to the blockage between the beacon and the mobile device receiver. Theoretically, the signal path loss obeys the distance power law as described below:

$$P_r(d) = P_r(d_0) - 10nlog\left(\frac{d}{d_0}\right) + X_\sigma \tag{3}$$

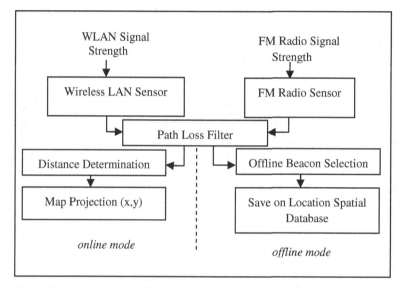

Fig. 1. Block Diagram of Offline Beacon Selection-based RSSI Fingerprinting

where Pr is the received power; $P_r(d_0)$ is the received power at the d_0(knownas the reference distance), and$n$ refers to the path loss exponent which indicates the rate of the path loss which increases with the distance. This depends on the surroundings, building type and other obstructions. In addition,d_0 is the close-in reference distance (1m) and d is the distance of the separation between the RF signal transmitter and the receiver (with the transmitter being the AP and the receiver is the mobile device receiver). The term X_σ is a zero mean Gaussian random variable with the standard deviation σ. Equation (3) is modified to incorporate the Wall Attenuation Factor (WAF) [20]. The modified distance power law is given as Equation (4):

$$P_r(d) = P_r(d_0) - 10nlog\left(\frac{d}{d_0}\right) - T * WAF \tag{4}$$

where, T is number of walls between the transmitter and the receiver.

$$d = e^{\left(\frac{\Pr(d_0)-\Pr(d)-T*WAF}{10}\right)} \tag{5}$$

The offline mode of this algorithm must be activated in order to locate a trade-off point between the number of beacons used and the accuracy they can achieve. The offline mode is used to collect spatial locations and signals to be saved into the spatial database. With this, the location information can be established during the online mode by comparing the current received signals with the signals within the spatial location database (that was collected during offline mode). The beacon is selected during the offline mode using the Information Gain-based beacon selection method (InfoGain for short) [21]. For example, in a grid-based location system where m is the number of grids and m is the total number of beacons detectable, each beacon $(AP_i, 1 \leq i \leq m)$ is considered a feature and each $\text{grid}(G_j, 1 \leq j \leq n)$ is expressed by these m features. For a particular grid G^*, signal samples from the beacons are accumulated offline and the average signal strength from $Beacon_i$ is taken as the value of the ith feature of G^*. There is also a possibility that some beacons may not be detected in G^*due to their physical locations as well as the characteristics of signal propagation. When this happens, the features of the corresponding missing beacons assume a default value, which is fixed at -95, i.e. the minimum strength of the signal received in the environment. The InfoGain carries out the beacon selection by the evaluating the worth of each feature i.e. beacon, in terms of their discriminative powers, following which the highest ones are selected. The discriminative power of the feature $Beacon_i$is measured by the information gained when its value is known. Specifically, it is calculated as the reduction in entropy as follows;

$$InfoGain(Beacon_i) = H(G) - H(G|Beacon_i) \tag{6}$$

where $H(G) = \sum_{j=1}^{n} \Pr(G_j) \log \Pr(G_j)$ is the entropy of the grids when the value of $Beacon_i$'s value is not known. Here, $\Pr(G_j)$ is the prior probability of gridG_j , which can be uniformly distributed if a user can be equally expected to be in any grid. $H(G|Beacon_i) = -\sum_v \sum_{j=1}^{n} \Pr(G_j, Beacon_i = v) \log \Pr(G_j, Beacon_i = v)$ computes the conditional entropy of grids given $Beacon_i$'s value. $v$is one possible value of signal strength from $Beacon_i$ and this summation is taken over all possible values of the beacon. The information gained for each $Beacon_i$,is computed using (6) and the top k beacons with the highest values are selected. Upon completion of the offline mode tasks, the online mode can be activated. As mentioned previously, the online mode is used to determine the location information by comparing the current received signals with the signals in the spatial location database. Whilst on the online mode, the distance of every device or node will be calculated by means of the Euclidean Distance equation (7) [22].

$$Distance = \sqrt{((X_1 - X_2)^2 + (Y_1 - Y_2)^2)} \tag{7}$$

The distance for every device in the network is calculated by the location server and all the distances are compared to determine the nearest device from the chosen mobile node. The method undertaken for the nearest computation is the nearest neighbour(s) in signal space (NNSS). The idea here is to calculate the distance (in signal space) between the observed set of SS measurements (ss_1, ss_2, ss_3) and the recorded SS (ss_1', ss_2', ss_3') at a fixed set of locations, and then selecting the location which minimises the distance. To carry out calculations based on three (3) measurements, Equation (7) can accede to become equation (8) with D being the distance between the observed signal and the recorded signal:

$$D = \sqrt{((ss_1 - ss_1')^2 + (ss_2 - ss_2')^2 + (ss_3 - ss_3')^2)} \tag{8}$$

3   Preliminary Result

This section discusses the preliminary results which are basically based on the WLAN data collected. The data were used to test the proposed system. In order to do that, the experiments need to carried out at Level 2, N28, at Universiti Teknologi, Malaysia, Johor in an area of approximately 120 m × 95 m (for additional details refer to Fig. 2),whereby WLAN and FM radio signal strength data were collected along the corridor in four (4) orientations. Data recording, processing and most of the positioning tests were completed using the HTC HD mini smart phones, with a Windows Mobile 6 platform, and equipped with Atheros AR5007EG wireless network adapters and stereo FM Radio. In Fig. 3, Fig. 4, Fig. 5 and Fig. 6, show the RSSI strength measured against the distance between the mobile node and the access points in different user orientations. As noted, the RSSI shows a normal trend where the value decreases if the distance increases. This trend is almost the same in Fig. 7, Fig. 8, Fig. 9 and Fig. 10.

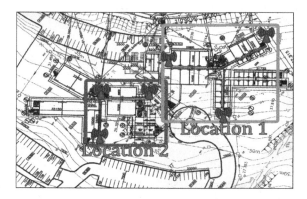

Fig. 2. Experiment Location (Location 1, Location 2). The icon of wireless transmitter refers to the location of the WLAN access point which is attached with the FM local transmitter.

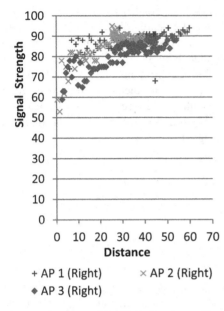

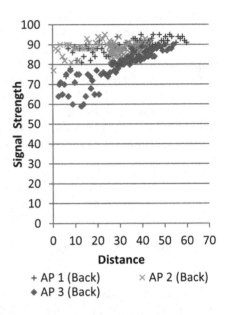

Fig. 3. Distance and RSSI during data collection in location 1 (User orientation: 0°). (Note: Signal strength in –dBm and distance in metres).

Fig. 4. Distance and RSSI during data collection in location 1 (User orientation: 90°). (Note: Signal strength in –dBm and distance in metres).

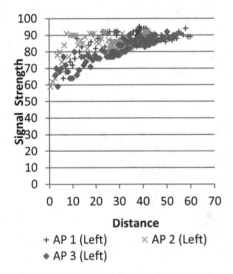

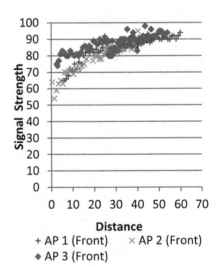

Fig. 5. Distance and RSSI during data collection in location 1 (User orientation: 180°). (Note: Signal strength in –dBm and distance in metres).

Fig. 6. Distance and RSSI during data collection in location 1 (User orientation: 270°). (Note: Signal strength in –dBm and distance in metres).

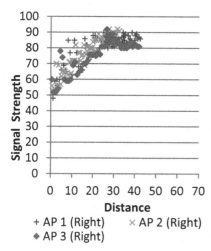

+ AP 1 (Right)   × AP 2 (Right)
◆ AP 3 (Right)

Fig. 7. Distance and RSSI during data collection in location 2 (User orientation: 0°) (Note: Signal strength in –dBm and distance in metres)

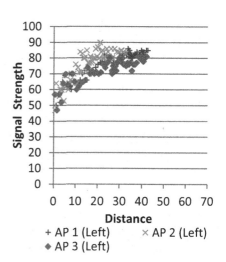

+ AP 1 (Left)   × AP 2 (Left)
◆ AP 3 (Left)

Fig. 8. Distance and RSSI during data collection in location 2 (User orientation: 180°). (Note: Signal strength in –dBm and distance in metres).

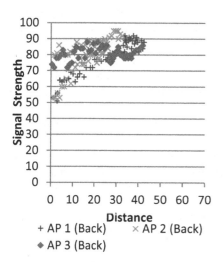

+ AP 1 (Back)   × AP 2 (Back)
◆ AP 3 (Back)

Fig. 9. Distance and RSSI during data collection in location 2 (User orientation: 90°) (Note: Signal strength in –dBm and Distance in metres)

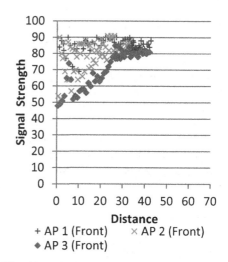

+ AP 1 (Front)   × AP 2 (Front)
◆ AP 3 (Front)

Fig. 10. Distance and RSSI during data collection in location 2 (User orientation: 270°). (Note: Signal strength in –dBm and Distance in metres).

The following explains the wide disparity in signal strengths between points at similar distances: the design of the rooms in the building, the assignment of base stations, and the position of the mobile user all affect the received signal. Particularly in Fig. 3, Fig. 4, Fig. 5 and Fig. 6, the signal strength value obtained by the mobile device varied according to the user's orientation in location 1. For instance, the signal strength value of AP2 recorded its highest during user orientation: 270° (see Fig. 6) compared with the signal strength obtained at 0° (see Fig. 3). At the same time, the signal strength value obtained by the mobile device in location 2 is portrayed in Fig. 7, Fig. 8, Fig. 9 and Fig. 10 and shows that although the distance between the mobile device receiver and AP is close, the signal strength also depends on the user's orientation. In this study, the signal strength value of AP3 was the highest during the user's orientation: 270° (see Fig. 10) compared with the signal strength obtained at 0° (in Fig. 7). The explanation for this is that the orientation of the user could have possibly contributed towards blockage and therefore reduced the signal strength obtained by the mobile device. The same rationale can be applied to the other values in the graph.

4   Conclusion and Future Works

This paper discussed the problem and solutions regarding the development of a location-aware shopping assistance, specifically in terms of its low computation, since this may lead large power consumptions. To do this, there needs to be a decrease in the beacons processed during location determination. Towards this end, the use of an offline beacon selection-based RSSI fingerprinting for location-aware shopping assistance has been proposed. The reduction of the number of beacons processed will also reduce the number of RSSI fingerprinting searches in the location spatial database. Preliminary results have shown that most signal strengths decreased when the distance was increased. It also showed that the signal strength was dependant on the user's orientation as the user's body acted as an obstacle between the mobile device and the WLAN access point. For future works, a continuation of experiments to ascertain the extent to which the approach can be more ubiquitous in other environments will be carried out, using these results combined with other mobile internal sensors, e.g. cameras. This part will undoubtedly be undertaken by the next researcher.

Acknowledgement. This work has been funded by the UTM Research University Grant (Title: The Prototype of Indoor Tracking System based on FM Radio) under project no. Q.130000.2628.1J10.

References

1. Othmane, L.B., Weffers, H., Ranchal, R., Angin, P., Bhargava, B., Mohamad, M.M.: A Case for Societal Digital Security Culture. In: Janczewski, L.J., Wolfe, H.B., Shenoi, S. (eds.) SEC 2013. IFIP Advances in Information and Communication Technology, vol. 405, pp. 391–404. Springer, Heidelberg (2013)

2. Kheirabadi, M.T., Mohamad, M.M.: Greedy Routing in Underwater Acoustic Sensor Networks: A Survey. International Journal of Distributed Sensor Networks, 2013, Article ID 701834, pages (2013)
3. Schougaard, K.R., Grønbæk, K., Scharling, T.: Indoor Pedestrian Navigation Based on Hybrid Route Planning and Location Modeling. In: Kay, J., Lukowicz, P., Tokuda, H., Olivier, P., Krüger, A. (eds.) Pervasive 2012. LNCS, vol. 7319, pp. 289–306. Springer, Heidelberg (2012)
4. Roth, S.P., Tuch, A.N., Mekler, E.D., Bargas-Avila, J.A., Opwis, K.: Location Matters, Especially for Non-Salient Features–An Eye-Tracking Study on the Effects of Web Object Placement on Different Types of Websites. International Journal Human-Computer Studies 71(3), 228–235 (2013)
5. Yang, X.M., Li, X.Y.: Research on Data Preprocessing Technology in Location Based Service. Advanced Materials Research 740, 134–139 (2013)
6. Kourogi, M., Sakata, N., Okuma, T., Kurata, T.: Indoor/Outdoor Pedestrian Navigation with an Embedded GPS/RFID/Self-contained Sensor System. In: Pan, Z., Cheok, D.A.D., Haller, M., Lau, R., Saito, H., Liang, R. (eds.) ICAT 2006. LNCS, vol. 4282, pp. 1310–1321. Springer, Heidelberg (2006)
7. Schmidt, A., Fularz, M., Kraft, M., Kasiński, A., Nowicki, M.: An Indoor RGB-D Dataset for the Evaluation of Robot Navigation Algorithms. In: Blanc-Talon, J., Kasinski, A., Philips, W., Popescu, D., Scheunders, P. (eds.) ACIVS 2013. LNCS, vol. 8192, pp. 321–329. Springer, Heidelberg (2013)
8. Evennou, F., Marx, F.: Advanced Integration of WiFi and Inertial Navigation Systems for Indoor Mobile Positioning. EURASIP Journal on Applied Signal Processing 2006, 164–164 (2006)
9. Popleteev, A.: Indoor positioning using FM radio signals. University of Trento (2011)
10. Wu, Y., Pan, X.: Velocity/Position Integration Formula Part I: Application to In-Flight Coarse Alignment. IEEE Transactions on Aerospace and Electronic Systems 49(2), 1006–1023 (2013)
11. Fang, S.-H., Wang, C.-H., Chiou, S.-M., Lin, P.: Calibration-Free Approaches for Robust Wi-Fi Positioning against Device Diversity: A Performance Comparison. In: 75th IEEE Vehicular Technology Conference, pp. 1–5. IEEE Press, New York (2012)
12. Bejuri, W.M.Y.W., Saidin, W.M.N.W.M., Bin Mohamad, M.M., Sapri, M., Lim, K.S.: Ubiquitous Positioning: Integrated GPS/Wireless LAN Positioning for Wheelchair Navigation System. In: Selamat, A., Nguyen, N.T., Haron, H. (eds.) ACIIDS 2013, Part I. LNCS, vol. 7802, pp. 394–403. Springer, Heidelberg (2013)
13. Bejuri, W.M.Y.W., Mohamad, M.M.: Performance Analysis of Grey-World-based Feature Detection and Matching for Mobile Positioning Systems. Sensing & Imaging 15(1), 1–24 (2014)
14. Bejuri, W.M.Y.W., Mohamad, M.M., Sapri, M., Rosly, M.A.: Ubiquitous WLAN/Camera Positioning using Inverse Intensity Chromaticity Space-based Feature Detection and Matching: A Preliminary Result. ArXiv Prepr. ArXiv12042294, 2012. In: International Conference on Man-Machine Systems, UniMAP, Penang (2012)
15. Bejuri, W.M.Y.W., Mohamad, M.M., Sapri, M., Rosly, M.A.: Performance Evaluation of Mobile U-Navigation Based on GPS/WLAN Hybridization. Journal of Convergence Information Technology 7(12), 235–246 (2012)
16. Bejuri, W.M.Y.W., Mohamad, M.M., Sapri, M., Rosly, M.A.: Investigation of color constancy for ubiquitous wireless LAN/Camera positioning: an initial outcome. International Journal of Advancements in Computing Technology 4(7), 269–280 (2012)

17. Bejuri, W.M.Y.W., Mohamad, M.M., Sapri, M., Rahim, M.S.M., Chaudry, J.A.: Performance Evaluation of Spatial Correlation-based Feature Detection and Matching for Automated Wheelchair Navigation System. International Journal of Intelligent Transportation Systems Research 12(1), 9–19 (2014)
18. Bejuri, W.M.Y.W., Mohamad, M.M., Sapri, M.: Ubiquitous positioning: A taxonomy for location determination on mobile navigation system. Signal & Image Processing: An International Journal (SIPIJ) 2(1), 24–34 (2011)
19. Bejuri, W.M.Y.W., Mohamad, M.M.: Wireless LAN/FM Radio-based Robust Mobile Indoor Positioning: An Initial Outcome. International Journal of Software Engineering & Its Applications 8(2) (2014)
20. Narzullaev, A., Park, Y., Yoo, K., Yu, J.: A Fast and Accurate Calibration Algo-rithm for Real-Time Locating Systems Based on The Received Signal Strength Indica-tion. AEU - International Journal of Electronics and Communications 65(4), 305–311 (2011)
21. Chen, Y., Yang, Q., Yin, J., Chai, X.: Power-efficient access-point selection for indoor location estimation. IEEE Transactions Knowledge Data Engineering 18(7), 877–888 (2006)
22. Laoudias, C., Michaelides, M.P., Panayiotou, C.G.: Fault detection and mitigation in WLAN RSS fingerprint-based positioning. Journal of Location Based Services 6(2), 101–116 (2012)

Web-Based Implementation
of Data Warehouse Schema Evolution

Jai W. Kang, Fnu Basrizal, Qi Yu, and Edward P. Holden

Rochester Institute of Technology
152 Lomb Memorial Drive, Rochester, NY, USA
{jai.kang,fxb3717,qi.yu,edward.holden}@rit.edu

Abstract. An organization collects current and historical data for a data warehouse from disparate sources across the organization to support management for making decisions. The data sources change their contents and structure dynamically to reflect business changes or organization requirements, which causes data warehouse evolution in order to provide consistent analytical results. This evolution may cause changes in contents or a schema of a data warehouse. This paper adapts a schema evolution method to address the data warehouse evolution given a data warehouse is built as a multidimensional schema. While existing works have identified and developed schema evolution operations based on conceptual models to enforce schema correctness, only a few have developed software tools to enforce schema correctness of those operations. They are also coupled with specific DBMSs (e.g., SQL Server) and provide limited GUI capability. This paper aims to develop a web-based implementation to support data warehouse schema evolution. It offers all the benefits that Internet browser-based applications provide, by allowing users to design, view, and modify data warehouse schema graphically. This work focuses on evolution operations in dimensional tables, which includes changes in levels, hierarchies, and paths. Schema correctness for each schema evolution operation is ensured by procedural codes implemented in PHP.

Keywords: data warehouse, schema evolution, web-based application, dimensional modelling, data mart, star schema.

1   Introduction

An organization develops and maintains a data warehouse to support management for making decisions. The data warehouse houses both current and historical transactional data from various data sources across the organization. A data warehouse can be a large normalized enterprise-wide relational database from which de-normalized multidimensional databases can be generated for end users to understand and write queries easily against.

While data marts can be generated from the enterprise data warehouse, modeling as a star schema for each business process can develop a data mart independently. The data mart as a dimensional model classifies data as facts and dimensions. Facts contain numeric data as a result of a business, called a measure. Dimensions contain descriptive

© Springer International Publishing Switzerland 2015   313
D. Barbucha et al. (eds.), *New Trends in Intelligent Information and Database Systems,*
Studies in Computational Intelligence 598, DOI: 10.1007/978-3-319-16211-9_32

information to filter and aggregate the facts. Kimball & Ross [4] advocate that an enterprise data warehouse is a union of data marts as long as all data marts are conformed to following the same convention and standards, which is called a bus architecture.

A dimension consists of mostly textual attributes that can be characterized as levels and hierarchies. Malinowski & Zimányi [5] state that the level allows data warehouse users to explore the measures from a different perspective of analysis. The levels may create a hierarchy so that users can access the measures in detailed or generalized information. The hierarchy is defined as a relationship between two levels. A path is defined as a set of hierarchies in a dimension table. A dimension can have multiple paths for alternative or different analysis criteria. A path represents a specific analysis criterion.

Data of a data warehouse come from many different data sources. These data sources tend to change their contents and structure dynamically to reflect business changes or organization requirements. Changes in contents and structures of data sources may cause schema changes in a data warehouse [3]. Therefore, data warehouse evolution can be caused by the dynamic nature of data sources. The evolution may cause changes in content or a schema of a data warehouse. Slowly changing dimension (SCD) as proposed by Kimball & Ross [4] handles changes that occur for the content of data warehouse. For schema changes, there are three different methods to address a data warehouse evolution: (1) schema evolution, (2) schema versioning when a data warehouse is built as a multidimensional schema and (3) view maintenance when a data warehouse is built as a collection of materialized views [6].

Schema change operations require taking into account hierarchies in a data warehouse in order to ensure that the schema changes do not violate existing hierarchies. Hierarchy has an important function to process information that allows users to view and explore data in a data warehouse at various levels of granularity. Specifically, using hierarchy, a user can view data from a general view to a detailed view through roll-up, drill-down, slicing and dicing operations.

This paper aims to handle schema evolutions that occur in dimension tables of a data warehouse. Schema evolution can be structure changes in levels, hierarchies, and paths of dimension tables. The proposed approach implements a web-based schema evolution that performs and validates schema evolution operations to achieve schema correctness in a data warehouse. A web-based application offers key benefits over platform specific ones for obvious reasons, as users of a web-based system do not require having their computers equipped with particular software and hardware, but an Internet browser.

The rest of the paper is organized as follows. Section 2 presents dimension hierarchical categorization and classification, and existing works on the implementation of data warehouse schema evolution. Web-based implementation is given by adopting a Multi-level Dictionary Definition (MDD) approach in Section 3. Finally, Section 4 concludes the paper and discusses future work.

2   Background, Related Works and Motivation

2.1   Dimensional Hierarchies

A data warehouse offers multidimensional models for users to analyze a large volume of data. A dimensional model classifies data as a fact and dimensions, which are modeled as a star schema. Users can then explore the facts in various perspectives using the hierarchical nature of dimension attributes. These hierarchies allow users to view and explore data at various levels of granularity. Using hierarchy, a user can view data from a general view to a detailed view through operations like roll-up and drill-down.

Talwar & Gosain [7] recognize the necessity of properly categorizing dimension hierarchies so as to properly model them during evolution. They define a hierarchy as a set of binary relationships existing between dimension levels, where a dimension level participating in a hierarchy is called a hierarchical level. Based on two consecutive levels of a hierarchy, the higher level is called *parent* and the lower level is called *child*. The *leaf* level is the finest level in a hierarchical path, which has no child; the last level, i.e., the one that has no parent level called *root* level. A hierarchy represents some organizational, geographical, or other type of structure that plays a key role for analysis [5].

Talwar & Gosain [7] performs a literature survey on hierarchical classification of data warehouse proposed by various researchers based on certain parameters from a wide range of business scenarios. Their classifications include ones by Malinowski & Zimányi [5] and Banerjee et al. [2] such as simple, multiple, parallel dependent and independent hierarchies. These hierarchies are described in Table 1, which also includes non-strict, non-onto and non-covering hierarchies defined by Talwar & Gosain [8].

2.2   Related Works

Existing works on the implementation of data warehousing schema evolution include Banerjee and Davis [2] and Talwar and Gosain [7]. Both works define a formal meta-model for data warehouse core features using Uni-Level Description (ULD) language and a Multi-level Dictionary Definition (MDD). The core features of a data warehouse conceptual model refers to a multidimensional model consisting of facts, dimensions, measures, levels and hierarchies that conform to: 1) a many-to-one relationship between a fact and a dimension, 2) a one-to-many relationship between two levels in a dimension, and 3) hierarchies in a dimension have a single path to roll-up or drill-down operations. While the ULD definition provides formal semantics and uniform representation of schema data, meta-model layers and their inter-dependencies, the MDD allows direct implementation in a relational database system. Banerjee and Davis [2] implemented such advanced features as multiple hierarchies, non-covering, non-onto, and non-strict hierarchies. Talwar and Gosain [7] implemented such extended hierarchies as multiple alternative, parallel dependent, and parallel independent hierarchies. Both works use stored procedures and triggers to enforce schema correctness in Microsoft SQL Server.

Table 1. Dimension hierarchies [7]

Hierarchy	Description
Simple	This hierarchy can be represented as tree. They use only one criterion for analysis
Multiple	This hierarchy contains several non-exclusive simple hierarchies and share some levels, but all of these hierarchies have the same analysis criterion
Parallel	Parallel hierarchies arise when there are multiple hierarchies, accounting for different analysis criteria. Two types: Parallel independent/dependent hierarchy depending sharing any level(s) or not
Non-strict	This exists when a dimension can have many-to-many relationships
Non-onto	This hierarchy exists when lower level can exist without a corresponding data in the higher level to roll-up to.
Non-covering	This exists when at least one member whose logical parent is not in the level immediately above the member

2.3   Motivation

Users of these existing applications perform data warehouse schema evolution on a specific DBMS, such as Microsoft SQL Server. The system offers a simple GUI that prompts users to supply arguments or using the SQL command line in order to execute stored procedures.

The paper aims to make a key extension of existing approaches by moving from platform specific applications to a web-based system. The benefits of a web-based system can be related to minimizing software installation, update and training among others. Google search returns over 140 million hits for the query "Benefits of web-based application" as of June 2014. Furthermore, [9] includes "Enterprise applications migrating to browsers" in Gartner's 2014 top ten technology trends. In fact, such a trend has already been observed during the past decades in the application development industry.

3   Web-Based Implementation of Schema Evolution Using MDD

3.1   Implementation Example

This project adopts the MDD approach of schema evolution developed by Atzeni et al. [1], which manages schema and describes its components. It follows the same approach for designing the data model constructs as [8] and [2]. It uses MDD to represent core features of a data warehouse and meta-constructs of ULD instead of the supermodel proposed by Atzeni et al. [1]. The constructs are implemented in a relational database system.

Fig. 1 shows an example of data warehouse schema to demonstrate schema evolution using the MDD approach. The example is taken from Banerjee & Davis [2] and Talwar & Gosain [7]. The schema shows four dimensions: product, customer, store and location. Location dimension here is used specifically to show an implementation of non-covering, non-strict and non-onto hierarchies. The tabular format of constructs includes Schema, Fact, Primary Key and Measure, but they are not shown here due to space restriction. Tabular constructs of Dimension, Level, Hierarchy, Path, Non-Covering, Non-Onto and Non-Strict are given in Tables 2 through 8 respectively.

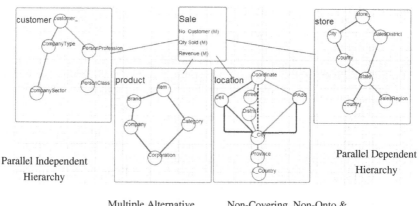

Fig. 1. Sales schema

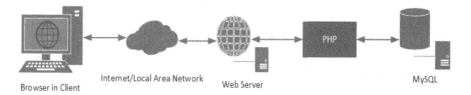

Fig. 2. Application architecture

Table 2. Dimension construct

Id	Name	DLevel	DPKey	DHierarchy	Dpath
d1	Product	L1,L2,L3,L4,L5	pk1	h1,h2,h3,h4,h5	p1,p2
d2	Customer	L6,L7,L8,L9,L10	pk2	h6,h7,h8,h9	p3,p4
d3	Store	L11,L12,L13, L14,L15,L16,L17	pk3	h10,h11,h12,h13, h14,h15,h16	p5,p6
d4	Location	L18,L19,L20,L21, L22,L23,L24,L25	pk4	h17,h18,h19,h20, h21,h22,h23,h24h25	p7,p8,p9

Table 3. Level Construct

Id	Name
L1	Item
L2	Brand
L3	Company
L4	Category
L5	Corporation
L6	Customer
L7	Company Type
L8	Company Sector
L9	Person Profession
L10	Person Class
L11	Store
L12	City
L13	County
L14	State
L15	Country
L16	Sales District
L17	Sales Region
L18	Coordinate
L19	Street
L20	District
L21	L_City
L22	Province
L23	L_Country
L24	IPAdd
L25	Cell

Table 4. Hierarchy Construct

Id	Parent Level	Child Level
h1	L2	L1
h2	L3	L2
h3	L5	L3
h4	L4	L1
h5	L5	L4
h6	L7	L6
h7	L8	L7
h8	L9	L6
h9	L10	L9
h10	L12	L11
h11	L13	L12
h12	L14	L13
h13	L15	L14
h14	L16	L11
h15	L14	L16
h16	L17	L14
h17	L19	L18
h18	L20	L19
h19	L21	L20
h20	L22	L21
h21	L23	L22
h22	L24	L18
h23	L21	L24
h24	L25	L18
h25	L21	L25

Table 5. Path Construct

Id	Phierarchy
p1	h1,h2,h3
p2	h4,h5
p3	h6,h7
p4	h8,h9
p5	h10,h11,h12,h13
p6	h14,h15,h16
p7	h17,h18,h19,h20,h21
p8	h22,h23,h20,h21
p9	h24,h25,h20,h21

Table 6. Non-Covering construct

Id	NCParent	NCChild	Conformance
nc1	L21	L18	p7

Table 7. Non-Onto construct

Id	NOParent	NOChild	Conformance
no1	L21	L24	h23

Table 8. Non-Strict construct

Id	NSParent	NSChild	Conformance
ns1	L21	L25	h25

3.2   Architecture

Schema evolution in this work is implemented as a web-based application to interact with a particular schema. Users perform schema evolution operations such as add/edit schema, dimension, level hierarchy, and path. Architecture of the application is shown in Fig. 2. It consists of a web interface as a front end/client side application and PHP as a back end/server side application that access MySQL.

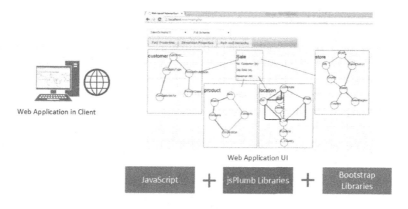

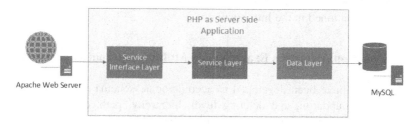

Fig. 3. Client-side application architecture

Fig. 4. Server-side application architecture

Client side application shows a user interface to allow users to view schema structure and perform schema evolution operations (Fig. 3). It is developed using JavaScript and API libraries of jspPlum and Bootstrap. Data exchange between client and server is done by using AJAX (Asynchronous JavaScript and XML) in JSON (JavaScript Object Notation) format. Then, backend processes in the server side validate the operations with associated constraints.

Fig. 4 shows the PHP/JSP server side application, which comprises a service interface, service and data layers. The service interface layer is responsible for routing client's requests to the service layer. It then calls a particular function in the service layer according to the request. The service layer contains all available methods the application has. The actual computation of the schema evolution operations happens in the data layer, which has access to the meta-schema database to query and modify the database. A return from the data layer can be data or status of the operations.

Data layer functions validate schema evolution operations with corresponding constraints. These evolution operations include multiple alternative, parallel dependent and parallel independent hierarchies [8] and core features and extended hierarchies (non-covering, non-onto and non-strict hierarchy), constraint functions [2].

3.3   Meta-schema Database

Meta-schema database stores metadata information of schemas in the form of relational tables. Schema evolution operations modify records in the tables. The meta-database consists of a schema construct table, which stores sets of dimension, fact and primary key constructs. The dimension construct has level, hierarchy and path constructs. The fact construct has measure constructs.

Conforming Dimensions

The application allows a user to create more than one schema to represent different data marts. The dimension, hierarchy, level, fact and measure constructs are available to all schemas, which means every change that occurs in the constructs are reflected to all existing schemas. The objective of this feature is to achieve conformity of all data marts, which emphasizes Kimball's data warehouse being a union of conformed data marts as mentioned in the Introductiuon

Schema Evolution Algorithm Example: add Multiple Alternative Hierarchy

Many functions have been developed to accommodate schema evolution operations such as adding, updating and deleting level, hierarchy, path, dimension, fact and measure in a schema. Below is an algorithm that adds a new multiple alternative (MA) hierarchy. The algorithm requires a dimension name, parent and child levels as input arguments. The parent level should be a level in an existing path. For the child level, the user must create a new attribute as the child level before creating a new MA hierarchy. The output is a new path added into the dimension. The new path created should converge to the same level in the existing path.

Input: dimension name, child level, and parent level
Output: the updated dimension with a new path

Step 1: Check if the dimension is valid.
Step 2: Check if the dimension has at least one existing path.
Step 3: Check if the input child and parent levels are valid levels.
Step 4: Check if the parent level is part of an existing path.
Step 5: Check if the parent level is not the leaf level of the existing path.
Step 6: Create a new path by calling the *addSimpleHierarchy* function to create two hierarchies: one between the leaf and child level and another one between the child and parent level. If the parent level is not the root level, call *addSimpleHierarchy* to add the rest of the hierarchies of the existing path into the new path.

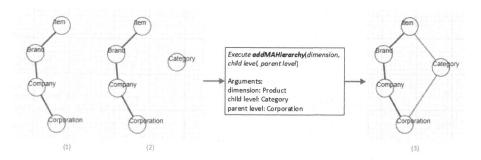

Fig. 5. Multiple Alternative hierarchy addition process (*addMAHierarchy*)

Fig. 5 illustrates the algorithm. There should be at least one existing path (Fig. 5(1)) and the new path must use the same level of the existing path as a parent level. In this case, the parent level is *Corporation*. For the child level, the user must add the new level, *Category*, before adding the new path (Fig. 5(2)). Then, the application executes the *addMAHierarchy* function, which requires three arguments namely, *Product* as the dimension name, *Category* as the child level and *Corporation* as the parent level. The function then creates a new path containing two simple hierarchies: 1) one between the leaf level of the existing path, which is *Item,* and the child level argument, which is *Category,* and 2) another one between the child level, *Category* and parent level arguments, *Corporation*. The *addMAHierarchy* function actually invokes the *addSimpleHierarchy* function to create these hierarchies and to add them into the path. As a result, Fig. 5(3) shows a newly created path as a Multiple Alternative hierarchy.

4   Conclusions

This paper presents a web-based application to implement data warehouse schema evolution, which allows the user to create, view and modify a schema of a multi-dimensional model based data warehouse. The application has objectives to allow the users to perform schema evolution operations in a user-friendly environment and to enforce schema correctness. The Internet browser-based system offers not only a platform independent environment but also broad user types beyond database experts. The application uses PHP functions to implement schema evolution operators to ensure schema correctness, and supports schema evolution over core features and extended hierarchies in multi-dimensional models. It also uses the MDD approach to not only implement such metadata constructs as dimension, level, hierarchy and path on a relational database, but also satisfies conforming dimensions among data mart schemas.

This work can be extended to include evolutions of more generalized dimensional hierarchies. Next level challenges would be extending this metadata-level conceptual schema evolution towards the physical level. The extension will work on modification of an underlying data warehouse schema and data in order to perform schema evolution operations.

References

1. Atzeni, P., Cappellari, P., Bernstein, P.A.: A multilevel dictionary for model management. In: Delcambre, L.M.L., Kop, C., Mayr, H.C., Mylopoulos, J., Pastor, Ó. (eds.) ER 2005. LNCS, vol. 3716, pp. 160–175. Springer, Heidelberg (2005)
2. Banerjee, S., Davis, K.C.: Modeling data warehouse schema evolution over extended hierarchy semantics. Journal on Data Semantics XIII 5530, 72–96 (2009)
3. Bebel, B., Eder, J., Koncilia, C., Morzy, T., Wrembel, R.: Creation and management of versions in multiversion data warehouse. In: Symposium on Applied Computing, pp. 717–723. ACM, New York (2004)
4. Kimball, R., Ross, M.: The data warehouse toolkit, 3rd edn. John Wiley & Sons, Indianapolis (2013)
5. Malinowski, E., Zimányi, E.: OLAP hierarchies: A conceptual perspective. In: Persson, A., Stirna, J. (eds.) CAiSE 2004. LNCS, vol. 3084, pp. 477–491. Springer, Heidelberg (2004)
6. Oueslati, W., Akaichi, J.: A survey on data warehouse evolution. International Journal of Database Management Systems 2(4), 11–24 (2010)
7. Talwar, K., Gosain, A.: Hierarchy classification for Data Warehouse: A Survey. Procedia Technology 6, 460–468 (2012)
8. Talwar, K., Gosain, A.: Implementing schema evolution in data warehouse through complex hierarchy semantics. International J of Scientific & Eng. Research. 3(7) (2012)
9. Gartner's Top 10 IT Trends of (2014), http://www.news-sap.com/gartners-top-10-trends-2014/

The Usage of the Agent Modeling Language for Modeling Complexity of the Immune System

Martina Husáková

University of Hradec Králové,
Faculty of Informatics and Management,
Hradec Králové,
Czech Republic
martina.husakova.2@uhk.cz

Abstract. Immune system is complex system which is composed of thousands entities interacting with each other. If we want to understand the immune system behavior, we can use the bottom-up approach investigating interactions occurring in the low-levels (e. g. molecular level) where particular biological entities exist. Multi-agent systems are bottom-up approach used for exploration of the immunity, but the complexity complicates clarifying immune processes. The paper investigates the Agent Modeling Language (AML) for conceptual modeling of particular immune properties and processes. T-cell dependent B-cell activation is used as the case study for finding out if the language can offer value added for conceptual modeling in computational immunology.

Keywords: AML, Conceptual modeling, Computational immunology, Agent, StarUML, Multi-agent system.

1   Introduction

Investigation of biological systems requires deep study of processes appearing in various scales, i. e. molecular, cellular, tissue or organ. Various approaches exploring these levels of organization exist. The paper deals with the computational immunology uncovering amazing world of the immune system (IS) behavior. Main attention is paid for transferring facts about domain of interest into more formal shape – diagrams which can ease communication with immunologists owning expert knowledge and the process of simulators development. Main aim of the paper is to investigate whether and how the Agent Modeling Language (AML) can be applied for conceptual models development in computational immunology. The paper is organized as follows. Section 2 introduces conceptual modeling approaches used in computational immunology. Structure of the AML is presented in section 3. Section 4 practically demonstrates the application of the AML for T-cell dependent B-cell activation. Section 5 describes the usefulness of AML-based diagrams. Section 6 concludes the paper and mentions future direction in the research of the topic.

© Springer International Publishing Switzerland 2015   323
D. Barbucha et al. (eds.), *New Trends in Intelligent Information and Database Systems*,
Studies in Computational Intelligence 598, DOI: 10.1007/978-3-319-16211-9_33

2 Conceptual Modeling in Computational Immunology

Petri nets (state-transition systems, [1]) has been already used for modeling chemical reactions, modeling catalyzed reactions [3] or dynamic behavior of infection caused by Mycobacterium marinum in tropical freshwater fish Zebrafish [4]. Petri nets are not a part of conceptual languages family, but dynamic behavior of the investigated system can be visualized and represented as graphs.

Entity-relationship modeling (ERM) is the approach for data modeling in software engineering [5]. Usage of ERM is mentioned e. g. in [6] where ERM is used for modeling relations between enzymes, properties of proteins and biopolymers.

Statecharts are the extensions of state-transition diagrams which suppose that all states of the system are known. This is true only for less complex and smaller systems. Statecharts are useful for complex and larger systems [7]. State diagrams were used for modeling dynamic behavior of immune cells in the lymph node where agent-based approach was used [8]. Humoral immune response was investigated in the lymph node together with statistical analysis of the occupancy of various compartments of the lymph node e. g. in [9].

The Biochart language was proposed as the extension of state diagrams for the possibility in modeling biological processes occurring in biological systems [10]. This approach was applied e. g. in modeling chemotaxis of Escherichia Coli.

Ontologies model generic or domain knowledge under the terms of ontological engineering [11]. Bio-ontologies are specific classes of ontologies offering common vocabulary for communication between people, machines and people with machines. The OBO Foundry (Open Biological and Biomedical Ontologies) is the collaborative experiment with the goal to develop biomedicine-based ontologies and provide the space where it is possible to find and share bio-ontologies. Gene ontology (GO) is one of the best known OBO foundry ontologies. It offers controlled vocabulary of terms (cellular components, molecular functions or biological processes) for annotation gene products without distinguishing animal species [12].

The SBML (Systems Biology Markup Language) is the XML-based and software-independent language for computational models representation in systems biology. It is perceived as the standard for modeling formal, quantitative and qualitative models of biological phenomena, i. e. metabolic networks, cell-signaling pathways, regulatory networks and biochemical reactions [13, 14]. SBML-based models can be perceived as conceptual models, because they can be depicted with the usage of various tools, e. g. CellDesigner [16] or Cytoscape [17].

The SBGN (Systems Biology Graphical Notation) offers limited and sufficient collection of symbols for unambiguous description of biochemical events with the usage of graphs at different levels of granularity [18].

The UML (Unified Modeling Language) is used for design and analysis of software systems. Several research studies advocate the usefulness of the UML for domain modeling of biological systems. Domain model of granuloma formation process in the liver is proposed as the initial step before simulation development in [19]. The UML-based domain model of autoimmune disease Experimental Autoimmune Encephalomyelitis is presented in [20].

3   The Agent Modeling Language

The AML (Agent Modeling Language) is semi-formal visual language for modeling systems in terms of concepts drawn from multi-agent systems (MAS) theory [21]. The main motivation behind the proposal of the AML is to offer conceptual language which could satisfy specific needs related to the MAS development. The AML introduces eleven new UML 2.0-based diagrams. These diagrams arise from the AML meta-model. The AML meta-model is composed of five packages proposing elements for modeling different aspects of MAS:

- Architecture: modeling entities (agents, resources, environments), deployment of MAS and social aspects of MAS (role types, social associations),
- Mental: representation of mental states (believes, goals, plans and mental relationships),
- Behaviors: modeling dynamic nature of entities (behavior decomposition, observations and effecting interactions, services), mobility and communicative interactions,
- Ontologies: modeling basic ontological hierarchy (classes and relations between them, merging and importing ontologies),
- Model management: situation-based modeling for representation of context.

The AML is investigated as the approach for domain knowledge capturing of selected immune characteristics and processes, because this conceptual language is not deeply investigated in relation to the computational immunology. Only these AML-based diagrams ensuring added value in the computational immunology are mentioned in the paper. The StarUML open-source tool is applied for domain modeling, because offers the AML profile with all AML-based diagrams.

4   Case Study

B-cells can be activated and differentiated into plasma or memory cells with the assistance of T_{helper}-cells (T-cells dependent B-cells activation (TDA) [2], [22]). TDA is selected as the case study for demonstration of the AML usefulness in computational immunology. TDA is based on the recognition of antigens by B-cells on the surface of antigen-presenting cells (APC). B-cell receptor (BCR) recognizes antigens on the surface of APC and receive first signals for their activation, but these signals are not enough for their proliferation and differentiation. Second signals (cytokines) are transmitted by T_{helper} 2-cells recognizing antigenic peptides on the surface of APC. Activation processes are mainly occurred in secondary lymphoid organs (e. g. lymph nodes, spleen, Peyer´s patches). Activated B-cell proliferates and differentiates into memory or plasmatic cell. Memory cell remembers antigens for faster reactions in the future and plasmatic cell produces antibodies. Mutations occur during differentiation of B-cells. Unproductive mutation often causes apoptosis of the cell. Productive mutation can increase the affinity of the cell and improve reactions of B-cells during the fight with pathogens.

4.1   External and Static View

Entity diagram proposes global view on the investigated system. The diagram extends
the UML class diagram in the view of distinguishing autonomous entity types (agents,
environments) and non-autonomous entity types (resources). Agent types are compact
units able to solve specific problem. Resources are information sources, outputs of
processes or objects of interest and environments specify living conditions for agents
and resources types. The most important entity types and relations of the TDA are
represented in the Fig. 1. Literature and consultations with expert were used for
identification crucial entity types in case of the TDA. Blood can be type of the
environment but it is not integrated into the diagram because of the readability of the
diagram. Entity diagram is simplified in the Fig. 1 – communication molecules
cytokines are not represented within the diagram.

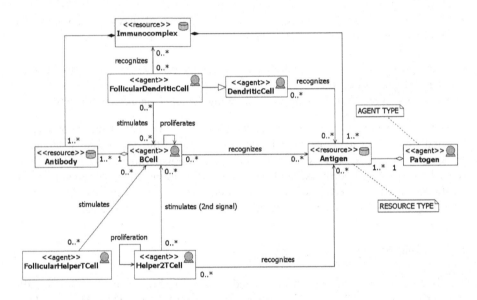

Fig. 1. Entity diagram

Group of immune cells can be represented as the society, because they can
coordinate and cooperate their activities. Social aspects are modeled by society
diagram representing organization units, social roles and relations, see Fig. 2. Lymph
node is perceived as the society – organization unit, where for example antibody plays
the role of an opsonin. The opsonin is perceived as the social role emphasizing
meaning and function of the entity type in social context. B-cells can play the role of
memory or plasmatic cells. T-cells can play the role of helper cells. Social relation
peer-to-peer represents equal relation between entity types, see Fig. 2 – antigen and
opsonin. Social relation superordinate-subordinate is represented between plasmatic
cell and antibody, because antibody is the output of the plasmatic cell.

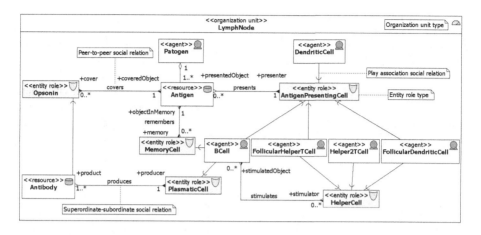

Fig. 2. Society diagram

4.2 External and Dynamic View

Protocol communication diagram is special case of the UML communication diagram. It models the information stream between entity types where the order of communication messages is important and highlighted in the diagram. Entity types or social roles can be used as "players" transmitting communication messages. Relations between "players" are modeled as UML association roles, but the UML message is concrete communication message. Simplified two-phase model of B-cell activation is depicted in the Fig. 3. The initial step in the B-cell activation is the interaction between APC (B-cell) and the pathogen (step 1 – 4). This interaction is not often enough for the B-cell stimulation. The second signal is provided mainly by the T_{helper} 2-cells (step 7 – 9) after stimulation of the T_{helper} 2-cells by the APC (step 5, 6).

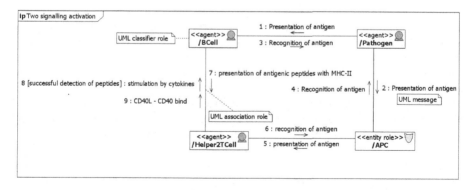

Fig. 3. Protocol communication diagram for the two-phase model of B-cell activation

4.3   Internal and Static View

Perceptor-effector diagram is the extension of the UML class diagram used for modeling perceptors and effectors of the entity type together with the dependencies between them, see Fig. 4. Perceptors are applied for sensing the environment by entity type (Fig. 4 - e. g. BCR, CD40). Entity type reacts on the inputs from the environment by effectors (Fig. 4 - e. g. MHC-II, CD40L). Relations between preceptors and effectors are modeled with UML dependencies – perceives and effects (Fig. 4 - BCR perceives Antigen. MHC-II stimulates BCR.).

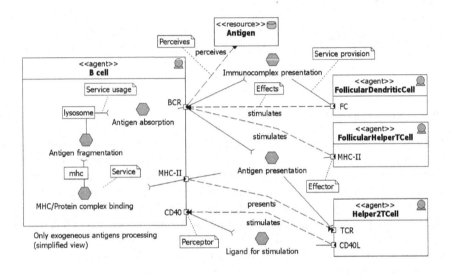

Fig. 4. Perceptor-effector diagram and service diagram

Service diagram is specialized composite structure diagram used for modeling services. Services are offered by entity types (service providers) and used by service clients. Biological entities (e. g. cells or their parts, pathogens, microorganisms, organs or proteins) can play a role of service providers or service clients. Simplified service diagram for the B-cell is mentioned in the Fig. 4. As the example, MHC molecule uses the service Antigen fragmentation which is offered by the lysosome. MHC molecule offers the service MHC/Protein complex binding process. This process is used by MHC-II molecule for presentation of antigenic fragments.

Mental diagram models mental attitudes of autonomous entity types. It is used for representation of goals, believes, plans, mental relationships (contributions, responsibilities), mental actions (committing or cancelling goals) and mental constraints. Individual cell does not have mental characteristics similar to humans. On the other hand, cells can be perceived as social entities living in communities and displaying pro-active or competitive behavior with the orientation to fulfill goals which are part of "their plans". Mental models of biological entities can be useful for development plausible models of immune system behavior or for the investigation of similarities and differences among social life of cells and humans. Simplified mental model of the B-cell is depicted in the Fig. 5. The B-cell has three plans: antigen presentation, antibodies production and antigen recognition. These plans are supported by believes that the cell is able to follow plans and fulfill goals (decidable, un-decidable) which are part of these plans. Responsibility for goals can be defined with the usage of the mental relationship responsibility.

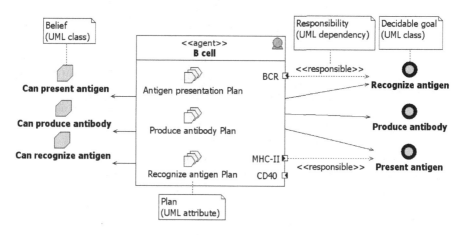

Fig. 5. Mental diagram for the B-cell

5   Contributiveness of the Agent Modeling Language

Conceptual modeling of particular immune processes is difficult because of the complexity of the immune system. Complexity of the immune system is caused by the existence of many "players" (cells, molecules, antigens, proteins, pathogens, etc.) and relations between them. It is inevitable to abstract for dealing with the complexity of the immunity, but towards correct and useful models. The UML language is often used for conceptual modeling in computational immunology (e. g. [15], [19, 20]), but it is not satisfactory for the MAS-based applications. The immune system can be perceived as the MAS. The AML is more precise in several aspects of the immune system modeling in comparison to the UML, see Table 1. The AML is promising approach for analysis and design of immune simulators if the immune system is represented as the MAS. On the other hand these claims are received on the basis of the only one case study. It is necessary to apply the AML for more case studies for verification its usefulness. The paper presents initial step in the research of conceptual modeling in computational immunology with the usage of the AML. Table 1 mentions beneficial AML-based diagrams with new elements extending the UML diagrams.

Table 1. Contribution of the AML in computational immunology

AML diagram (Type of diagram)	New elements	Contribution for modeling in computational immunology
Entity diagram (Class diagram)	Agent type Environment type Resource type	Entity types ensure structuring conceptual models in terms of the MAS. The UML does not deal with the agent, resource and environment, but these ones are modeled as UML classes.
Society diagram (Class diagram)	Organization unit type Entity role type Social association Play association	The diagram can model social aspects of the biological system. The UML is able to represent only roles as a part of the associations. Social role is the UML class in the AML.
Protocol communication diagram (Communication diagram)	Create attribute Destroy attribute Multi-message Join Subset	Integration of social roles into interactions and their changes (create/destroy attribute) is the added value of this diagram. Communication between multiple participants is solved with the AML (multi-message, join, subset). The UML does not offer this possibility.
Perceptor-effector diagram (Class diagram)	Perceptor (type) Effector (type) Perceives dependency Effects dependency	The diagram represents sensing and effecting interactions (perceives and effects dependency). It can be used e. g. for modeling interactions between immune cells or for representation of surface structures for immune cells. The UML does not offer these characteristics, but preceptors/effectors are represented as the UML ports.
Service diagram (Composite structure diagram)	Service specification Service provision dependency Service usage dependency	The diagram models abilities (services) of biological systems, i. e. what these systems offer (service provision dependency) and use (service usage dependency). The UML can represent web services.
Mental diagram (Class diagram)	Belief Goal Commit goal action Cancel goal action Plan Mental association Mental responsibility	The diagram is useful only if biological entities are perceived as systems having mental states (believes, goals, plans). Relations between entity types and mental states are represented by mental associations or mental responsibilities. Mental states are not offered by the UML, but they are represented as UML classes.

6   Conclusion

The paper investigates the usefulness of the Agent Modeling Language for conceptual models development in computational immunology. The main purpose of these diagrams is to transfer domain knowledge about specific immune process into more formal shape which can ease immune simulators development. T-cell dependent B-cell activation is chosen as the case study where the AML is tested. Immune system is considered as the multi-agent system in the paper. The six AML-based diagrams are applied for this case study, i. e. entity diagram, society diagram, protocol communication diagram, perceptor-effector diagram, service diagram and mental diagram. Not all "players" of immune processes are taken into account, but key players are included in these diagrams thanks to discussion with computer scientists and immunologist. The future research is going to be focused on the usage of the AML for modeling interactions between immune cells occurring in the lymph node of a mouse. Simulator of interactions between dendritic cells, B-cells and T-cells is going to be developed. The AML is going to be used for domain modeling of related immune processes.

Acknowledgements. This article was supported by the project No. CZ.1.07/2.3.00/30.0052 - Development of Research at the University of Hradec Králové with the participation of postdocs, financed from EU (ESF project).

References

1. Petri, C.A.: Kommunikation mit Automaten, Dissertation Thesis, Institut für Instrumentelle Mathematik, Schriften des IIM, nr. 2, 68 (1964)
2. Murphy, K.: Janeway´s Immunobiology. Garland Science (2014)
3. Blätke, M.A.: Tutorial Petri Nets in Systems Biology, http://www.regulationsbiologie.de/pdf/BlaetkeTutorial.pdf
4. Carvalho, R.V., et al.: Modeling Innate Immune Response to Early Mycobacterium Infection. J. Computational and Mathematical Methods in Medicine (2012), http://www.hindawi.com/journals/cmmm/2012/790482/
5. Chen, P.: The Entity-Relationship Model - Toward a Unified View of Data. J. ACM Transactions on Database Systems 1, 9–36 (1976)
6. Bornberg-Bauer, E., Paton, N.W.: Conceptual data modeling for bioinformatics. J. Briefings in Bioinformatics 3, 166–180 (2002)
7. Harel, D.: Statecharts: A visual formalism for complex systems. J. Science of Computer Programming 8, 231–274 (1987)
8. Swerdlin, N., Cohen, I.R., Harel, D.: The Lymph Node B Cell Immune Response: Dynamic Analysis In-Silico. In: Proc. of The IEEE, pp. 1421–1442. IEEE Press, New York (2008)
9. Belkacem, K.: Foudil, Ch.: An Anylogic Agent Based Model for the Lymph Node Lymphocytes First Humoral Immune Response. In: The International Conference on Bioinformatics and Computational Biology, pp. 163–169. IACSIT Press, Singapore (2012)
10. Kugler, H., Larjo, A., Harel, D.: Biocharts – a visual formalism for modeling biological systems. J. of The Royal Society Interface 7, 1015–1024 (2010), http://research.microsoft.com/pubs/115444/rsif20090457.pdf

11. Gruber, T.R.: A Translation Approach to Portable Ontology Specification. J. Knowledge Acquisition 5, 199–220 (1993)
12. Ashburner, M., et al.: Gene Ontology: tool for the unification of biology. J. Nat. Genet. 25, 25–29 (2000)
13. Hucka, M., et al.: Systems Biology Markup Language (SBML) Level 1 - Structures and Facilities for Basic Model Definitions, http://sbml.org/Special/specifications/sbml-level-1/version-1/sbml-level-1.pdf
14. Hucka, M., et al.: The Systems Biology Markup Language (SBML): Language Specification for level 3 version 1 Core, http://sbml.org/Special/specifications/sbml-level-3/version-1/core/sbml-level-3-version-1-core.pdf
15. Bersini, H.: Immune System Modeling: The OO Way. In: Bersini, H., Carneiro, J. (eds.) ICARIS 2006. LNCS, vol. 4163, pp. 150–163. Springer, Heidelberg (2006)
16. Funahashi, A., Matsuoka, Y., Jouraku, A., Morohashi, M., Kikuchi, N., Kitano, H.: CellDesigner 3.5: A Versatile Modeling Tool for Biochemical Networks. In: Proc. of The IEEE, pp. 1254–1265. IEEE Press, New York (2008)
17. König, M., Drager, A., Holzhutter, H.-G.: CySBML: a Cytoscape plugin for SBML. J. of Bioinformatics 28, 2402–2403 (2012), http://bioinformatics.oxfordjournals.org/content/early/2012/07/05/bioinformatics.bts432.full.pdf
18. Le Novère, N., et al.: The Systems Biology Graphical Notation. J. Nature Biotechnology 27, 735–741 (2009)
19. Flugge, J., Timmis, P., Andrews, J.: Moore and P. Kaye: Modelling and Simulation of Granuloma Formation in Visceral Leishmaniasis. In: Congress on Evolutionary Computation, pp. 3052–3059. IEEE Press, New York (2009)
20. Read, M., Timmis, J., Andrews, P., Kumar, V.: Domain Model of Experimental Autoimmune Encephalomyelitis. In: Proc. of the 2009 Workshop on Complex Systems Modelling and Simulation, pp. 9–44 (2009)
21. Červenka, R., Trenčanský, I.: The Agent Modeling Language – AML: A Comprehensive Approach to Modeling Multi-Agent Systems. Springer Science & Business Media, New York (2007)
22. Abbas, A.K., Lichtman, A., Pillai, S.: Cellular and Molecular Immunology. Saunders, United States (2011)

A Healthcare Companies' Performance View through OSN

Camelia Delcea[1], Ioana Bradea[1], Ramona Paun[2], and Alexandru Friptu[2]

[1] Bucharest University of Economic Studies, Bucharest, Romania
camelia.delcea@csie.ase.ro, alexbradea1304@yahoo.com
[2] Webster University, Bangkok, Thailand
{paunrm,friptua}@webster.ac.th

Abstract. In the new economic context, the on-line social networks (OSN) are gathering among their users a whole array of persons, with a diversified background, culture, opinions and needs. Here is the meeting point for persons looking for information on different domains and areas of interest and also here is the exchange land for thoughts, experiences, news, etc. As the companies have sensed the global phenomenon of the OSN, they have paid more attention to the on-line social environment, investing more in the social media campaigns. But what about the companies in the healthcare system? Are their customers seeking for information in OSN? And if so, to what extent the information provided here is useful? Namely, is the OSN affecting their performance?

Keywords: healthcare system, grey incidence analysis, Deng's degree of incidence, on-line social networks.

1 Introduction

Along with the adherence at European Union, the Romanian healthcare system has undergone many changes in order to align to the external reference frame. The health status of a country is determined by: age, hereditary factors, education, living conditions, geographic areas, agriculture, water supply, hygiene and health services. Besides these factors, social networks started to impose lately as through it, population gets information about health and health services.

With the help of social networks, it can be provided information on healthcare regarding: the updates, the prices charged by different hospitals, the quality of the medical staff of a particular hospital unit and so on. Also, the on-line social networks (OSN) allow the dialogue between physician and patient on one side and between patients on the other side, providing the necessary feedback for the best decision on a man's health. The information available through the Internet and social networks allow the development of efficient, cheaper and better understood healthcare. The main key performance indicator of the health system is the degree of patient satisfaction, with influence on that company's image and reputation.

The principal problem of the Romanian healthcare system is the lack of performance. In Romania the focus is not on the individual needs, which results in major disruptions in various aspects of the medical system functionality. Medical services must focus on information technology and knowledge. A knowledge management enables the development and the use of knowledge to the maximum. The poor performance of the Romanian health system is determined by: the lack of hospitals in rural areas, inadequate financing, poor management of human resources, migration of health professionals in the EU countries, the lack of integration of health services for continuity of care, lack of medical equipment and the distrust in the medical system integrity [1].

The financing of the health system affects its functionality and performance. Because the notion that health is an unproductive sector, Romania awarded annually up to 5% of GDP to this sector, the lowest percentage in the EU. Also, contrary to the policies of other countries, Romania made the allocated proportion of GDP to decrease with the increasing of GDP. That is why, the low level of performance is given by the lack of transparency in the allocation of resources and their inefficient use.

The inadequate funding and the poor management have generated bankruptcy to several hospitals in rural areas and small towns, the access to medical services being made extremely difficult, especially for the residents of North-Eastern Romania. In this area, there is 1 doctor for 2778 inhabitants, 98 of villages remaining without medical staff. In recent years were closed 67 public hospitals, situated in smaller cities, which were not profitable. The patients were moved to hospitals in big cities, and the buildings were either returned or refurbished as homes for the elderly. Following these decisions were recorded more deaths among people who could not be saved due to the excessive distance to the first functional hospital. Currently, the Romanian health system is comprised of 459 hospitals, of which only 109 are private. People who benefit from the services of private hospitals are few, due to high prices. The obsolete medical equipment is another weakness of the health system. As a result of this, many interventions are made only in a few cities [2].

The medical staff is the key to the efficient workout of the medical system, in accordance with the population needs. According to WHO (World Health Organization), Romania is on the ante-penultimate place in the EU, concerning the density of physicians, being 1.9 physicians per 1,000 inhabitants.

Unchanged in many respects, the Romanian health system gravitates around the Ministry of Public Health and not around the patient, as in other EU countries. Currently there are several gaps in the quality of care, informational systems and risk management [3]. These shortcomings are made by the central authority, which takes all the decisions for the entire health system, not taking in consideration that in this way the decisions are made without knowing the problems from the level where the health services are offered and used.

2   Non-financial Performance

Performance can be classified in several ways. Considering the content, or the way of approach, can be distinguished the financial and non-financial performance. The financial performance quantitatively captures its level using the financial indicators, while the non-financial performance is considered less valuable in the eyes of managers, even if it can be used as an indicator that leads to financial performance, especially for the financial performance that is not yet contained in the accounting measurements. Non-financial information enable a rapid response of the decision makers to unexpected changes in the business environment, as operational managers use more qualitative data than monetary data. The key indicators for non-financial performance are: the consumer's satisfaction, the company's reputation and image, the employee's satisfaction and the quality of products or services [4].

A high level of consumer satisfaction can generate the following benefits: the customer loyalty, reduced price elasticity and new customers [5, 6]. The quality is the key for obtaining the customer loyalty, in terms of products offered for sale and available services, while the reputation allows the launch of new products and services on the market, by reducing the risk of rejection. Also, a good reputation enables to maintain positive relationships with key suppliers.

Obtaining a desired level of non-financial performance will drive to a high level of financial performance. There are two variables that change in direct proportion.

For this, the present paper focuses on the non-financial performance of the healthcare companies in Romania, in order to determine some of its main triggers in the now-a-days economy.

3   On-line Social Networks (OSN) Overview

With the passing of the years, the OSN has become more and more important from an economic point of view, some of the researchers in the field even speaking about the appearance of a global phenomenon [7]. In this context, it has been proven that the OSN has a potential business value along the value chain from multiple perspectives: for research and development [8, 9]; for marketing and sales [10, 11, 12]; for customer services [12, 13]; for human resources and internal applications   [14] etc.

Among all, consumer feedback is one on the most valuable information that can be extracted from the OSN. For example, in a study conducted by Synthesio for Nissan concerning the barriers that stopped the customers to buy Nissan Leaf (their electric car), it has been discovered some of the key elements that were stopping from buying, such as: running costs, battery recharge problems, public infrastructure, drive quality, etc. which were easily found out by analyzing these customers' conversations.

Starting from here, improvements have been made based on the high importance that have been given to the Web 2.0 technologies. In order to obtain higher interest from customers, the companies have rethink completely their business model, being aware on the fact that social media is responsible for achieving both promotion and brand consideration, but also for information searching and building a tighter relationship with customers [15]. For this, the investments in social media at company level have been recently increased.

Moreover, a company's reputation can be shaped through social networks [16] and its products can be analyzed in terms of susceptibility in on-line environments based on their customers' reviews [17]. As a recent study shows, [18] the users of the OSN are responsible persons, as it has been proven that before they are sharing the information with others, they normally try to judge the outcome of their actions. For this, the information sharing is a conscious act in two stages: considering both the personal and social consequences of sharing.

Another interesting approach was the measurement of media strategies impact on companies' financial performance [19]. As a result, the authors have underlined that there is a positive correlation among the considered variables.

4   Grey Incidence Analysis

One of the main problems encountered in this situations, that are involving OSN, is the limited data access, mainly the impossibility of gathering the necessarily amount of data or the inexistence of these data due to different situations [20, 21].

This is also one of the grey systems covering situations as this theory is treating the deficiencies generated by the inexistence of big data sets through some adequate methods and techniques [22, 23].

The second advantage offered by grey systems theory is that the relatively reduced amount of calculations necessarily for an analysis, offering in the same time useful information to the analyst, without generating pointless situations among different approaches such as the quantitative and the qualitative one.

The main idea of grey incidence is that the similitude between two situations can be analyzed through their graphics' tracking and evolution's monitoring. But, a necessarily step before conducting such a study is the choosing of the main factors that are best describing such a behavior.

For this, in the following, the Deng's degree of incidence calculus [24, 25, 26] will be described in order to be used in section 5

Among all these for the present analysis it has been chosen the Deng's degree of grey incidence due to its easiness of use and demonstrated practical applicability.

For this, it shall be considered the following:

- a sequence of systems' characteristics:

$$X_0 = \left(x_{1,0}, x_{2,0}, \dots, x_{t,0}\right) \tag{1}$$

- a number of relevant factors sequences for this system, given by:

$$X_j = \left(x_{1,j}, x_{2,j}, \dots, x_{t,j}\right) \tag{2}$$

where: $j = 1, 2, \dots, n$, t = the period of time and n = the considered variables and: $x_{k,j} > 0, k = 1, \dots, t$;

- the calculated values of the incidence degree of a relevant factors sequence on its main characteristics will be noted as: $\gamma\left(X_0, X_j\right)$.

Deng's degree of grey incidence [26] is calculated as follows:

$$\gamma\left(X_0, X_j\right) = \frac{1}{t} \sum_{k=1}^{t} \gamma\left(x_{k,0}, x_{k,j}\right) \tag{3}$$

where:

$$\gamma\left(x_{k,0}, x_{k,j}\right) = \frac{\min\limits_{k} \min\limits_{j} |x_{k,0} - x_{k,j}| + \rho \max\limits_{k} \max\limits_{j} |x_{k,0} - x_{k,j}|}{|x_{k,0} - x_{k,j}| + \rho \max\limits_{k} \max\limits_{j} |x_{k,0} - x_{k,j}|} \tag{4}$$

with: $\rho \in (0, 1)$

Both the sequence generation methods and Deng's grey incidence analysis will be used in the next section for modelling the OSN's influence on the healthcare companies' performance in the next section.

5   Case Study on Romanian Healthcare Companies

A questionnaire was applied in order to determine whether there is an influence from the usage of the OSN on the medical services consumers' preferences. For assuring a great number of answers from the respondents, one question from each set of questions could be skipped, the answer to this being filled in the final data records using the sequence generation offered by the grey systems theory. Beside the personal questions (such as: gender, age, last school graduated, etc.), the questionnaire was divided into two major parts: the customers' overall medical service satisfaction (COMSS) and the customers' on-line social network activity (COSNA).

As a result, a number of 301 respondents have successfully completed the questionnaire, 59.47% of them being female and 40.53% male. The age distribution was 20.93% under 30 years old, 33.89% - between 30 and 45, 29.24% - between 46 and 60, and 15.95% above 60 years old.

5.1   Medical Services Satisfaction and OSN Activity

The two main categories analyzed as being the triggers for the healthcare companies' non-financial performance (measured through image and reputation), namely COMSS and COSNA were determined based on a series of questions, evaluated through a 5-point Likert scale.

For the COMSS construction, the respondents were asked to give their opinion on the physicians they have been treated by in such healthcare companies -PH (e.g. their attitude PH_1, attention PH_2, explanations PH_3, empathy PH_4), on the medical facilities offered –MF (e.g. location MF_1, furniture MF_2, hygiene MF_3, equipment adequacy MF_4) and on medical appointments –MA (e.g. easiness of getting an appointment MA_1, waiting time MA_2, equity MA_3).

After conducting the confirmatory factor analysis, the number of the items in COMSS was reduced to nine, as it can be seen in Fig. 1.

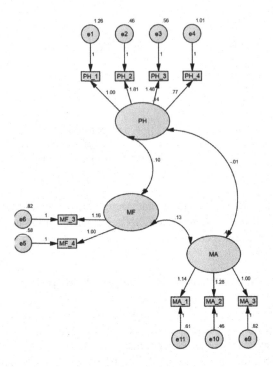

Fig. 1. The COMSS construction in AMOS 22

During the confirmatory factor analysis, the values obtained for medical facilities MF_1 and MF_2, location and furniture, had very low factor loadings and, therefore, they have been eliminated from the final analysis.

As for the COSNA construction, three main categories have been considered: interaction with other users on OSN regarding the usage of the medical services – IU (e.g. reading comments and reviews IU_1, asking opinions IU_2, and making a decision IU_3), commercials –CM and news –NW (e.g. watching commercials in OSN CM_1, reading short advertising in OSN CM_2, accessing the healthcare companies websites NW_1, reading on-line news reports NW_2,).

After conducting the confirmatory factor analysis with AMOS 22, COSNA construction retained only five elements (see Fig. 2), validating through this analysis the construct convergence, validity and reliability.

Having these constructions and using grey systems theory method in the next section the grey incidence analysis will be proceeded for establishing the influence of each construction on firm's non-financial performance, also measured based on the applied questionnaire, with a 5-point Likert scale, where the respondents were asked to underline their opinion on the image and reputation they have over their medical healthcare provider.

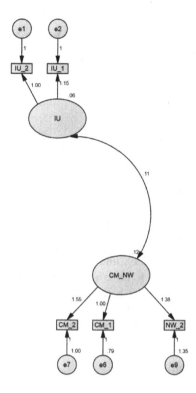

Fig. 2. The COSNA construction in AMOS 22

5.2   OSN's Grey Influence on Non-financial Performance

First of all, the influence of COMSS and COSNA on non-financial performance was determined using Deng's degree of grey incidence as presented in the above sections.

As it is stipulated in the literature, this degree can take values between 0 and 1, the closer the value is to 1 the stronger is the influence of that variable on the result. Moreover, a value above 0.5 underlines a significant influence of one of the observed variables on the other.

Therefore, for the COMSS and COSNA the Deng's degree of grey incidence was of calculated using the GSMS 6.0 Software and the values are presented in Fig. 3. Also, we considered to be interesting to view the each component's incidence on non-financial performance, which is also presented in the following figure:

As a result, it can be said that the consumer's overall medical service satisfaction has a great impact on the non-financial performance (as the value of the Deng's degree of grey incidence 0.8467 is between 0.5 and 1), the component regarding the physicians' characteristics being the most important component, with a incidence degree of 0.9197, approaching the maximum value of 1, followed by the medical facilities offered by the healthcare companies (Deng's degree of 0.7285).

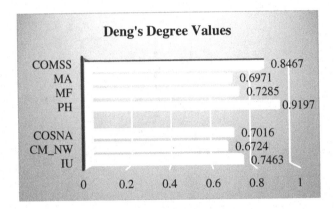

Fig. 3. Deng's Degree Values – GSTM 6.0

As for the customers' on-line social network activity regarding the healthcare service providers, it seems to have a significant influence (Deng's degree of 0.7016), the most important component being the one related to the interaction of the customers with other users in the on-line environment in the process of seeking information regarding a specific healthcare company provider.

6 Conclusions

Along with the passing of the years, the OSN has become more and more important from an economic point of view, some of the researchers in the field even speaking about the appearance of a global phenomenon. In this context, the consumers' feedback is one on the most valuable information that can be extracted from the OSN, with direct influence of the companies' evolution on the market.

In this context, the present paper underlines the influence OSN can have on the performance of the Romanian healthcare companies as the health domain is one of the most fragile in this country.

For this, by applying a questionnaire, the respondents were asked to express their opinion on several elements regarding the medical life there are experiencing. Two main components have been extracted: one for measuring the overall medical service satisfaction and other measuring the activity on consumers' in OSN in order to find the best healthcare provider.

The questionnaire validation was performed through a confirmatory factor analysis using AMOS 22, which results in the reduction of the implied variables from both of the considered constructions.

Next, grey systems theory was used in order to determine the influence of each of these components on the non-financial performance, measured through image and reputation evaluation. As a result, the component regarding the overall customers' satisfaction has obtained higher values for Deng's degree.

Even though, the component regarding the users' activity on OSN has reached a good value for the Deng's degree, which can be translated through a significant influence of the OSN users' activity on the non-financial performance.

Moreover, among the considered elements of COSNA, the users' interaction with the other users on their attempt of finding the best healthcare provider seems to have the greater influence.

Furthermore, the research can be extended by including a financial analysis as done by [19] in parallel with the non-financial one, for getting a better picture of what the OSN influence is on the healthcare system and how this can be used for both customers and healthcare providers.

Acknowledgments. This paper was co-financed from the European Social Fund, through the Sectorial Operational Programmee Human Resources Development 2007-2013, project number POSDRU/159/1.5/S/138907 "Excellence in scientific interdisciplinary research, doctoral and postdoctoral, in the economic, social and medical fields -EXCELIS", coordinator The Bucharest University of Economic Studies. Also, this work was co-financed from the European Social Fund through Sectorial Operational Programme Human Resources Development 2007-2013, project number POSDRU/159/1.5/S/134197 „Performance and excellence in doctoral and postdoctoral research in Romanian economics science domain". Moreover, the authors gratefully acknowledge partial support of this research by Webster University Thailand.

References

1. Sturmberg, J.P., Martin, C.M.: Handbook of Systems and Complexity in Heatlh. Springer, London (2013)
2. Bradea, I.A., Delcea, C., Scarlat, E., Boloş, M.: KRIs in Hospitals - Network, Correlations and Influences. Economic Computation and Economic Cybernetics Studies and Research 48, 81–94 (2014)
3. Cotfas, L.A., Roxin, I.: Alternate reality game for learning. In: 12th International Conference on Informatics in Economy, Education, Research & Business Technologies, pp. 121–125 (2013)
4. Prieto, I., Revilla, E.: Learning Capability and Business Performance: a Non-financial and Financial Assessment. The Learning Organization 13(2), 166–185 (2006)
5. Fornell, S.: A National Customer Satisfaction Barometer: The Swedish Experience. Journal of Marketing 55, 6–21 (1992)
6. Cotfas, N., Cotfas, L.A.: Hypergeometric type operators and their super-symmetric partners. J. Math. Phys. 52, 11–21 (2011)
7. Heidemann, J., Klier, M., Probst, F.: On-line social networks: A survey of a global phenomenon. Computer Networks 56, 3866–3878 (2012)
8. Casteleyn, J., Mottart, A., Rutten, K.: How to use Facebook in your market research. International Journal of Market Research 51(4), 439–447 (2009)
9. Kettles, D., David, S.: The business value of social network technologies: a framework for identifying opportunities for business value and an emerging research. In: Proceedings of the Americas Conference on Information Systems, pp. 14–17 (2008)

10. Ermecke, R., Mayrhofer, P., Wagner, S.: Agents of diffusion – insights from a survey of Facebook users. In: Proceedings of the Hawaii International Conference on System Sciences, pp. 1–10 (2009)
11. Bernoff, J., Li, C.: Harnessing the power of the oh-so-social web. MIT Sloan Management Review 43(3), 36–42 (2008)
12. Libai, B., Bolton, R., Bugel, M.S., Gotz, O., Risselada, H., Stephen, A.T.: Customer-to-customer interactions: broadening the scope of word of mouth research. Journal of Service Research 13(3), 267–282 (2010)
13. Ye, C., Hui, Z.: The impact of customer to customer interaction on service company-customer relationship quality. In: Proceedings of 8th International Conference on Service System and Service Management, pp. 1–4 (2011)
14. Richter, D., Riemer, K., vom Brocke, J.: Internet social networking – research state of the art and implications for enterprise 2.0. Business and Information System Engineering 53(2), 89–103 (2011)
15. Parveen, F., Jaafar, N.I., Ainin, S.: Social media usage and organizational performance: Reflections of Malaysian social media managers. Telematics and Informatics 31 (2014), http://dx.doi.org/10.1016/j.tele.2014.03.001
16. Bonchi, F., Castillo, C., Gionis, A.: Social Network Analysis and Mining for Business Applications. ACM Transactions on Intelligent Systems and Technology 2(3), 1–38 (2011)
17. Xu, Y., Zhang, C., Xue, L.: Measuring product susceptibility in on-line product review social network. Decision Support Systems, pp. 22–31. Elsevier (2013)
18. Sohn, D.: Coping with information in social media: The effects of network structure and knowledge on perception of information value. Computers in Human Behavior 32, 145–151 (2014)
19. Schniederjans, D., Cao, E., Schniederjans, M.: Enhancing financial performance with social media: An impression management perspective. Decision Support Systems 55(4), 911–918 (2013)
20. Cotfas, L.A.: A finite-dimensional quantum model for the stock market. Physica A: Statistical Mechanics and its Applications 392(2), 371–380 (2013)
21. Delcea, C.: Not Black. Not even White. Definitively Grey Economic Systems. The Journal of Grey System 26(1), 11–25 (2014)
22. Andrew, A.: Why the world is grey, Keynote speech. In: The 3th International Conference IEEE GSIS, Nanjing, China (2011)
23. Forrest, J.: A Systemic Perspective on Cognition and Mathematics. CRC Press (2013)
24. Delcea, C., Cotfas, L.A., Paun, R.: Grey Social Networks – a Facebook Case Study. In: Proceedings of the 6th International Conference on Computational Collective Intelligence, Technologies and Applications, Seoul, Korea, pp. 125–134 (2014)
25. Liu, S.F., Lin, Y.: Grey Systems – Theory and Applications. Understanding Complex Systems Series. Springer, Heidelberg (2010)
26. Deng, J.L.: Theory basis. Huanzhong University of Science and Technology Publishing House (2002)

Production Unit Supply Management Solution in Agent-Based Computational Economics Model

Petr Tucnik and Zuzana Nemcova

University of Hradec Kralove, Hradec Kralove, Czech Republic
{petr.tucnik,zuzana.nemcova}@uhk.cz

Abstract. This paper is focused on the description of (supply) management of a production unit (represented by multi-agent sub-system) in the model of agent-based computational economy. The concept is described both from internal and external perspective. Internal view describes internal processes like supplies management, production line processes and workforce allocation. External view is focused on interactions with other agents in the model environment, which provide necessary resources, workforce, and products required for continuous production. Both sides of communication are handled by production unit interface agent. Production of the unit is planned according to production contracts traded on the common market via broker agents. In order to function properly, behavior of agents in the model is designed according to common microeconomic principles. Internal architecture of the production unit is of large importance in the model, which is the main topic we would like to present and discuss in this paper.

Keywords: Agent, multi-agent system, SCM.

1 Introduction

The agent-based computational economies represent bottom-up approach for developing economic systems, where the desired behavior is achieved through the interaction of its individual components – agents [9]. In this paper, we would like to focus on description of agent-based economic sandbox environment, currently under the development at the Faculty of Informatics and Management of University of Hradec Kralove.

The idea of agent-based economy is not novel, since there exist a large number of agent-based economic simulations, models and applications. But in general, this approach is not very often used for large-scale (sand-box) applications. Typically, specialized model focused on singular problem is more common approach for experiments of this kind. However, our intention is to create environment where different approaches, methods and algorithms from areas of artificial intelligence, decision-making, planning, logistics, adaptive production, and supply chain management may be tested and verified. In case of several agent colonies (i.e. multiple agent communities), different strategies and high-level decision making may even be tested against

© Springer International Publishing Switzerland 2015
D. Barbucha et al. (eds.), *New Trends in Intelligent Information and Database Systems,*
Studies in Computational Intelligence 598, DOI: 10.1007/978-3-319-16211-9_35

each other. Our intention is to develop an agent-based environment that provides foundations for long term research based on above mentioned areas.

Since extensive and thorough description of any large-scale economic system is a very complex task, in this paper we will focus specifically on problem of production unit (i.e. factory) supply management. The concept will be described both from internal and external perspective. Internal view describes internal processes like supplies management, production line processes and workforce allocation. External view is focused on interactions with other agents in the model environment, which provide necessary resources, workforce, and products required for continuous production. Both sides of communication are handled by production unit interface agent.

2   State of the Art

Some of the typical features often mentioned when describing agent-based approach are (among others) adaptation and proactivity. Ability of agents to act autonomously and adapt to changing environment is useful for modelling economic entities because in real-world economic systems, individuals and enterprises also act in a similar manner – they adapt to changes and are able to act independently on each other. In general, this can be considered manifestation of rational behavior as long as every entity in the system is trying to maximize its benefit, whatever it may be (this phrasing is used since it is not easy to explicitly describe what lies behind notion of "benefit" when it comes to individual parties involved in economic transactions). In order to do so, economic entity tries to adapt to changes on markets where price volatility reflects changes in supply/demand relationships.

For example, price of goods, as one of the most important aspects of economic systems, constitutes a fundamental economic attribute and is basically aggregated, consisting of a large number of effects influencing each other to certain level. It is a matter of model complexity how precisely are these effects described. However, certain level of abstraction is generally required due to overwhelming complexity of real world functioning, and this is the reason why economic models therefore often work under assumption of *ceteris paribus*, resulting in abstraction from other phenomena, not included in the model. As a consequence to this, model ignoring important aspects of economic environment is not entirely valid and results of experiment conducted within such constraints have limited viability. This creates obstacles for real world application – model has to be sufficiently complex but - on the other hand - abstraction is also necessary. Finding the appropriate equilibrium for the model complexity is one of the most important goals of proper design of such models, if they are to be applicable in the real world scenarios.

Excellent general overview of the agent-oriented approach application in the economic modeling is well-described in Tesfatsion`s paper [9]. The notion of "*agent-based computational economy*" (ACE) is used here to refer to economic models created on computational (i.e. digital) rather than technical level. This approach is particularly useful for modeling on macro-level scale where implementation issues of technical nature may be avoided and attention is pointed at function of production units as whole. Our own approach is close to this view on economic modeling since questions of technical implementation are also left out of account in our case at this moment.

More general view on agent-based economic system modeling also offer Whitestein Technologies`s series of books, in our context especially Zimmermann`s [11] and van Aart`s [10]. While van Aart [10] is focusing on the general organizational principles of the work division and the effective work processes using agent-based approach, Zimmermann [11] describes supply network event management where "events" represent various kinds of events of disruptive nature, disturbing or delaying the supply processes. For optimal performance, the supply network is propagating information about such events, allowing participating entities to take the necessary actions in adequate time, minimizing risk and negative effects of such disruptive events.

Among other - more general - resources could be mentioned especially Chaib`s multiagent-based supply chain management [5] and Schuldt`s work on the autonomous logistics, using multiagent systems [8].

Recent development in areas such as robotics, control, communications, etc. allowed development of adaptive production systems, for relevant and more specialized research papers see e.g. [4], [2] or [3].

Hamichi [4] focuses on the phenomena of emergence of industrial clusters and each production firm in the environment is represented by an agent. The agents representing production units are responsible for management of B2B relations, making the investment decisions, reducing costs of inputs and fulfillment of orders. Simple decision making mechanisms at the micro level leads to emergence of meta-level clusters, creating global production system.

Brusaferri [2] describes the "*automation solution based on modular distributed approach for agile factory integration and reconfiguration, integrating a knowledge based cooperation policy providing self-adaptation to endogenous as well as exogenous events*". Scope of this paper is more focused, moving from global perspective of Hamichi`s paper to more individual view of production. Adaptation of the factories production to reflect the market demands should allow customization and personalization of production to the level of adaptation in the real-time.

Denkena [3] adopted more holistic approach, and is focused more on internal production processes. Goal here is to create adaptive process chain optimization, with emphasis on existing interdependencies between processes, resulting in more effective performance on global level. This leads to more adaptive behavior within production units and adaptive dimensioning of process parameters.

Last three mentioned papers are focused on the manufacturing processes and the production unit management, but in order to create more complex system, areas of adaptive logistics and supply chain management (SCM) are closely related as well. Useful review of the performance models analysis of SCM offers [1]. Agrell points out some of the best practice techniques for information integration and implementation across the supply chain. Another relevant work of Ivanov [6] is also focused on SCM, describing integration of the static supply chain design and planning models towards execution dynamics, creating correspondence between static and dynamic models of SCM.

3   Model Description

In this section the model from the perspective of single production unit and both its external and internal interactions/processes will be described. Effort will be made to describe such production unit (production agent) in the context of the whole economic system.

Our agent-based computational economics model does not aspire to be the all-inclusive real world simulation. Some areas important for real world economies have been deliberatively omitted, like currency markets, stock market trading, many financial services, etc., in order to keep model viable and comprehensible. Although some level of abstraction was necessary, there has been shown considerable effort to utilize the real statistical data (from sources like Czech Statistical Office, see www.czso.cz webpage) as much as it has been relevant and useful for the model. Also (although simplified for the same reasons as mentioned above) official classifications were used to classify model components/activities plausibly (e.g. CZ-NACE classification of economic activities or RAMON – Eurostat`s Classification Servers). This helped to clarify the model while maintaining reasonable level of detail at the same time.

The description of the production unit provided in the following subsections is general. In the model, the production unit (production agent) is a general concept consisting of several common parts (warehouse, production lines, etc.), universally describing production processes within such unit. From this perspective it is not so important **what** kind of product is produced but rather **how** (e.g. internal processes of production), **when** (e.g. effective storage management) and with what level of **efficiency** (e.g. necessary workforce, materials or energy allocation). Emphasis is on the efficient functioning of the production unit where there should be no delays in the production and optimal fulfillment of the orders represented by the **contracts** is expected. Contracts and respective trading system for contract trading is not described in this paper.

We will also intentionally omit the service-providing agents from the following description since it would be beyond the scope of this paper, but they function on similar principles as production agents. There are, however, some differences (services cannot be stored, transported, etc.), but in principle, their production is generally similar to production of goods.

3.1   Factory Models from External Perspective

The external perspective describes the production unit in the context of other entities in its economic environment. There are following important types of agents implemented in the model:

- Set of production agents P – consisting of harvesting agents (H) and factories (F), where $H \cup F = P$. Harvesting agents provide the basic materials (atomic inputs) for production and processing in factories. Factories transform the inputs through the production process into the goods (for consumption) and semi-products (used in next level(s) of production). All agents included in P forms together *production chain* where the agents are positioned in levels 0-9, where 0 consist exclusively of *H* agents and levels 1-9 are *F* agents of respectively increasing technical complexity of production. Higher level in the production chain reflects higher technical level of the production and the outputs of any agents from lower levels 0 to n-1 can be used as inputs in any combination for agents at level n. The production is divided into segments of specialized production areas according to the area they belong to – agriculture, machinery, electronics, etc.

- Set of transport agents *T*. They provide the transportation services for both population and goods/materials/resources. Transport agents` services are contracted (i.e. there exists a formal contract) and cost of the transportation is included in the order of materials and goods when purchase is made.
- Set of consumer agents *C*. Consumer agents represent the population in the model and they serve both as a consumption pool for generated goods and as a workforce. These agents work at companies providing services or manufacturing goods, they receive salary for their work, and consume goods and services in ratio according to their salary level. The detailed description of internal functioning of *C* agents is again beyond the scope of this paper, but in general the higher the *C* agent`s salary is, the more luxurious goods/services are consumed. This results in changes of  the consumer`s basket composition and as income level of population rises, they become more demanding for higher-level goods (i.e. produced on higher levels of production chain).
- Facilitator agents – these agents form a special group of agents responsible for efficient communication within the system. They constitute a framework for market functioning because they allow conclusion of contracts between agents. Specifically, agents for indirect communication are implemented using mainly matchmaker and broker architectures.

P agent communicates with other entities in the system via the interface agent. The *P* agent is viewed as a single entity from external perspective, but consists of other agents handling internal processes. This encapsulation reflects the functioning of companies in the real world – customers and other organizations are unable to see internal processes, but they can communicate via intermediary (company representatives) with the company environment, placing orders, exchanging information, etc. as needed. The basic task of interface agent is to provide necessary resources, goods, workforce and services for company to function, while respecting constraints are given by the negotiation protocol within the system.

3.2   Factory Models from Internal Perspective

Figure 1 demonstrates the universal factory model. In general, each factory can produce *M* final products, the index *m* takes values $m = \{1, 2, ..., M\}$.

The factory manufactures these products by one or several - say *N* production lines (depicted as a number in the circle), which can be interconnected as a combination of the parallel and/or the series manner according to the needs of individual factory type (the index *n* takes values $n = \{1, 2, ..., N\}$).

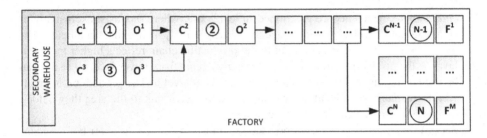

Fig. 1. The general model of the factory

Let us denote by C^n the set of the components (inputs) and by O^n the set of the outputs of n-th production line. The final product, marked as F^m, is the final, desired product of the factory. It is easily seen that the product F^m is a special type of output of a production line. The output O^n is so called semi-product. It is used for further processing in order to complete or manufacture the final product. The example of such product in the figure is output O^3, which is utilized as one of the components C^2 in a production line 2.

In general, it is supposed that there is a main warehouse and halls with the production lines in the factory area. Each hall contains also (small) secondary warehouse with limited capacity shared by production lines situated there. Employers in factories work in shifts. For simplicity we suppose that before the system starts working (e.g. initiation phase of the model), the components sufficient for production of at least one work shift are prepared in the secondary warehouse. Moreover, we suppose that necessary components for production are prepared at production lines.

Following part describes the production line of the factory in more detail. We will use explanatory example here, which is intentionally simplified in some details, since its main purpose is to provide comprehensible explanation of described processes.

The factory has a description of its final products (data file input). The final products are described in terms of their components in a table containing individual ratios of raw materials needed to produce it. For example, the bakery produces several types of products, including bread. The bread contains one kilogram of flour and 0.6 liter of water (see the Table 1.)

Table 1. The example of components quantities of final products

Bakery		
product	component	quantity
bread	flour	1,00 kg
	water	0,60 l
roll	flour	0,50 kg
	wood	0,25 l
baguette	flour	0,60 kg
	wood	0,35 l
	grain	0,10 kg

Each production line i has three service places for processing the production (see Figure 2). First, assembler, denoted by A, puts together the corresponding quantity of each of components required for manufacturing of the output O^i. In general, there can be K components at each production line, although due to the limitations given by our implementation platform, each assembler has maximal capacity of five components preset (however, this limitation can be solved by concatenation of several assemblers if needed). The components are then denoted by C_k^i, the index $k = \{1, 2, \ldots, K\}$. The second service place is marked by P – the production. Here, the assembled components are processed and then the product falls into the third service place responsible for packing the output denoted by B (by word "batch").

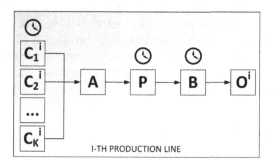

Fig. 2. The general model of the production line

We can distinguish two types of packing. The first is the packing of a semi-product – when the output of the production line is needed for further production (in other words it is not the final product), the packing consists of preparation of the product for transportation to another line. The second type (packing of the final product) is preparing the output of the factory for exportation outside of the factory (to market). Partial example of model interface for factories is shown at the Fig. 3.

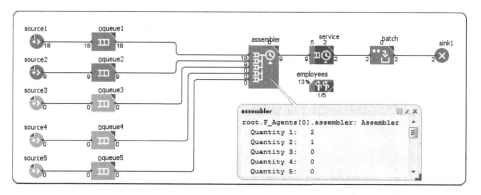

Fig. 3. Screenshot of the factory production line from the model interface (implemented in Anylogic multi-agent modeling platform)

The Production Time. For each factory we want to compute the time of the production of the final product in order to decide whether to accept the contract or not. For this purpose, the manager agent needs to know if the terms of delivery are achievable and/or the shortest time for which the supply is able to meet expected requirements.

We distinguish three types of time-consuming activities at each production line i. The total time needed for preparation of the components denoted by T_C^i, the production time denoted by T_P^i and the time for packing the product denoted by T_B^i.

The variable T_C^i is the total time for delivering components from the secondary warehouse to the ith production line. The time for preparation of particular component varies according to the type – it is denoted by t_k^i. The assembler can start the work if and only if all components are available at a time, so the time the components are prepared for assembling is maximum of all t_k^i. The need to prepare subsequent components for assembling will arise during the production of the just-assembled components. Because this can be done in parallel (in order to prepare the next components there is no need to wait for the production to be done), the T_C^i during the production is computed by following formula:

$$T_C^i = \max\left(\max_{1 \leq k \leq K}(t_k^i) - T_P^i; 0\right) \tag{1}$$

The total time of production of the output O^i at ith production line, T^i, is then computed as

$$T^i = T_C^i + T_P^i + T_B^i \tag{2}$$

To compute the time to produce the final product of the factory we suggest using well-known Critical path method described, for example, in [7]. Note that in case of concatenated lines the possibility of parallel processing of the semi-products has to be considered.

The formulas can be verified on a following short example (see Figure 4). The times needed for preparation of the components, production and packing are shown in the brackets. Patterned fields show the time lines of the activities according to the above-described rules. It is easily seen that the total production time for the first run of the line is computed as $T^i = \max(\max(2,1,3) - 0; 0) + 5 + 2 = 10$. The time period needed for production of the next outputs is computed as $T^i = \max(\max(2,1,3) - 5; 0) + 5 + 2 = 7$.

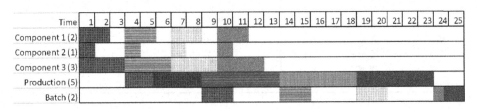

Fig. 4. The production of the line in time

The Secondary Warehouse. The demand for raw materials within the considered time period is given – we can say it has the deterministic character. As mentioned above, before the shift starts, necessary materials and components are prepared in the secondary warehouse and at the lines. Each component has assigned some given space. The secondary warehouse has certain limit of maximal capacity denoted by $CapSW$. This capacity is given in units of volume. It holds that the capacity in time t is less or equal then the maximal capacity: $CapSW(t) \leq CapSW_{max}$.

We will first express the relation to the volume of the particular component denoted by C_l, $l = \{1, 2, ..., L\}$. Let us denote this volume by $VolC_l$. For the volume of the component C_l in time t it holds that

$$VolC_l(t) = VolC_l(t - 1) - \sum_{\substack{i=1...N \\ k \in W}} \overline{C_k^i}(t) + VolC_l'(t) \qquad (3)$$

where W is a set of components of kind C_l that have been sent in time (t) to some of the production lines i and $VolC_l'(t)$ is the amount of just delivered components C_l (ordered from the main warehourse). Note that one particular component C_l can appear in the model under the one or more designations C_k^i.

In the model, there is defined minimal amount for each of the components in the secondary warehouse. If the stock level falls to the minimal level, the preset amount of the component is ordered from the main warehouse.

For the capacity of the secondary warehouse it holds that the actual capacity is dependent on the capacity in time $(t - 1)$, the volume of the components that have been sent on the basis of demand to some production line in time and the volume of components that have been delivered into the warehouse (from the main warehouse), i.e.

$$CapSW(t) = CapSW(t - 1) - \sum_{\substack{i=1...N \\ k \in U}} \overline{C_k^i}(t) + \sum_{l \in V} VolC_l'(t) \qquad (4)$$

where U is a set of components that have been sent to the production line i, $\overline{C_k^i}$ is the amount of particular component converted into units of volume, V is the set of just delivered components C_l.

Pollution and Waste. With the final products some given amount of the pollution (have impact on the environment surrounding respective production unit) and waste (can be processed further - recycled) is produced. These parameters do not have a great value for the run of the system, but they could have an impact on the prices of products and also the pollution of the environment considered in the model.

4   Conclusions

In this paper, we have provided the detailed description of one of the components of agent-based computational economy model, a production unit. The model is intended to be a sandbox solution, allowing experimentation with different approaches, strategies and decision making methods. This paper was focused mainly on the description of the internal processes of production unit, which allow it to manage the production autonomously, taking the advantage of the multi-agent approach to the full extent. Formal description of the production unit concept that was presented here is not related to a specific product. It is general and re-applicable to all production units in the model in a similar manner as it was shown on provided example. Future work will be focused on decision making and planning of production units (factories) on strategic (long-term) level.

Acknowledgements. The support of the Czech Science Foundation projects Nr. 14-02424S and Nr. 15-11724S is gratefully acknowledged.

References

1. Agrell, P.J., Hatami-Marbini, A.: Frontier-based performance analysis models for supply chain management: State of the art and research directions. Computers & Industrial Engineering 66, 567–583 (2013)
2. Brusaferri, A., Ballarino, A., Carpanzano, E.: Distributed Intelligent Automation Solutions for Self-adaptive Manufacturing Plants. In: Ortiz, Á., Franco, R.D., Gasquet, P.G. (eds.) BASYS 2010. IFIP AICT, vol. 322, pp. 205–213. Springer, Heidelberg (2010)
3. Denkena, B., Henjes, J., Lorenzen, L.E.: Adaptive Process Chain Optimisation of Manufacturing Systems. In: ElMaraghy, H.A. (ed.) Enabling Manufacturing Competitiveness and Economic Sustainability, pp. 184–188. Springer, Heidelberg (2012)
4. Hamichi, S., Brée, D., Guessoum, Z., Mangalagiu, D.: A Multi-Agent System for Adaptive Production Networks. In: Di Tosto, G., Van Dyke Parunak, H. (eds.) MABS 2009. LNCS, vol. 5683, pp. 49–60. Springer, Heidelberg (2010)
5. Chaib-Draa, B., Müller, J.P.: Multiagent based supply chain management. Springer (2006)
6. Ivanov, D., Sokolov, B., Kaeschel, J.: Integrated Adaptive Design and Planning of Supply Networks. In: Dangelmaier, W., Blecken, A., Delius, R., Klöpfer, S. (eds.) IHNS 2010. LNBIP, vol. 46, pp. 152–163. Springer, Heidelberg (2010)
7. Lawrence, J.A., Pasternack, B.A.: Applied management science: modeling, spreadsheet analysis, and communication for decision making. Wiley, New York (2002)
8. Schuldt, A.: Multiagent coordination enabling autonomous logistics. KI-Künstliche Intelligenz 26, 91–94 (2012)
9. Tesfatsion, L.: Agent-based computational economics: modeling economies as complex adaptive systems. Information Sciences 149, 262–268 (2003)
10. van Aart, C.: Organizational Principles for Multi-Agent Architectures. Springer (2005)
11. Zimmermann, R.: Agent-based Supply Network Event Management. Springer (2006)

Data Flow Processing Framework
for Multimodal Data Environment Software

Mateusz Janiak, Marek Kulbacki, Wojciech Knieć,
Jerzy Paweł Nowacki, and Aldona Drabik

Polish-Japanese Academy of Information Technology, Koszykowa 86,
02-008 Warszawa, Poland
{mjaniak,kulbacki,wkniec,nowacki,adrabik}@pjwstk.edu.pl

Abstract. Presented is a general purpose, very efficient data processing platform Multimodal Data Environment (MMDE), that includes a new framework for flexible data analysis and processing - Data Flow Programming Framework (DFPF). It was designed to unify already developed solutions, integrate them together and reuse as much of them in the simplest possible way. At the lowest level DFPF provides wrappers for current code to run it within a new framework and at the higher level it is possible to visually construct new algorithms for data processing from previously created or provided processing elements according to visual programming paradigm. DFPF is implemented as a dedicated service for MMDE in C++, but we expect that such solution can be easily implemented in any modern programming language.

Keywords: software architecture, multithreading, parallel data processing, data flow, visual programming, framework, pipeline, graph.

1   Introduction

Most often, for new research problems there do not exist any dedicated tools or software supporting research work. General purpose commercial solutions are enough for a reasonable price, but sometimes, despite pure research work, also some software development and implementation must be done. Usually not all research team members have programming skills required to implement particular solutions or they don't know different technologies that need to be integrated together. This unfortunately slows down the research process and increases its costs. In result, developed software is very similar in its general functionality to existing applications, only data types and operations are different. New application still loads the data, process it, optionally visualizes and saves the results. What usually software for data processing does with the data is: **load**, **process**, **visualize** and **save**.

Generally data processing scheme is always the same, only data types and operations on them change. Despite this, every time a research team is starting a new project and decides to create custom software for this purpose, a similar software is developed from its general functionality perspective. Some code might be reused, but

© Springer International Publishing Switzerland 2015

D. Barbucha et al. (eds.), *New Trends in Intelligent Information and Database Systems,*
Studies in Computational Intelligence 598, DOI: 10.1007/978-3-319-16211-9_36

new data types require generally completely different handling. This makes that plenty of time that could be spend on real research work is wasted on another redundant software implementation. To address this problem Multimodal Data Environment (MMDE) software was developed at the Research and Development Center of Polish-Japanese Academy of Information Technology (PJWSTK) in Poland. MMDE is a more general software and a successor of Motion Data Editor (MDE) [1]. MDE was oriented on supporting physicians in analysis of motion kinematic and kinetic parameters from motion data with advanced tools [2,3,4] based on quaternions and lifting scheme [5,6,7].

MMDE can support any data types because of its unique and simple architecture and a novel concept of various data types handling in an uniform manner for strongly typed C++ programming language.

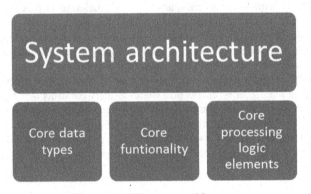

Fig. 1. MMDE system architecture

MMDE architecture (Fig. 1) is based on a simple conceptual model for data processing (core processing logic elements), covering three steps:

- load
- modify
- store

Fig. 2 presents MMDE core processing logic elements used for modeling of those operations. The **load step** covers data browsing (searching local file system, ftp, ...), reading various file formats, connecting to other devices or querying database. It also involves data normalization and conversion. The **store step** performs inverse operations, serializing data to known formats and saving them for further use. The **modify step** is the most complex part of application logic. It is the heaviest computational part of such data processing pipeline and covers not only data processing but also results validation by different kind of visualizations. Keeping in mind the presented data operations, it was decided to design a high level MMDE core with dedicated processing logic. Each stage was wrapped with a corresponding functional elements.

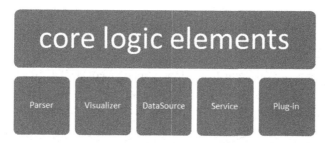

Fig. 2. MMDE core processing logic elements

Data browsing and loading are performed by Source elements. They are supported with *Parsers*, which main task is to deliver of normalized data from various file formats and streams. Data is saved with the help of *Sink Objects*. As MMDE should support various functionalities, it was decided to represent them as Services responsible for particular tasks, possibly cooperating together and managing application resources and behavior. To visualize data *Visualizer* elements were proposed. Loaded data is available through central storage - *Memory Manager*. As MMDE is implemented in C++, it was required to introduce an equivalent Object type concept, known from higher level programming languages, being a common base type for all other types. New approach for variant type in strongly typed C++ was developed for this reason. It allows to unify data management and exchange. It is based on meta-programming techniques and policies [8]. The garbage collection mechanism was introduced, based on this solution. It effectively reduces overall processing platform memory consumption. It is based on data reference counting and time of last data access. MMDE delivers also many other functionalities, like threads management, jobs scheduler and file system manipulation, which unify those common data processing tasks.

2   Data Flow Programming Framework

Data Flow Programming Framework (DFPF) module is implemented as a service for MMDE. It provides a functionality for creating and running data processing pipelines. Its concept is based on graph structures. Describing DFPF, two layers can be pointed out:

- model structure,
- execution logic and control.

2.1   Model Structure

Model structure defines basic DFPF elements and their connections. Together they define how data would be processed by particular operations within DFPF. Any data processing (load, save, modify) is done within three types of nodes (Fig. 3):

- sources,
- processors,
- sinks.

Those elements are different from already mentioned MMDE functional core elements.

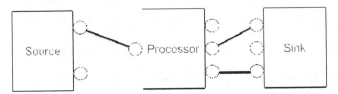

Fig. 3. DFPF model structure elements

Nodes are defined within DFPF service scope, independently from general MMDE architecture. Each node, despite data operations it performs, is described with inputs and outputs, called *pins*. Each pin represents one data type. Nodes can be connected through pins of different types (input-output), representing compatible data types. Such structure creates a directed graph, where pin types define the direction (from output to input). One important difference between input and output pin is that output pin can create many connections to other nodes in relation one to many and input pin can be a part of at most one connection. This can be compared to a function signatures in programming languages. Each input pin is an argument for particular function (node), lack of input pins means function with no arguments. For each argument only one value (variable, constant, default value) can be passed. Function can also return some values after it has finished computations. In general, functions return one or none values. As nodes can have more output pins they can be interpreted as a fixed structure, which is returned by a function and its fields are mapped to particular output pins. Such data flow model might be more complex. Some input pins might be required by node to operate properly and some not. Also output pins might have limited connectivity capabilities (limited number of connections). In general node might require some specific number of connections. These are model details specific to a problem realized with data flow and they can be defined with various policies. The model structure layer is responsible for model structure validation and integrity. Improperly defined models cannot process the data. DFPF model has one additional and valuable property. It allows to group properly connected nodes to create new, more general nodes. Such aggregates can be used to create completely new data operations from already available functions. This dynamical behavior allows to save time and does not require any additional programming. Such nodes can be stored and reused in the future, instead of creating such structures from scratch each time they are needed.

2.2 Execution Logic

When data processing model is ready it is time to describe how such processing could be done. At this layer nodes and pins have some additional functionality. Nodes produce, process and consume the data. To do the work they rely on attached pins functionalities, where input pins deliver new data for processing and output pins are used to propagate produced data. Pins themselves are responsible for effective data exchange. Based on such assumption there are generally two possible execution schemes for DFPF:

- serial,
- parallel.

In the serial case, there is a trivial, iterative algorithm, starting from source nodes and going step by step deeper with data processing (through connected successor nodes) as long as possible.

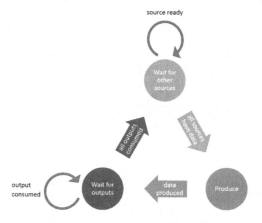

Fig. 4. Source node execution logic

When at some stage (node) not all required data are available for processing, algorithm moves back to the previous stages and tries to continue there, until processing continues within next nodes or data flow is finished.

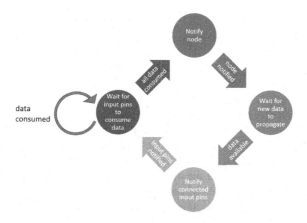

Fig. 5. Output pin execution logic

More interesting is parallel approach, allowing to utilize effectively computational Central Processing Unit (CPU) resources. In this case each node execution logic is processed by a single thread. Processing starts with source nodes. Fig. 4 presents source node processing logic.

Each node prepares data for propagation through DFPF. Data is set up to output pins. Fig. 5 presents output pin execution logic.

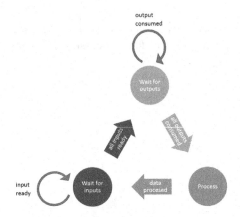

Fig. 6. Processor node execution logic

Output pins notify attached input pins through connections about new available data for processing. Source nodes wait for notifications until data from their all connected output pins are consumed to produce and propagate next data portion.

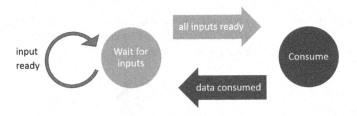

Fig. 7. Sink node execution logic

Data from output pin is assumed to be consumed, when all attached input pins have copied it. After this stage processing or sink nodes can operate. Fig. 6 and Fig. 7 presents processor node and sink node execution logic respectively. When their input pins are notified about available data to consume, they immediately copy the data to attached node. Fig. 8 presents complete input pin execution logic. When all input pins connected to a node have copied the data such node can perform its specific data operations. Sink nodes simply store the data and wait until new data is available. Processor nodes produce new data, pass them to their connected output pins and wait until this data is consumed by the following nodes. If so they wait once again for new input data for next processing stage. Parallel execution approach allows to fill model with as much data as possible. After some transient time, where there are large enough data sets to process, DFPF reaches 100% efficiency. It means all nodes are loaded with required data for processing and with each processing stage some data leaves DFPF (through sink or processor nodes) and new data are loaded (with source nodes).

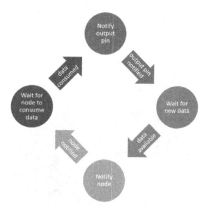

Fig. 8. Input pin execution logic

This layer is also responsible for controlling when to stop processing of data. There are possible two approaches:

- process as long any of source nodes provide new data,
- process as long all source nodes provide new data.

In the first case DFPF will try to process as much data as possible, possibly stopping its processing somewhere in the middle of the graph structure. This might happen due to a lack of required data from paths leading from other sources which might be empty at this stage. All sources work independently to each other. In the second approach new data is loaded only if all source nodes can provide more data. When one of them becomes empty none new data is loaded to the model. With this approach source nodes are additionally synchronized together to verify their data capabilities. In both cases execution logic must track data processing status within the model. This is required to define if some data processing is in progress or DFPF has finished its work. While executing processing in parallel scheme, it has to be mentioned that more complex graphs will lower overall performance of DFPF, as number of required threads servicing particular nodes might exceed available physical CPU resources [9]. To overcome this problem it is suggested to apply *Job Manager* pattern, where threads controlling nodes execution will schedule data computations in form of independent Jobs to Job Manager. Job Manager is responsible for efficient utilization of computational resources, so that overall performance would be kept at highest possible rate.

3   Visual Programming with DFPF

Visual programming is a paradigm used for data processing simplification in tools like Blender (http://www.blender.org) or LabView (http://www.ni.com/labview). It allows users to compose graphically processing pipelines with the use of simple building blocks performing particular operations. Users can connect blocks together to create more complex processing algorithms. Similar solution was developed for DFPF. An abstraction layer above DFPF model was prepared to allow a user graphically compose processing structures based on delivered nodes.

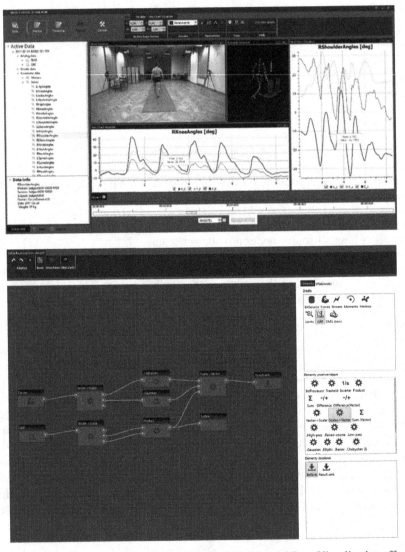

Fig. 9. Visual layer for DFPF with data flow and Multimodal Data Visualization effect

User utilizes simple drag and drop mechanism to create instances of particular nodes on the scene, representing processing pipeline. To create connection user simply clicks on visual representation of particular pins. Additionally, user is assisted when creating connections by highlighting compatible pins matching set up connectivity rules for the model. Visual layer prevents incompatible and improper connections. Additionally, it validates whole DFPF model structure before launching execution logic. User can move nodes around the scene, group them and create more abstract structures for future use. The whole graph structure can be serialized and loaded when required to save time required for custom pipelines designing. Fig. 9 presents an example of graphical layer for DFPF. With the help of visual layer user can control execution of DFPF: **run**, **pause** and **stop**.

4   Applications and Conclusion

In the paper we presented assumptions and implementation of Data Flow Processing Framework. DFPF can be easily adopted not only for processing data on a local computer, utilizing all computational power (GPU and CPU cores), but also for controlling cluster computing or distributed computations. Designed data flow graph might be translated to fit known cluster topology or be distributed over some network. What is more, it does not require to implement all processing elements in C++. With such approach user could distribute Java, C or any other data processing application and just control its parallel execution, deliver required data and collect the results. If node data operations (applications) support checkpoints (serialization), such processing could easily migrate around various nodes and computers, allowing better resources utilization and energy consumption management. Presented DFPF was successfully adopted in such research projects and solutions as [10, 11, 12]. Those demonstrations show that presented DFPF is not only a theoretical concept but can be used to solve problems of many research groups by unifying their current solutions.

Acknowledgment. This work was supported by a project UOD-DEM-1-183/001 from the Polish National Centre for Research and Development.

References

1. Kulbacki, M., Janiak, M., Knieć, W.: Motion Data Editor Software Architecture Oriented on Efficient and General Purpose Data Analysis. In: Nguyen, N.T., Attachoo, B., Trawiński, B., Somboonviwat, K. (eds.) ACIIDS 2014, Part II. LNCS, vol. 8398, pp. 545–554. Springer, Heidelberg (2014)
2. Świtoński, A., Polański, A., Wojciechowski, K.: Human Identification Based on Gait Paths. In: Blanc-Talon, J., Kleihorst, R., Philips, W., Popescu, D., Scheunders, P. (eds.) ACIVS 2011. LNCS, vol. 6915, pp. 531–542. Springer, Heidelberg (2011)
3. Josinski, H., Switonski, A., Jedrasiak, K., Kostrzewa, D.: Human identification based on gait motion capture data. In: Proceeding of the International MultiConference of Engineers and Computer Scientists, Hong Kong, vol. 1 (2012)

4. Switonski, A., Mucha, R., Danowski, D., Mucha, M., Cieslar, G., Wojciechowski, K.: Sieron. A.: Human identification based on a kinematical data of a gait. In: Electrical Review, R. 87, NR 12b/2011, pp. 33–2097 (2011) ISSN 0033-2097
5. Szczesna, A., Slupik, J., Janiak, M.: Motion data denoising based on the quaternion lifting scheme multiresolution transform. Machine Graphics & Vision 20(3), 238–249 (2011)
6. Szczęsna, A., Słupik, J., Janiak, M.: The smooth quaternion lifting scheme transform for multi-resolution motion analysis. In: Bolc, L., Tadeusiewicz, R., Chmielewski, L.J., Wojciechowski, K. (eds.) ICCVG 2012. LNCS, vol. 7594, pp. 657–668. Springer, Heidelberg (2012)
7. Szczesna, A., Slupik, J., Janiak, M.: Quaternion lifting scheme for multi-resolution wavelet-based motion analysis. In: The Seventh International Conference on Systems, ICONS 2012, pp. 223–228 (2012)
8. Alexandrescu, A.: Modern C++ Design: Generic Programming and Design Patterns Applied, 1st edn. Addison-Wesley Professional (2001)
9. Williams, A.: C++ Concurrency in Action: Practical Multithreading, 1st edn. Manning Publications (2012)
10. Janiak, M., Szczęsna, A., Słupik, J.: Implementation of quaternion based lifting scheme for motion data editor software. In: Nguyen, N.T., Attachoo, B., Trawiński, B., Somboonviwat, K. (eds.) ACIIDS 2014, Part II. LNCS, vol. 8398, pp. 515–524. Springer, Heidelberg (2014)
11. Kulbacki, M., Segen, J., Nowacki, J.: 4GAIT: Synchronized Mocap, Video, GRF and EMG datasets: Acquisition, Management and Applications. In: Nguyen, N.T., Attachoo, B., Trawiński, B., Somboonviwat, K. (eds.) ACIIDS 2014, Part II. LNCS, vol. 8398, pp. 555–564. Springer, Heidelberg (2014)
12. Kulbacki, M., Koteras, R., Szczęsna, A., Daniec, K., Bieda, R., Słupik, J., Segen, J., Nawrat, A., Polański, A., Wojciechowski, K.: Scalable, Wearable, Unobtrusive Sensor Network for Multimodal Human Monitoring with Distributed Control. In: Lacković, I., Vasić, D. (eds.) IFMBE Proceedings, pp. 914–917. Springer (2015)

Simulation Modeling Approach to Environmental Issues in Supply Chain with Remanufacturing

Paulina Golinska-Dawson and Pawel Pawlewski

Poznan University of Technology, Poznan, Poland
{paulina.golinska,pawel.pawlewski}@put.poznan.pl

Abstract. The aim of the paper is to present the modeling approach using simulation tool to assess environmental issues in supply chain with remanufacturing. Remanufacturing is usually multi-optional and multi-operations process, not as transparent as the manufacturing process itself. The supply chain with remanufacturing has to take in consideration the reverse materials flow and additional activities like for example collection of used product, disassembly and reprocessing. The simulation tools seem to have a big potential to model such environmental indicators like for example energy consumption, water usage and sewages control. The paper examines the potential of application of simulation model in a field of remanufacturing. Authors elaborated generic simulation modeling approach which might be used in global and local level of a supply chain.

Keywords: Remanufacturing, simulation, sustainability, environment, indicators.

1 Introduction

Environmental problems gain a lot of interest in the framework of the supply chain management. The environmental regulations, customers demand pattern and economics make company to redesign their supply chains and to include the reverse flows. Companies have faced the need to seek for a new way of costs reduction and appropriate products returns management. The effort to improve supply chains using environmental friendly management approaches results in performance improvements. In order to measure the improvements (or to identify lack of sufficient improvement) reliable tools are needed. Companies in supply chain need to define and systematically measure their common environmental objectives to reduce adverse impacts over environment. The materials flows in supply chain with product recovery are not so transparent as in a primary (traditionally) supply chain.

The paper examines the potential of application of simulation model in area of supply chain with remanufacturing. Authors elaborated generic simulation modeling approach which might be used in global and local level of the supply chain.

The paper is organized as follows in the second section authors discuss the characteristics of supply chain with product recovery. The focus in placed on remanufacturing.

© Springer International Publishing Switzerland 2015
D. Barbucha et al. (eds.), *New Trends in Intelligent Information and Database Systems,*
Studies in Computational Intelligence 598, DOI: 10.1007/978-3-319-16211-9_37

In the third section are described the green simulation aspects. Section 4 presents the elaborated by the authors approach to simulation of "green" indicators at a local and global level in the supply chain. Final conclusions are stated in Section 5.

2   Environmental Problems in Supply Chain with Remanufacturing

The nature of reverse flows in a supply chain depends on stage of product life cycle and source of its collection. Reverse materials management (re-supply) can be perceived as an opportunity to lower the costs of raw materials acquisition, as well as a way to gain benefits of decreasing cost of energy or waste disposal. "Green" supply chain might be also a driver for gaining more customers. Legal regulations put the pressure on companies participating in a supply chain to change from non-sustainable to more environmentally friendly operations.

According to the circular economy concept the most common scenario for closing loop in the supply chain are, as followed   [1]:

- Recycling - the process of reclaiming materials from used products or materials from their manufacturing and using them in the manufacturing of new products.
- Remanufacturing - an industrial process in which new product is reassembly from an old one and, where necessary, new parts are used to produce fully equivalent in performance to original new product.
- Refurbishment is the reprocessing of product in order to provide required functionality. Usually is it connected with some improvement in existing product parameters like improvement of product's existing operating software.
- Repair is related to actions taken to return a product's functioning condition after detected failure, particularly at the service. Repairs are not equivalent to return its state to as good as new.
- Reuse as it is, mainly for spare parts.

The most preferable options for product recovery are those in which products are not destroyed and the residual value can be saved. Fleischman [2] has grouped the reverse flows into five categories: end-of-life returns, commercial returns, warranty returns, production scrap and by-products, reusable packaging material. In this paper we focus on returns which are suitable for remanufacturing like: end-of-life returns, and warranty returns. Such products usually have still relatively high residual value. Remanufacturing is preferable option for products with high residual value like cars, automotive parts or computers. Remanufacturing allows the recovery of value added during primary processes in supply chain. The supply chain with product's recovery is presented in Fig. 1.

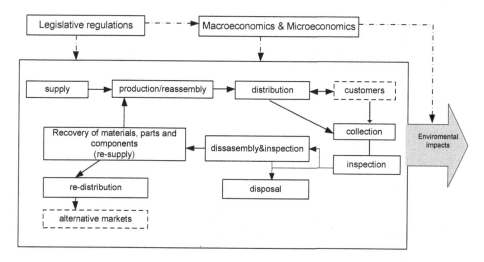

Fig. 1. Supply chain with product recovery

Sundin and Lee [3] provided an overview on the assessment of the environmental performance of remanufacturing. They identified a number of case studies from re-manufacturing companies and discussed the environmental indicators, which might be used for an environmental impact assessment. They classified the environmental measures used for assessment of the remanufacturing process, as following [3]:

- Direct: consumption of materials, energy and waste generated, which translates this directly to resource savings,
- Indirect: Life Cycle Assessment methods, which calculates eco-points to assess the environmental impact, they assess the long-term potential environmental impact.

Both direct and indirect measures require a rather big scope of data. In case of supply chain with recovery options the indicators should be measured locally (at single operation, single company) and at global level (group of companies).

3   Green Simulation and Simulation Tools in Remanufacturing

Nowadays some of the simulation software allow to perform, so called green simulation. In previous research [4, 5] analyzed five simulators for "green" features, including Anylogic, ExtendSim, Arena, Witness and Quest. The conclusions was that the green simulation options are rather limited in standard simulation software. Among analyzed software only Witness had built in green simulation tools. The traditional approach to green simulation is based on local measurements. Below is example (Fig. 2) of measuring "green" parameters as the "built- in" in the Witness simulation software. The main disadvantage of such approach is that the measurement can be only done for one "object" each time and then individual (local level) assessments have to be summarized according to some predefined by the researcher algorithm.

Such approach is not sufficient for supply chain with remanufacturing   where material flows are overlapping for example: returned products is transferred after recovery
to the primary supply chain as "re-supply" (see Fig. 1).

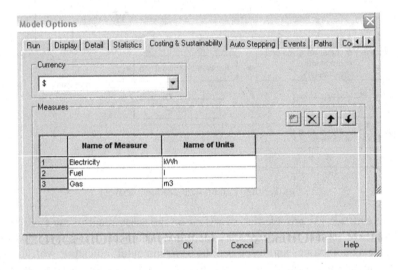

Fig. 2. Menu for selection of green measures in Witness [4]

After search of Google Scholar and Scopus under search criteria "green simulation" and "remanufacturing" satisfactory matches were not found. The problem of
"green" simulation in remanufacturing or supply chain with remanufacturing is not
well described in the literature. There limited amount of papers on application of
simulation in remanufacturing, on the following topics:

- behavior of the reverse supply chain with focus on inventory and bullwhip effect
  reduction [e.g. 6,7],
- pricing and demand management [e.g. 8],
- planning and control in remanufacturing [e.g.9],
- simulation of lot sizing problem with remanufacturing   [e.g.10]
- reverse logistics network design [e.g. 11, 12].

The most relevant paper seems to be work by Ismail et al. [13]. The requirements
for the development of „green" simulation in remanufacturing are presented in Fig.3.

Taking in consideration the research gap regarding the application of simulation tools
for green simulation in remanufacturing we have elaborated approach which allow to
monitor the green indicators in a supply chain with remanufacturing.

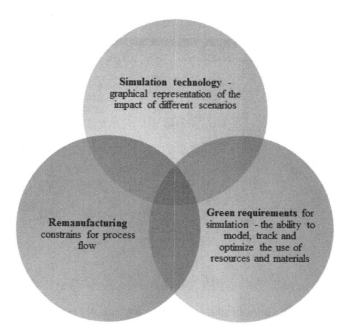

Fig. 3. Green simulation in remanufacturing requirements

4   Environmental Indicators and Simulation Approach – Example of an Application

Authors have defined a set of indicators which allow to assess the environmental impact of operations in supply chain with remanufacturing, they are as followed:

- ECL – Energy Consumption leve l,
- WCL - Water Consumption Level,
- GEL co2 – Generated Emission Level,
- GEL$_{sludge}$ – Genereted leve of sludge,
- PCL - Pressure Consumption Level – which measures compressed air consumption where relevant .

The indictors were implemented into FlexSim modeling software. The example of elaborated by the authors simulation model is presented in Fig. 4.

SIRO_ECO_Library

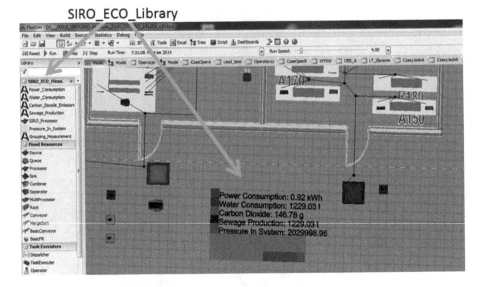

Fig. 4. Green simulation indicators implemented in FlexSim

We define in the simulation software a special ECO-object, which is a universal measure (universal "plug") and which can be connected to any objects in the supply chain model by combining information. It performs the measurement for one or more indicators (from the list above). The Eco-object has modules which allow to measure any of the listed indicators individually or joint with other indicators. It is also possible to measure all indicators at the same time. The Eco-object is used according to the following algorithm:

Step 1 – select the machine/operation for which is measured the environmental indicator.

Step 2 – in table „Operations" – insert the relevant information about active consumption of the machine/object (during operation) and passive consumption of the machine/object during operation (when it is turned on but not working) array operations involving this consumption. It will be saved in the labeling of this machines/object.

Step3 – connect the object /machines for which should be collected measurements to the module of Eco-object. If information is gathered locally for the machine, then connect the Eco-object to this machine, if to collect together for several machines connect them all to the Eco-object.

Step 4 – define the time of simulation and simulation scenario.

Step 5 – run the simulation experiment.

Step 6 – analyze the indicator values in the „Results' table".

The „Results table" is built automatically during the execution of simulation experiment. It contents as many items as there are the days of simulation. It shows the amount of machines multiplied by values in two columns. In the first column is the energy consumption during the working day, the second column is the total cumulative consumption. This Table can be used to more sophisticated analysis by combining it with other results for other indicators. Fig. 5 presents the example of Eco-object usage for measurement of modules: WCL (water consumption) and GEL$_{sludge}$ (waste water generation) for group of machines/objects. After connecting the meter (Eco-object) to at least one machine through a central port, and after setting all the labels it runs the model and makes measurements. In case of water consumption every machine/object has two labels:

- Water active – parameter of machine – how many water is used during work,
- Water passive – parameter of machine – how many water is used when machine doesn't work (does not perform the operations).

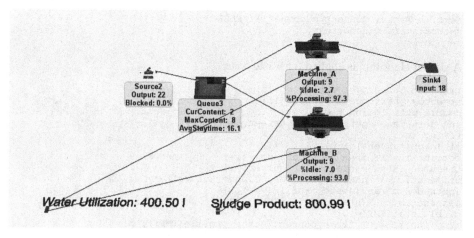

Fig. 5. Example – system with 2 machines/objects

Analyzed following machine states:

- 12 - STATE_SCHEDULED_DOWN – e.g. when a planned lunch break takes place,
- 20 - STATE_DOWN – when it doesn't work, for example it is waiting for material to arrive to a work station,
- 2 - STATE_PROCESSING
- 21 - STATE_SETUP
- The example of water calculation is presented below.

```
treenode current = ownerobject(c);
treenode  textnode = parnode(1);
double utilization = 0;
int setuping = getlabelnum(current,"setup");  // indicator - calculation
during setup
if  (nrcp(current) > 0)  {   //Number of connected machines
for(int i=1; i <= nrcp(current); i++)  {   // loop by machines
treenode TMP = centerobject(current,i);   // getting address of machines
// calculation of Water Active
```

```
utilization += getlabelnum(TMP,"Water_Active") *
(getnodenum(rank(state_profile(TMP),2)) +
getnodenum(rank(state_profile(TMP),21)) * setuping);
// calculation during setup
// calculation of Water passive
double stateTMP = 0;
for (int ii=0; ii <= content(state_profile(TMP)); ii++){
if (ii+1 != 12 && ii+1 != 20 && ii+1 != 2 && ii+1 != 21){
stateTMP += getnodenum(rank( state_profile(TMP),ii+1));}}
   utilization += getlabelnum(TMP,"Water Passive") * stateTMP; } //FOR1
double czas = getlabelnum(current,"unit");
if (czas <= 0 ) {
text = "Water: Set - unit";}else {
utilization *= czas;
text= concat("Water Utilize: ",numtostring(utilization,0,2), " l");}}
//IF1 else  { //ELSE1
text = "Water: Connect machine";}//ELSE
setnodestr(textnode,text);
return 1;
```

A similar algorithm is applied for waste water

```
treenode current = ownerobject(c);
treenode  textnode = parnode(1);
double utilization = 0;
int setuping = getlabelnum(current,"setup");
string text ;
if (nrcp(current) > 0)  {    //IF1
for(int i = 1; i <= nrcp(current); i++)  {    //FOR1
treenode TMP = centerobject(current,i);
utilization += getlabelnum(TMP,"Sludge_Active") *
(getnodenum(rank(state_profile(TMP),2)) +
getnodenum(rank(state_profile(TMP),21)) * setuping);
double stateTMP = 0;
for (int ii = 0; ii <= content(state_profile(TMP)); ii++){
if (ii+1 != 12 && ii+1 != 20 && ii+1 != 2 && ii+1 != 21) {
stateTMP += getnodenum(rank( state_profile(TMP),ii+1));}}
utilization += getlabelnum(TMP,"Sludge_Passive") * stateTMP;} //FOR1
double czas = getlabelnum(current,"unit");
if (czas <= 0 )  {
text = " Sludge:  Set unit";  } else  {
utilization *= czas;
text= concat("Sludge product: ",numtostring(utilization,0,2), "l"); }}
//IF1 else  {  //ELSE1
text = " Sludge:  Connect machine ";}  //ELSE
setnodestr(textnode,text);
return 1;
```

We can perform the measurements of the ecological indicators in the supply chain locally, for example in the enterprise where the Eco-object is connected to one or more machines (including mean of transportation). For many machines, when applying many meters or bulk meter for a wide range of machines (see fig 6). The same thing can be performed on a global level in a supply chain, where it is possible to connect the meter to the resources of different enterprises.

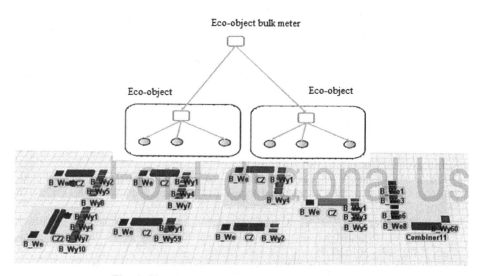

Fig. 6. Global environmental indicators measurements

5   Conclusions

The paper presents work on the simulation approach to the assessment of the environmental issues in a supply chain with remanufacturing. The proposed approach is different to the existing build-in "green" simulation tools, which are available on a very limited scale. Main advantage of the elaborated solution, it is the universality of the tool which might be "plug in" to unlimited number of object like machine, means of transportation ect. Furthermore it can measure one or more environmental indicator.

Acknowledgements. This work was financed by the Narodowe Centrum Badan i Rozwoju (National Centre for Research and Development) in the framework of the German-Polish cooperation for sustainable development, project "Sustainability in remanufacturing operations (SIRO)", grant no WPN/2/2012

References

1. Golinska, P., Kosacka, M.: Environmental Friendly Practices in the Automotive Industry. In: Golinska, P. (ed.) Environmental Issues in Automotive Industry, pp. 3–22. Springer, Heidelberg (2014)
2. Fleischmann, M.: Quantitative Models for Reverse Logistics. Springer, Berlin (2001)
3. Sundin, E., Lee, H.M.: In what way is remanufacturing good for the environment? In: Proceedings of the 7th International Symposium on Environmentally Conscious Design and Inverse Manufacturing (EcoDesign-2011), Kyoto, Japan, November 30 – December 2, pp. 551–556 (2011)

4. Pawlewski, P., Borucki, J.: Green Possibilities of Simulation Software for Production and Logistics: A Survey. In: Golinska, P., Fertsch, M., Marx Gomez, J. (eds.) Information Technologies in Environmental Engineering, pp. 675–688. Springer, Heidelberg (2011)
5. Pawlewski, P., Otamendi, F.J.: Simulation Software and Technologies for "Green" Eco-Production. In: Golinska, P. (ed.) EcoProduction and Logistics, pp. 239–259. Springer, Heidelberg (2013)
6. Qingli, D., Hao, S., Hui, Z.: Simulation of remanufacturing in reverse supply chain based on system dynamics. In: 2008 International Conference Service Systems and Service Management, pp. 1–6. IEEE Press, New York (2008)
7. Zolfagharinia, H., Hafezi, M., Farahani, R.Z., Fahimnia, B.: A hybrid two-stock inventory control model for a reverse supply chain. Transportation Research Part E: Logistics and Transportation Review 67, 141–161 (2014)
8. Pokharel, S., Liang, Y.: A model to evaluate acquisition price and quantity of used products for remanufacturing. International Journal of Production Economics 138(1), 170–176 (2012)
9. Li, J., González, M., Zhu, Y.: A hybrid simulation optimization method for production planning of dedicated remanufacturing. International Journal of Production Economics 117(2), 286–301 (2009)
10. Andrew-Munot, M., Ibrahim, R.N.: Development and analysis of mathematical and simulation models of decision-making tools for remanufacturing. Production Planning & Control 24(12), 1081–1100 (2013)
11. Suyabatmaz, A.Ç., Altekin, F.T., Şahin, G.: Hybrid simulation-analytical modeling approaches for the reverse logistics network design of a third-party logistics provider. Computers & Industrial Engineering 70, 74–89 (2014)
12. Golinska, P., Kawa, A.: Remanufacturing in automotive industry: Challenges and limitations. Journal of Industrial Engineering and Management 4(3), 453–466 (2011)
13. Ismail, N.H., Mandil, G., Zwolinski, P.: A remanufacturing process library for environmental impact simulations. Journal of Remanufacturing 4(1), 1–14 (2014)

Development of Interactive Mobile-Learning Application in Distance Education via Learning Objects Approach

Sheng Hung Chung and Ean Teng Khor

School of Science and Technology, Wawasan Open University, Malaysia
{shchung,etkhor}@wou.edu.my

Abstract. This paper presents the implementation and development of Mobile-Learning Initiatives in Wawasan Open University in Open Distance Learning (ODL) environment. The Mobile-Learning Initiative was introduced by Mobile Learning Research Group (MLRG) of WOU aiming to promote interactive mobile-learning experiences as an alternative learning for distance learners and carry out studies on mobile-learning applications and developments in the context of distance learning from ODL institutions. The components in this study intend to help educators in the implementation of mobile-learning and stimulate further research on mobile applications. This initiative is also hope to assist open and distance learning practitioners to better understand the concepts of m-Learning and how to more effectively incorporate mobile technologies into the teaching and learning environment.

Keywords: Mobile-learning, distance education, interactive learning.

1 Introduction

The high penetration rates of mobile phone subscriptions of 142.5 per 100 people (Quarter 2, 2012) and 143.3 per 100 people (Quarter 2, 2013) in Malaysia [1] and the rapid growing of mobile users show that it is viable for making inroads towards the usage of mobile devices as an alternative learning mode for Malaysian distance learners. According to Hisham [2] and Flowers [3], the study reveals that distance learners experienced isolation due to lack of interaction and communication with fellow learners, tutors, and the university as compared to those in traditional universities. Zirkle [4] and Vrasidas & Glass [5] reported the learner's support should be established to facilitate distance learning and strategies are frequently suggested in relation to promoting active and collaborative learning using learning technologies [6]. The study of Chou [7] indicates that the positive impact of mobile-learning integration on student learning includes active engagement and increased time for project involvement activities. Based on challenges stated above, the mobile-learning initiative is proposed to encourage learning and interactions in WOU distance learning communities aiming to bridge the transactional distances faced by the learners and adopt mobility as the key tool in course delivery. This study serves as the guidelines of mobile-learning initiatives carried out by Mobile Learning Research Group (MLRG) in

© Springer International Publishing Switzerland 2015

D. Barbucha et al. (eds.), *New Trends in Intelligent Information and Database Systems,*
Studies in Computational Intelligence 598, DOI: 10.1007/978-3-319-16211-9_38

WOU and prepare the review of several important aspects including: (i) The understanding of mobile-learning design, requirements and specifications of ODL learners, (ii) Design and development of mobile-learning App, (iii) Case studies of mobile-learning in ODL institutions, (iv) Collaborative Webinar and Web-conference sessions on mobile application development. The preliminary studies on mobile-learning implementation and focus areas are reported in Section 2. Section 3 describes the system architecture of WOU mLearning and Section 4 illustrates the WOU mLearning prototype design.

2   Mobile-Learning Implementation and Focus Areas

The mobile-learning implementation strategies undertaken are aimed to provide mobile-learning as an alternative way to accommodate the major needs of interactive learning by using mobile devices. The initiative also demonstrates and promotes the use of handheld devices to view the course modules, learning materials, attempt quizzes anywhere and anytime, allowing learners in engaging academic activities without the time and location barriers.

The five main objectives of the WOU mobile-learning initiative includes:

- To support u-learning (ubiquitous-learning)
- To encourage interactive learning by providing WOU Education Learning App via handheld devices
- To increase flexibility of learning by accessing course modules and materials offered to distance learners
- To promote the use of mobile-learning in learning and teaching environment
- To conduct studies of mobile-learning implementation and development in distance learning context

The following section describes the implementation and focus areas for mobile-learning initiative mainly:

1. **The Understanding of mLearning Designs, Requirements and Specifications of ODL Learners**
2. **Design and development of mobile-learning App**
3. **Case studies of mobile-learning in ODL institutions**
4. **Collaborative Webinar and Web-conference sessions on mobile application development**

2.1   The Understanding of mLearning Designs, Requirements and Specifications of ODL Learners

This section explores the factors, navigations and design requirements of mobile-learning environment in WOU. The relevant design, requirements and specifications are studied during the initial study stage of mobile-learning initiative. In considering the mobile-learning design and specifications, the preferable learning contents and the receivable features are taken into considerations as one of the important components to engage learners by providing useful and interactive learning experience. The mobile-learning pre-development was also evaluated by asking the learners on time spent using the mobile-learning app. The readiness study covers the determinacy factors of mobile learners and the readiness to be a mobile learner in ODL context. The readiness studies focus on the following aspects and data collections:

i.   Mobile devices platforms
ii.   Tablet PC platforms
iii.   Time spent using the mobile-learning app
iv.   Requirements, navigations and specifications of mobile user interface
v.   Preferable learning content in mobile (audio, video, course materials, simulations)
vi.   Additional receivable features (quizzes, reminder, notifications, announcements)

The preliminary study [8] of the requirements highlights the mobile-learning support features for mobile learners and indicated that repositories search, interactive multimedia support, content ranking, location based services, help documentation and cloud storage are having the highest votes and play important role in mobile-learning for the study of learning content accessibility (Fig. 1). The research also uses quantitative data which were obtained at the end of the post-test performance after learners completed the mobile-learning apps. The post-test performance is also followed by a satisfaction with the mobile-learning experience upon successful completion of mobile-learning module.

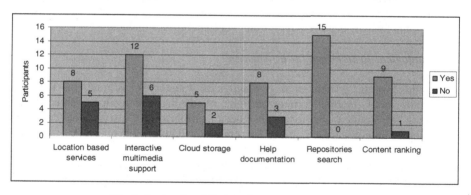

Fig. 1. Study of mobile-learning support features [8]

2.2   Design and Development of Mobile-Learning App

The **WOU Education Learning Apps** are constructed based on mobile-learning specifications with appropriate matching of mobile-learning features obtained from *Readiness Study* stage to suit learners' experience and expectations in mobile learning environment. The development of the mobile-learning apps was conducted by MLRG core team members, instructional designers and mobile apps developers. The output of the project includes the development of the following applications:

WOU Education Learning Mobile App (iOS)
Mobile app that enable learners to engage interactive learning from iPhone devices (iOS version 5.2 or higher), screen size (320 by 480 pixels)

WOU Education Learning Tablet App (iOS)
Tablet app that enable learners to engage interactive learning or consume learning materials from iPad devices (iOS version 5.2 or higher), screen size (1024 by 768 pixels).

2.3   Case Studies of Mobile-Learning in ODL Institutions

A review of mobile-learning literature reveals a lack of study and findings conducted pertaining to mobile-learning related activities in ODL institutions. The use of flexible learning via mobile technology provides opportunities for researchers to conduct research for the use of mobile technology in teaching and learning that exists nowadays. This presents the opportunity for conducting a study based on five aspects as illustrated in the section follows:

(I) Exemplary m-learning Projects

The study of exemplary projects implemented in various ODL institutions such as designs and implementation describes the mobile-learning initiatives in multiple operating systems (OS) adopted by ODL institutions project spearheaded by the respective universities.   The demonstration of mobile Apps, use case models, scenarios and designs in carried out through collaborative approach.

(II) Pedagogical Framework for m-learning

In mobile-learning environment, pedagogical framework is one of the key considerations of mobile-learning design. In the recent reports published by Yeonjeong [9], mobile technologies have been incorporated by instructional designers in developing effective teaching and learning in pedagogical framework of mobile-learning by focusing on technological attributes such as individuality and interactivity. Pedagogical affordances such as usefulness and concentration are also highly contributed towards designing an effective learning tool.

(III) Educational Standards for m-learning

According to Judy et al. [10], the study of educational standards for mobile-learning application development is mainly used to address the portability and development of learning materials. The aspects which influence the role of educational standards include the ubiquity of mobile devices and the opportunity to leverage the portability and connectivity of learning.

(IV) Media Formats and Technologies for m-learning Platform

According to Uther [11], successful mobile applications incorporate various media objects and play an important role in the learner's experience. One of the commonly stated characteristics of mobile-learning content is that it should be delivered in short 'nuggets' rather than large units of information, with the appropriate implementation of media presentation such as video lectures, audio, images and text. Therefore the mobile media types have to be chosen with care and should support content appropriately.

(V) Examining the Challenges and Issues Related to Teaching and Learning Aspects in m-learning

The limitations and weaknesses faced by mobile-learning in ODL environment is taken into considerations as important assessment in the study. The investigation covers both teaching and learning aspects conducted in mobile-learning applications. The analysis framework proposed by Koole [12] evaluates three different aspects of mobile-learning such as device, learner and social environment. This model investigates the issues arises while interactions take place through each of the three aspects.

2.4   Capacity Building on Mobile Application Development

This section describes the organizing of webinar, training sessions and demonstrations of mobile application in Android or iOS to interested participants in ODL environment who are interested in Apps development. The relevant details concerning the web-conference series are shown in Table 1:

Table 1. Mobile-learning related topics and scope in promoting mobile development

Title of web-conference sessions	Scope
i) Instructional Design for the mobile-learning and Fundamentals of effective m-Learning interface design for learners	Design, requirements and specifications
ii) Transition from e-learning to m-Learning design and support	Design, learner supports
iii) Effective e-learning to for m-Learning design	Design, learning experiences
iv) m-Learning: Increasing learners engagement and interactivity	Interactive learning
v) Case Study of m-Learning development and application in ODL support environment (Participation of ODL institutions in Asian perspective)	Case studies, exemplary projects
vi) Mobile Application Development in Android and iOS platform	Design and development

3   System Architecture

The mLearning framework is designed to be implemented on WOU-level involving learners across all regional centres. This project determines the mLearning framework include aspects such as the process and message flow, system architecture. The learning experiences and reviews are also derived from the pilot study of WOU learners to track downloads, reviews and statistics on respective WOU courses. Acceptance test and experimental studies aims to gauge the acceptance level of the WOU learners towards mLearning and determine the preferable contents and framework for WOU App. The following Fig. 2 depicts the work flow of tester, iOS mobile agent and program portal.

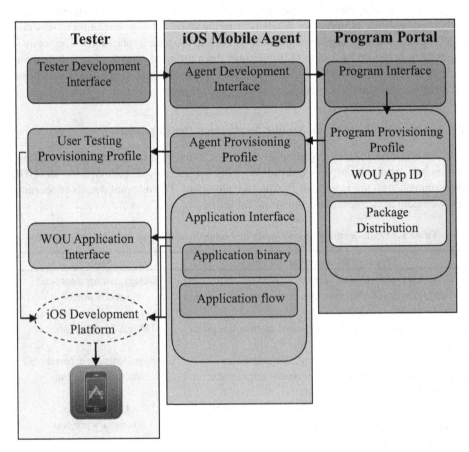

Fig. 2. System Architecture for WOU mLearning

4   WOU mLearning Prototype

In Fig. 3, the WOU mLearning App interface that deployed in iOS such as iPhone (iOS version 5.2 or higher), screen size (320 by 480 pixels) enable learners to engage interactive learning which takes place via automated, real-time access of courses such as Programming Fundamentals with Java in WOU. The WOU mLearning App prototype provides the alternative to course materials for learners in Java Programming important concepts such as Arrays, Sorting Arrays, Binary Search, Sequential Search, Bubble Sort and Insertion Sort in learning objects environment.

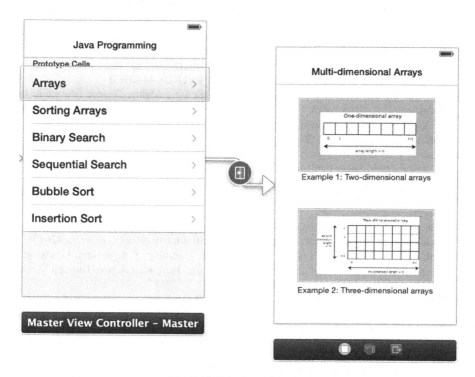

Fig. 3. WOU mLearning App

5   Discussion and Conclusion

The mobile learning initiative focuses on the initial study of mobile-learning applications and explains the technologies involved and development of mobile-learning roadmap. This study provides the prospects and strategies employed in mobile-learning strategies. Considering mobile-learning as a supplementary option to enhance learners' learning experience and encourage interactivity in distance learning environment, this paper present the use of WOU Education Learning App as the alternative delivery platform in WOU to boost the learners' motivation for lifelong learning and cater as an additional learning mode in ODL environment. The WOU Mobile-Learning initiative also promotes mobile-learning related development and collaborative research works with various ODL practitioners.

References

1. Communications and Multimedia: Pocket Book of Statistics, Q2 2013, Malaysian Communications and Multimedia Commission (2013)
2. Dzakiria, H.: The role of learning support in open and distance learning: Learners' experiences and perspectives. Turkish Online Journal of Distance Education (TOJDE) 6(2) (2005)
3. Flowers, J.: Online learning needs in technology education. Journal of Technology Education 13(1), 17–30 (2001)
4. Zirkle, C.: Identification of distance education barriers for trade and industrial teacher education. Journal of Industrial Teacher Education 40(1), 20–44 (2002)
5. Vrasidas, C., Glass, G.V.: A conceptual framework for studying distance education. In: Vrasidas, C., Glass, G.V. (eds.) Current Perspectives in Applied Information Technologies: Distance Education and Distributed Learning, pp. 31–56. Information Age Publishing, Inc, Greenwich (2002)
6. Stanley, C., Porter, E. (eds.): Engaging Large Classes: Strategies And Techniques For College Faculty. Anker Publishing Company, Bolton (2002)
7. Chou, C.C., Block, L., Jesness, R.: A case study of mobile learning pilot project in K-12 schools. Journal of Educational Technology Development and Exchange 5(2), 11–26 (2012)
8. Hung Chung, S.: Implementation of M-OER Initiative (Mobile OER): Prospect and Challenges. In: Asian Association of Open Universities International Conference (2013)
9. Park, Y.: A Pedagogical Framework for Mobile Learning: Categorising Educational Applications of Mobile Technologies into Four Types. Increasing Access Through Mobile Learning, pp. 27–48 (2014)
10. Brown, J., Hruska, M., Johnson, A., Poltrack, J.: Educational Standards for Mobile Learning and Mobile Application Development. Increasing Access Through Mobile Learning, pp. 17–25 (2014)
11. Uther, M.: Mobile Internet Usability: What Can 'Mobile Learning' Learn From the Past. In: IEEE International Workshop on Wireless and Mobile Technologies in Education (WMTE 2002), pp. 174–176. Växjö, Sweden (2002)
12. Koole, M.L.: A model for framing mobile learning. In: Ally, M. (ed.) Mobile learning: Transforming the Delivery of Education and Training, pp. 25–47. University Press, Edmonton (2009)

Author Index

Printed in the United States
By Bookmasters